͵cture Notes in Computer Scien

T0237954

Commenced Publication in 1973
Founding and Former Series Editors:
Gerhard Goos, Juris Hartmanis, and Jan van Leeuwen

Editorial Board

David Hutchison
Lancaster University, UK

Takeo Kanade
Carnegie Mellon University, Pittsburgh, PA, USA

Josef Kittler
University of Surrey, Guildford, UK

Jon M. Kleinberg
Cornell University, Ithaca, NY, USA

Alfred Kobsa
University of California, Irvine, CA, USA

Friedemann Mattern
ETH Zurich, Switzerland

John C. Mitchell
Stanford University, CA, USA

Moni Naor
Weizmann Institute of Science, Rehovot, Israel

Oscar Nierstrasz
University of Bern, Switzerland

C. Pandu Rangan
Indian Institute of Technology, Madras, India

Bernhard Steffen
University of Dortmund, Germany

Madhu Sudan
Microsoft Research, Cambridge, MA, USA

Demetri Terzopoulos
University of California, Los Angeles, CA, USA

Doug Tygar
University of California, Berkeley, CA, USA

Gerhard Weikum
Max-Planck Institute of Computer Science, Saarbruecken, Germany

Niall M. Adams   Céline Robardet
Arno Siebes   Jean-François Boulicaut (Eds.)

Advances in Intelligent Data Analysis VIII

8th International Symposium
on Intelligent Data Analysis, IDA 2009
Lyon, France, August 31 – September 2, 2009
Proceedings

 Springer

Volume Editors

Niall M. Adams
Imperial College London
Department of Mathematics
South Kensington Campus, London SW7 2PG, UK
E-mail: n.adams@imperial.ac.uk

Céline Robardet
Jean-François Boulicaut
University of Lyon, INSA Lyon, LIRIS CNRS UMR 5205
Bâtiment Blaise Pascal, F-69621 Villeurbanne, France
E-mail: {celine.robardet,jean-francois.boulicaut}@insa-lyon.fr

Arno Siebes
University of Utrecht
Department of Information and Computer Science
Utrecht, The Netherlands
E-mail: Arno.Siebes@cs.uu.nl

Library of Congress Control Number: 2009933040

CR Subject Classification (1998): H.3, I.5, G.3, J.1, J.3, H.3.3, I.5.3

LNCS Sublibrary: SL 3 – Information Systems and Application, incl. Internet/Web
and HCI

ISSN   0302-9743
ISBN-10   3-642-03914-6 Springer Berlin Heidelberg New York
ISBN-13   978-3-642-03914-0 Springer Berlin Heidelberg New York

This work is subject to copyright. All rights are reserved, whether the whole or part of the material is concerned, specifically the rights of translation, reprinting, re-use of illustrations, recitation, broadcasting, reproduction on microfilms or in any other way, and storage in data banks. Duplication of this publication or parts thereof is permitted only under the provisions of the German Copyright Law of September 9, 1965, in its current version, and permission for use must always be obtained from Springer. Violations are liable to prosecution under the German Copyright Law.

springer.com

© Springer-Verlag Berlin Heidelberg 2009
Printed in Germany

Typesetting: Camera-ready by author, data conversion by Scientific Publishing Services, Chennai, India
Printed on acid-free paper   SPIN: 12745715   06/3180   5 4 3 2 1 0

Preface

The general theme of the Intelligent Data Analysis (IDA) Symposia is the intelligent use of computers in complex data analysis problems. The field has matured sufficiently that some re-consideration of our objectives was required in order to retain the distinctiveness of IDA. Thus, in addition to the more traditional algorithm- and application-oriented submissions, we sought submissions that specifically focus on aspects of the data analysis process. For example, interactive tools to guide and support data analysis in complex scenarios. With the increasing availability of automatically collected data, tools that intelligently support and assist human analysts are becoming important.

IDA-09, the 8th International Symposium on Intelligent Data Analysis, took place in Lyon from August 31 to September 2, 2009. The invited speakers were Paul Cohen (University of Arizona, USA) and Pablo Jensen (ENS Lyon, France). The meeting received more than 80 submissions. The Programme Committee selected 33 submissions for publication: 18 for full oral presentation, and 15 for poster and short oral presentation. Each contribution was evaluated by three experts and has been allocated 12 pages in the proceedings. The accepted papers cover a broad range of topics and applications, and include contributions on the refined focus of IDA.

The symposium was supported by INSA Lyon (Institut National des Sciences Appliquées de Lyon) and by IXXI (Rhône-Alpes region Complex Systems Institute). We also thank our generous sponsors, La Région Rhône-Alpes and Le Ministère de l'Enseignement Supérieur et de la Recherche. We would like to express our gratitude to the many people involved in the organization of the symposium and the reviewing of submissions. The local Organizing Committee was co-chaired by Guillaume Beslon and Céline Robardet, and the committee members were Loïc Cerf, Serge Fenet, Pierre-Nicolas Mougel and David Parsons (all at INSA Lyon, France). We thank them for making the conference an unforgettable event. We are grateful to Joaquina Labro from Insavalor s.a., who managed the registration process. We are thankful for the support of the IDA council, especially for the advice of Michael Berthold, Joost Kok, Xiaohui Liu and José-Maria Peña. Finally we would like to thank Richard van de Stadt and Springer for preparing the proceedings.

August 2009

Niall Adams
Céline Robardet
Arno Siebes
Jean-François Boulicaut

Conference Organization

Conference Chair

Jean-François Boulicaut University of Lyon, France

Programme Committee

Niall Adams (PC Co-chair)	Imperial College London, UK
Fabrizio Angiulli	University of Calabria, Italy
Alexandre Aussem	University of Lyon, France
Tony Bagnall	University of East Anglia, UK
Riccardo Bellazzi	University of Pavia, Italy
Bettina Berendt	KU Leuven, Belgium
Daniel Berrar	Systems Biology Institute, Tokyo, Japan
Michael Berthold	University of Konstanz, Germany
Klemens Böhm	University of Karlsruhe, Germany
Christian Borgelt	European Centre for Soft Computing, Spain
Elizabeth Bradley	University of Colorado, USA
Pavel Brazdil	University of Porto, Portugal
Bruno Crémilleux	University of Caen, France
Werner Dubitzky	University of Ulster, UK
Sašo Džeroski	Jozef Stefan Institute, Slovenia
Fazel Famili	IIT - NRC, Canada
Jason Farquhar	University of Nijmegen, The Netherlands
Ad Feelders	University of Utrecht, The Netherlands
Ingrid Fischer	University of Konstanz, Germany
Eibe Frank	University of Waikato, New Zealand
Elisa Fromont	University of Saint-Etienne, France
Johannes Fürnkranz	TU Darmstadt, Germany
Alex Gammerman	University of London, UK
Gérard Govaert	TU Compiègne, France
Pilar Herrero	Polytechnic University of Madrid, Spain
Alexander Hinneburg	University of Halle, Germany
Frank Höppner	University of Applied Sciences, Germany
Jaakko Hollmén	Helsinki University of Technology, Finland
Eyke Hüllermeier	University of Marburg, Germany
Daniel Keim	University of Konstanz, Germany
Frank Klawonn	University of Wolfenbuettel, Germany
Jiri Klema	Czech Technical University, Czech Republic
Arno Knobbe	Leiden University, The Netherlands

Trajectory Voting and Classification Based on Spatiotemporal
Similarity in Moving Object Databases 131
   Costas Panagiotakis, Nikos Pelekis, and Ioannis Kopanakis

Leveraging Call Center Logs for Customer Behavior Prediction 143
   Anju G. Parvathy, Bintu G. Vasudevan, Abhishek Kumar, and
   Rajesh Balakrishnan

Condensed Representation of Sequential Patterns According to
Frequency-Based Measures 155
   Marc Plantevit and Bruno Crémilleux

ART-Based Neural Networks for Multi-label Classification 167
   Elena P. Sapozhnikova

Two-Way Grouping by One-Way Topic Models 178
   Eerika Savia, Kai Puolamäki, and Samuel Kaski

Selecting and Weighting Data for Building Consensus Gene Regulatory
Networks 190
   Emma Steele and Allan Tucker

Incremental Bayesian Network Learning for Scalable Feature
Selection 202
   Grégory Thibault, Alex Aussem, and Stéphane Bonnevay

Feature Extraction and Selection from Vibration Measurements for
Structural Health Monitoring 213
   Janne Toivola and Jaakko Hollmén

Zero-Inflated Boosted Ensembles for Rare Event Counts 225
   Alexander Borisov, George Runger, Eugene Tuv, and
   Nuttha Lurponglukana-Strand

Selected Contributions 2 (Short Talks)

Mining the Temporal Dimension of the Information Propagation 237
   Michele Berlingerio, Michele Coscia, and Fosca Giannotti

Adaptive Learning from Evolving Data Streams 249
   Albert Bifet and Ricard Gavaldà

An Application of Intelligent Data Analysis Techniques to a Large
Software Engineering Dataset...................................... 261
   James Cain, Steve Counsell, Stephen Swift, and Allan Tucker

Which Distance for the Identification and the Differentiation of
Cell-Cycle Expressed Genes? 273
   Alpha Diallo, Ahlame Douzal-Chouakria, and Francoise Giroud

Ontology-Driven KDD Process Composition 285
  Claudia Diamantini, Domenico Potena, and Emanuele Storti

Mining Frequent Gradual Itemsets from Large Databases 297
  Lisa Di-Jorio, Anne Laurent, and Maguelonne Teisseire

Selecting Computer Architectures by Means of Control-Flow-Graph
Mining 309
  Frank Eichinger and Klemens Böhm

Visualization-Driven Structural and Statistical Analysis of Turbulent
Flows 321
  *Kenny Gruchalla, Mark Rast, Elizabeth Bradley, John Clyne, and
  Pablo Mininni*

Distributed Algorithm for Computing Formal Concepts Using
Map-Reduce Framework 333
  Petr Krajca and Vilem Vychodil

Multi-Optimisation Consensus Clustering 345
  Jian Li, Stephen Swift, and Xiaohui Liu

Improving Time Series Forecasting by Discovering Frequent Episodes
in Sequences 357
  Francisco Martínez-Álvarez, Alicia Troncoso, and José C. Riquelme

Measure of Similarity and Compactness in Competitive Space 369
  Nikolay Zagoruiko

Bayesian Solutions to the Label Switching Problem 381
  Kai Puolamäki and Samuel Kaski

Efficient Vertical Mining of Frequent Closures and Generators 393
  Laszlo Szathmary, Petko Valtchev, Amedeo Napoli, and Robert Godin

Isotonic Classification Trees 405
  Rémon van de Kamp, Ad Feelders, and Nicola Barile

Author Index 417

Intelligent Data Analysis in the 21st Century

Paul Cohen[1] and Niall Adams[2]

[1] University of Arizona
cohen@cs.arizona.edu
[2] Imperial College London
n.adams@imperial.ac.uk

Abstract. When IDA began, data sets were small and clean, data provenance and management were not significant issues, workflows and grid
computing and cloud computing didn't exist, and the world was not
populated with billions of cellphone and computer users. The original
conception of intelligent data analysis — automating some of the reasoning of skilled data analysts — has not been updated to account for
the dramatic changes in what skilled data analysis means, today. IDA
might update its mission to address pressing problems in areas such as
climate change, habitat loss, education, and medicine. It might anticipate
data analysis opportunities five to ten years out, such as customizing educational trajectories to individual students, and personalizing medical
protocols. Such developments will elevate the conference and our community by shifting our focus from arbitrary measures of the performance
of isolated algorithms to the practical, societal value of intelligent data
analysis systems.

Each time we hold an IDA conference, a distinguished conference committee
thinks hard about a theme and a distinguished researcher writes a keynote lecture
about what Intelligent Data Analysis is or might be. We suspect that all this
hard thinking does not influence the kinds of papers we receive. Every conference
season we review and accept roughly the same kinds of papers as appear at the
data mining and machine learning conferences.

The subject of the conference should not be a fifteen year old vision of intelligent data analysis, nor should the subject default to a sample of current work
in data mining and machine learning. The conference should provide a venue for
future interpretations of Intelligent Data Analysis. We should start publishing
in areas that are developing now and will reach full bloom in five years. At the
same time, we should stay true to the traditional goals of the IDA conference.

The first symposium on Intelligent Data Analysis was organized by Xiaohui
Liu and held in Baden-Baden in 1995, the same year as the first International
Conference on Knowledge Discovery and Data Mining. Professor Liu's idea was
that data analysis, like other kinds of human expert problem solving, could be
done by computers:

> [Computers] should also be able to perform complex and laborious
> operations using their computational power so that the analysts can

N. Adams et al. (Eds.): IDA 2009, LNCS 5772, pp. 1–9, 2009.
© Springer-Verlag Berlin Heidelberg 2009

focus on the more creative part of the data analysis using knowledge and experience. Relevant issues include how to divide up work between human and computer; how to ensure that the computer and human stay "in synch" as they work on parts of a data analysis problem; how to seamlessly integrate human domain and common sense knowledge to inform otherwise stupid search procedures such as stepwise regression; how to present data so human eyes can see patterns; how to develop an integrated data analysis environment...[6]

To a remarkable extent, these issues have been addressed and Liu's vision of automated data processing has been achieved. Computers do perform complex and laborious operations, we have integrated data analysis environments (such as R [7]) and packages of algorithms (such as WEKA [9]). The community has settled on a small collection of common "generic tasks," [1] such as prediction, classification, clustering, model selection, and various kinds of estimation. These tasks are more specific than "exploring a dataset," and yet are general enough to cover data from disparate domains such as finance and marketing, biology and ecology, psychology and education. Less progress has been made toward integrated, knowledge-intensive, human-computer systems, but, as we suggest later in this paper, this goal might not be as desirable as we once thought. If the IDA community has achieved most of its goals and abandoned those it cannot achieve, what remains to be done? IDA is today a data mining conference, and data mining has achieved the kind of maturity that produces only incremental progress. What will our next challenges be?

We will address two perennial answers to this question before turning to some new challenges.

1   Autonomous Expert Data Analysis

The kinds of autonomy we see in data analysis today are not the kinds we anticipated in 1995. Following the "knowledge revolution" and the widespread commercialization of expert systems, we expected intelligent data analysis systems to attack data sets with the strategies of expert human data analysts. Few such systems were built. It is worth reviewing one of them, built by Rob St. Amant as part of his PhD research, to understand why there are not more systems like it.

AIDE was a mixed-initiative planner for data exploration, meaning it and a human user could explore a dataset together, with AIDE sometimes following the user's lead and sometimes striking out on its own [2,3]. Its knowledge about data analysis was stored in plans and control knowledge. Plans contain sequences of data-processing actions, as well as preconditions and postconditions. In general, preconditions specify when a plan *can* be executed, not when it *should* be. When more than one plan applies, control knowledge ranks them. It is the job of control knowledge to make the data analysis follow a coherent path, rather than jumping around (which would be hard for users to follow). Of course, a human user can direct AIDE to do anything he or she pleases, so some control knowledge pertains to inferring or anticipating the user's focus.

At the time, it made sense for AIDE to be a domain-independent data analyst, so its plans and control knowledge referred to the form of data, not to its content or what it represents. AIDE was based on the assumption that the random variables in a dataset were actually connected by causal or functional relations, and the job of data analysis is to build one or more models of the variables and their relations — one or more graphs, if you will, in which nodes represent variables and arcs represent relations. AIDE was not a system for analyzing financial data, or intelligence analysis, or phenotyping, all of which are today aided by domain-specific tools. It was instead a system for general purpose data analysis. It was shown to improve data exploration, in the sense that users of AIDE were able to explore more of a dataset and figure out the relationships between variables more thoroughly and accurately than users without AIDE. This test was appropriate for a domain-independent data analysis tool, but it was not realistic: Generally one approaches a dataset with particular questions in mind, not looking for all significant relations between variables.

Although AIDE did a lot of things right — mixed initiative exploration, explicit plans and control knowledge, a clear user interface, and plenty of data analysis functionality — it could never work in practice. Its understanding of data (random variables in functional or causal relations) and the user's intentions was too weak to be a basis for really focused and intelligent analysis. If intelligent data analysis means something like expert systems for data exploration, then these systems will have to be a lot more expert. They will have to specialize in financial fraud data, or climate change data, and so on.

Another reason that programs like AIDE did not start a revolution might be that they did little to help human analysts. We have been here before: At first blush, stepwise multiple regression seems like an ideal use of computer power, but in practice, instead of letting the machine explore huge numbers of regression models, most analysts prefer to build them by hand. Ignorant of what is being modeled, stepwise multiple regression blunders through the hunt and rarely brings home a tasty model; or it brings home too many, overlapping, similarly performing models because it hasn't the analyst's knowledge to rank them. AIDE was more intelligent than stepwise multiple regression, but it probably wasn't intelligent enough to make analysts relinquish the creative parts of data analysis they find relatively easy and enjoyable. Much of its knowledge and intelligence was for planning analyses and working alongside analysts. The closest thing we have seen to this functionality in commercial systems is the graphical language in SPSS's PASW Modeler, which allows analysts — but not the system — to plan and program workflows for their analyses. By analogy, most cooks are happy for assistance with chopping, stirring, mashing, scrubbing and cleaning, but few cooks want help with inventing, menu planning, tasting, or observing the pleasure their food brings to others; especially if the "help" results in an inferior meal.

If past is prologue, we should not expect the original vision of IDA, as exemplified by AIDE, to be productive in future. Modern data mining software can chop, mash and scrub data, but we should not expect sophisticated analysis — beautiful models, new discoveries, finding parallel phenomena in disparate datasets, clever workflows, detection of semantic anomalies, and the like — until we make semantical reasoning about data a priority. By semantical reasoning, we mean reasoning about the phenomena that data represents. Although it has been a theme of every IDA conference, the IDA community evidently reserves semantical reasoning about data for human data analysts, and very little has been done to make machines that are capable of reasoning deeply about the content of data. Of course, this attention to form instead of content characterizes virtually all of Artificial Intelligence.

2   Challenge Problems

Periodically, IDA considers creating a community around challenge problems. This has worked well in some fields, particularly robotic soccer and robotic autonomous vehicles, and less well in others, notably the KDD Cup. Why are some challenges successful and others less so? The organization that runs the annual robotic soccer competitions has a fifty year goal: *By the year 2050, develop a team of fully autonomous, humanoid robots that can win against the human world soccer champion team.* [8] Progress toward this goal is steered by an expert committee that changes the rules of the competition and adds new intermediate challenges every year. These changes are monotonic in the sense that researchers can base next year's work on last year's work, and new competitors can join relatively easy by adopting state of the art technology. Robotic soccer has a relatively low cost of admission. It is enormously popular and captures the hearts and minds of participants and the general public. Bragging rights go to individual teams, but the biggest winner is the community as a whole, which gains new members and makes steady, impressive progress every year.

One methodological strength of robotic soccer is that individual algorithms matter less than complete soccer-playing systems. The same was true in St. Amant's AIDE system. While it is desirable to include the fastest and most accurate algorithms available, the marginal benefits of better algorithms will often be negligible, especially over many datasets. Systems like AIDE introduce some pragmatism into the evaluation of new algorithms. Perhaps your changepoint analyzer, or decision tree inducer, or association rule miner is slightly faster or more accurate than last year's model on datasets of your own choosing, but would a data analyst notice a difference if your new algorithm was substituted for an old one in AIDE? Would the analyst be more productive? Admittedly, this question might be harder to answer than a simple evaluation of speed or accuracy, but it is the real usability question, whereas speed and accuracy are only proxies for usability.

The KDD Cup does not encourage the development of complete systems, nor does it steer the community toward any long-term goal [5]. The problems it poses each year are important (several have been medical problems), but each is narrowly defined by a data set and performance targets. These problems emphasize high performance for individual algorithms, and do not require the development of systems that are complete in any sense. Nor is a progression of data mining capabilities apparent in the choice of KDD Cup problems. One has the sense that the problems could have been offered to the community in a different order with essentially the same results. The state of the art isn't being steered.

3 New Challenges for Intelligent Data Analysis

The IDA community should pick one or more significant problems that depend on intelligent data analysis, set a goal for a decade or longer in the future, and steer ourselves to achieve it. The choice of problems should mirror our aspirations for intelligent data analysis. Good problems will be bigger than individuals or small groups of human analysts can manage. They will feature every aspect of data analysis: acquisition, cleaning, storage, markup, analysis, visualization, and archiving and dissemination of results. They will require every kind of reasoning about data and the algorithms that process it: reasoning about provenance, design of workflows, and interpretation of results. Most importantly, good challenge problems will require machines to think about the phenomena that data represent — to think about the content or semantics of data. Here are some examples:

The Scientific Discovery Challenge. By the year 2030, a computer program will make a significant scientific discovery (indicated by publication in Science or Nature, or a comparable venue, or by the granting of an important patent). To qualify, the program will have to formulate a theory, direct the search for evidence, analyze the data, and explain the theory and supporting evidence in a formal language. Natural language understanding is not a necessary part of the challenge problem, but the ability to find and reason about relevant knowledge in the literature might be valuable. To demonstrate that it is more than an assistant to a human scientist (who does the hard intellectual work) the program will have to pass some tests that any human scientist is expected to pass; for example, the program should be able to say whether and why a hypothetical result is consistent or inconsistent with its theory.

A specific scientific discovery challenge could be to automatically construct a gene regulatory network for an important cascade or developmental process. Increasingly, biologists turn to modeling techniques that are familiar to us in the IDA community: Stochastic processes, Bayesian networks, the Viterbi algorithm and related methods. Today, there are only a few examples of automated construction of gene regulatory networks, and those are single algorithms, rather than systems that hypothesize a model by integrating the results of multiple data mining algorithms, and gather data to support or contradict it.

The Global Quality-of-Life Monitoring System. By 2060, every living human will have some kind of communication device that is at least as powerful, computationally, as today's most advanced cell phones. Each human will be an intelligent source of data. IDA can contribute to the infrastructure and algorithms necessary to provide real-time, high-resolution fusion of the data in service of research and policy. Challenges are to monitor habitat loss, species redistribution, erosion, water quality and management, epidemics and pandemics, and other consequences of climate change and population growth.

A specific, relatively near-term challenge problem might go like this: Model the population dynamics and food webs of foxes in London given data from a dedicated web site where residents can report sightings, indicators (e.g., scat, prints), and behaviors of foxes and other predators and prey. To qualify, a computer program would have to demonstrate autonomy in several areas of intelligent data analysis: Knowing what to do with new data (e.g., where to put it in a data set, cleaning it, handling missing values, flagging anomalies); directing data-gathering resources in an efficient way; planning a workflow of operations to build a specified kind of model (e.g., a food web) and estimate its parameters; flagging parameter values, data values, or model predictions that are unusual or anomalous (e.g., if the model predicted a huge jump in the fox population of Islington).

Similar similar challenges might involve automated modeling of human population dynamics, modeling effects of climate change through widely distributed monitoring of spring blooming and densities of species, and modeling evaporation (see below).

The Personalized Protocol. Some of the most important activities in life are controlled by protocols and standards that make little if any provision for differences between individuals. Procedures in intensive care, chemotherapy protocols, other kinds of drug therapy, primary and secondary education, even the mutual funds in which we invest, are quite generic, and rarely are tailored to particular individuals. However, there are hints that personalized protocols are both effective and profitable. Designer drugs, gene therapies, and "lifecycle" investments (which adjust portfolio parameters according to one's age and closeness to retirement) are examples of protocols that are personalized to some degree. Personalization requires data, of course, and the whole idea of personalizing medical care or education might fail because of legal and ethical challenges. But let us imagine that information technologies will permit us to both gather data and protect the rights of the people who provide it. Then we can envision a Personalized Protocol Challenge: To personalize any high-value procedure to maximize its utility to any individual.

As stated, this sounds like a planning problem or a sequential decision problem to be optimized by policy iteration, reinforcement learning, or a related optimization method. However, the difficult work is not to run one of these algorithms but to design the state space and objective function to be optimized. In terms of the old dichotomy between model specification and model estimation, data mining is pretty good at the latter, but the former is still in the realm of intelligent data analysis, and, thus, is a fit challenge for us.

These challenges have several things in common: Each is significant and could affect the survival of species, including our own. Each assumes enormous amounts of data generated by distributed sources. The Large Hadron Collider is a point-source of high-quality data, whereas the world's citizens are a distributed source of variable-quality data. Consequently, we will need new research in gathering, cleaning, and fusing data, all of which arrives asynchronously, yet must sometimes be integrated to construct a temporal or developmental story. Each of the challenges stresses our ability to model systems of dependencies, whether they are gene regulatory systems, ecologies, or social systems. Survival in the modern world will require a new science of complex systems, and statistical methods for discovering and estimating the parameters of these systems. The intelligent data analysis community should not sit this one out, but should take advantage of new opportunities for research and development.

Each of the challenges requires the IDA community to be more outward-looking, less concerned with the arcana of algorithms, more concerned with helping scientists ensure our future well-being. To respond effectively to any of the challenges, we will have to think more about data provenance, metadata and search for data, reasoning about the content or meaning of the data, user interfaces, visualizations of results, privacy and other ethical issues, in addition to the algorithmic research we usually do.

If we adopt a challenge that uses humans as data sources, such as the Global Quality-of-Life Monitoring System, then we should recognize that humans are both producers of raw data and consumers of knowledge, and the same communication infrastructure that supports data capture can support knowledge distribution. Said differently, there will be opportunities to engage people in science and social science, to educate them, to empower them to influence policy, and to create a sense that communities cross national borders. At the University of Arizona this reciprocal relationship between scientists and citizens is called *citizen science*. The science of evaporation provides a nice illustration. Evaporation matters in Arizona and in the arid lands that comprise 30% of the Earth's land surface. (If current models of climate change are correct, this proportion will increase dramatically.) More than 90% of the water that reaches the Sonoran Desert in Arizona evaporates back into the atmosphere, but scientists don't know how plants affect this process. There are no good analytical models of evaporation. At the Biosphere 2 facility, scientists are studying empirically how plant density and configurations affect evaporation, but it is slow work, and the number of factors that affect the results are daunting. Recently, our colleague Clayton Morrison, working with the Biosphere 2 scientists, built a version of their experiment to be run by children in classrooms in Tucson [4]. The kids benefit by being engaged in real science, run by local scientists, on locally important issues. The scientists benefit from data collected by the kids. And those of us who live in arid lands benefit from the resulting science.

How can IDA participate in citizen science? A natural role for IDA is to help scientists transform data into knowledge. In the evaporation experiment, for instance, this means transforming spatial and temporal data into models of evaporation that account for complex dependencies between types and distributions of plants and soils. Another natural role for IDA is to optimize the tradeoff between the quality and quantity of data. Having monitored the generation every data point in the Tucson classrooms, we know that children produce lower-quality data than trained scientists do. But there are many children and relatively few trained scientists. We look forward to new methods for cleaning, censoring, and otherwise editing enormous data sets that are a bit grubbier than most scientists are used to.

Whether IDA runs challenge problems of the kind we described earlier, or remains a conventional conference, it should expand its view of the field. What has been an algorithms conference should become a systems conference, where the systems typically will have components for data gathering, data processing, and disseminating results, and are built to solve problems that matter to society. We should encourage papers on crowd sourcing, social network analysis, experimental economics, new data markup schemas, mobile education, and other topics in the general areas of data gathering, data processing and disseminating results. We might accept papers on the ethics of semiautomated decision-making (if a credit scoring system misclassifies you, who is responsible, and what are the legal and ethical considerations?). We should particularly value papers that demonstrate reciprocity between citizens and scientists. The conference should recognize that systems may be harder to evaluate than algorithms, especially when these systems include humans as data sources or expert data processors, and it should adjust reviewing criteria accordingly. New criteria should reward autonomous and mixed-initiative analysis; integration of analysis with data management, workflow management, and new ways to present and justify results; integration of multiple data sets from different sources within analyses; and automated adjustment of algorithm parameters, so they don't have to be tuned by hand.

In conclusion, IDA can look forward to success if it organizes itself around problems that both matter to society and afford opportunities for basic research. These problems are not proxies for important problems (as robot soccer is a proxy for more important things to do with teams of mobile robots) but are themselves important. Good problems can be defined with a few, nontechnical words, and they have clear criteria and metrics for success. The IDA community could adopt one or a few, and organize annual challenges around them, using them to steer research toward major, long-term goals. As a practical matter, these problems should generate research funding for some years to come. Finally, we will find that all good, important problems already have people working on them: biologists, sociologists, economists, ecologists and so on. We should not be scared away but should remember our heritage: The job of intelligent data analysis is not to create more data analysis algorithms, but to make sense of data, nearly all of which is generated by experts in fields other than our own. These people need intelligent data analysis, and we need colleagues and intellectual challenges, and something more substantial than a half-point improvement in classification accuracy to demonstrate what we're worth.

References

1. Chandrasekaran, B.: Generic tasks in knowledge-based reasoning: High-level building blocks for expert systems design. IEEE Expert 1(3), 23–30 (1986)
2. St. Amant, R., Cohen, P.R.: Interaction With a Mixed-Initiative System for Exploratory Data Analysis. Knowledge-Based Systems 10(5), 265–273 (1998)
3. St. Amant, R., Cohen, P.R.: Intelligent Support for Exploratory Data Analysis. The Journal of Computational and Graphical Statistics (1998)
4. http://www.cs.arizona.edu/~clayton/evap-web/

5. http://www.sigkdd.org/kddcup/index.php
6. http://people.brunel.ac.uk/~csstxhl/IDA/IDA_1995.pdf
7. The R Development Core Team. R: A Language and Environment for Statistical Computing. R Foundation for Statistical Computing, Vienna, Austria (2009), http://www.R-project.org
8. http://www.robocup.org/
9. Witten, I.H., Frank, E.: Data Mining: Practical machine learning tools and techniques, 2nd edn. Morgan Kaufmann, San Francisco (2005)

Analyzing the Localization of Retail Stores with Complex Systems Tools

Pablo Jensen

Institut des Systèmes Complexes Rhône-Alpes, IXXI-CNRS,
Laboratoire de Physique, Ecole Normale Supérieure de Lyon and LET-CNRS,
Université Lyon-2, 69007 Lyon, France

Abstract. Measuring the spatial distribution of locations of many entities (trees, atoms, economic activities, . . .), and, more precisely, the deviations from purely random configurations, is a powerful method to unravel their underlying interactions. I study here the spatial organization of retail commercial activities. From pure location data, network analysis leads to a community structure that closely follows the commercial classification of the US Department of Labor. The interaction network allows to build a 'quality' index of optimal location niches for stores, which has been empirically tested.

1 Introduction

Walking in any big city reveals the extreme diversity of retail store location patterns. Fig. 1 shows a map of the city of Lyon (France) including all the drugstores, shoes stores and furniture stores. A qualitative commercial organisation is visible in this map: shoe stores aggregate at the town shopping center, while furniture stores are partially dispersed on secondary poles and drugstores are strongly dispersed across the whole town. Understanding this kind of features and, more generally, the commercial logics of the spatial distribution of retail stores, seems a complex task. Many factors could play important roles, arising from the distincts characteristics of the stores or the location sites. Stores differ by product sold, surface, number of employees, total sales per month or inauguration date. Locations differ by price of space, local consumer characteristics, visibility (corner locations for example) or accessibility. One could reasonably think that to understand the logics of store commercial strategies, it is essential to take into account most of these complex features. This seems even more necessary for finding potentially interesting locations for new businesses.

However, in this paper, I show that location data suffices to reveal many important facts about the commercial organisation of retail trade[1]. First, I quantify the interactions among activities and group them using network analysis tools. I find a few homogeneous commercial categories for the 55 trades in Lyon, which closely match the usual commercial categories: personal services, home

---

[1] C. Baume and F. Miribel (commerce chamber, Lyon) have kindly provided extensive location data for 8500 stores of the city of Lyon.

N. Adams et al. (Eds.): IDA 2009, LNCS 5772, pp. 10–20, 2009.
© Springer-Verlag Berlin Heidelberg 2009

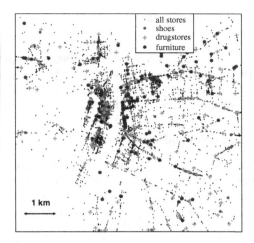

Fig. 1. Map of Lyon showing the location of all the retail stores, shoe stores, furniture dealers and drugstores

furniture, food stores and apparel stores. Second, I introduce a quality indicator for the location of a given activity and empirically test its relevance. These results, obtained from solely *location* data, agree with the retailing "mantra": *the three points that matter most in a retailer's world are: location, location and ... location.*

2   Quantifying Interactions between Activities

Measuring the spatial distribution of industries [1], atoms [2], trees [3] or retail stores [4,5] is a powerful method to understand the underlying mechanisms of their interactions. Several methods have been developed in the past to quantify the deviations of the empirical distributions from *purely random distributions*, supposed to correspond to the non-interacting case [6,7,8,9]. Recently, a method originally developed by G. Duranton and H. Overman [10], later modified by Marcon and Puech [11] has been proposed. Its main interest is that it takes as reference for the underlying space not a homogeneous one as for the former methods [6,7,8,9], but the overall spatial distribution of sites, thus automatically taking into account the many inhomogeneities of the actual geographical space. For instance, retail stores are inhomogeneously distributed because of rivers, mountains or specific town regulations (parks, pure residential zones, ...). Therefore, it is interesting to take this inhomogeneous distribution as the reference when testing the random distribution of, for instance, bakeries, in town. Furthermore, by using precise location data (x and y coordinates), this method avoids all the well-known contiguity problems, summarized in the 'modifiable areal unit problem' [12,13,14,15]. However, the method has two main drawbacks:

1. the need of precise location data (i.e. x and y coordinates, and not only knowing that a site belongs to a given geographical area),

2. the need for Monte Carlo simulations in order to compute the statistical significance of the deviations from a random distribution.

Point (1) is probably going to be less crucial as precisely spatialized data becomes more common. Moreover, it can be argued that, when only region-type data exists, it can be more convenient to locate all the sites at the region centroid and then apply the 'continuous' method, thus avoiding contiguity problems.

2.1   Definitions of the Spatial Indicators

The indicators that are studied here deal with the problem of quantifying deviations of empirical distribution of points from purely random and non-interacting distributions. One can be interested in the interaction of a set of points between themselves, or with some other set of points. From now on we shall work with two different types of points: A and B. We define two indicators, refered to as respectively the *intra* and *inter* coefficients [11], to characterize the (cumulative) spatial interaction between sites closer than a distance r. The *intra* coefficient is intended to measure the independence between points of type A, whereas the *inter* coefficient describes the type of interactions of fixed A points with random B points. One can also work with indicators characterizing the (differential) spatial distributions between distances r and $r + dr$ (with $dr \ll r$) [10]. Those last coefficients are potentially more sensitive to spatial variations of the distributions because they do not integrate features from 0 to r. We shall start by calculating the variance of the cumulative coefficient and then extend our results to other quantifiers of spatial distributions.

We shall use the following definitions and notations:

- one has N_t sites, of which N_A sites are of type A, and N_B sites are of type B,
- for any site S, one denotes by $N_t(S, r)$, $N_A(S, r)$ and $N_B(S, r)$ the number of respectively total, A and B sites that are at a distance lesser than r of site S, where site S is *not* counted, whichever its state.

The notation $N_A(D)$ (resp. $N_B(D)$) will denote the number of A (resp. B) sites in a subset D of T, T being the set of all the points.

In this discrete model, the locations of stores A and B are distributed over the total number of possible sites, with mutual exclusion at a same site. Therefore, the geographical characteristics of the studied area are carried by the actual locations of those possible N_t sites.

The coefficients that we introduce depend on the reference distance r, however we shall drop this dependency in the notations, unless when strictly necessary.

2.2   *Intra* Coefficient

Let us assume that we are interested in the distribution of N_A points in the set T, represented by the subset $\{A_i, i = 1 \ldots N_A\} \subset T$. The reference law for this set, called *pure random distribution*, is that this subset is uniformly chosen at random from the set of all subsets of cardinal N_A of T: this is equivalent to an urn model with N_A draws with no replacement in an urn of cardinal N_t.

Intuitively, under this (random) reference law, the local concentration represented by the ratio $N_A(A_i, r)/N_t(A_i, r)$ of stores of type A around a given store of type A should, in average, not depend on the presence of this last store, and should thus be (almost) equal to the global concentration N_A/N_t, this leads us to introduce the following *intra* coefficient:

$$M_{AA} = \frac{N_t - 1}{N_A(N_A - 1)} \sum_{i=1}^{N_A} \frac{N_A(A_i, r)}{N_t(A_i, r)} \tag{1}$$

In this definition, the fraction $0/0$ is taken as equal to 1 in the right hand term. Under the *pure randomness hypothesis*, it is straightforward to check that the average of this coefficient is equal to 1: for all $r > 0$, we have $E[M_{AA}] = 1$.

We deduce a qualitative behaviour in the following sense: if the observed value of the *intra* coefficient is greater than 1, we may deduce that A stores tend to aggregate, whereas lower values indicate a dispersion tendency.

2.3 *Inter* Coefficient

In order to quantify the dependency between two different types of points, we set the following context: the set T has a fixed subset of N_A stores of type A, and the distribution of the subset $\{B_i, i = 1 \ldots N_B\}$ of type B stores is assumed to be uniform on the set of subsets of cardinal N_B of $T \setminus \{A_1, \ldots, A_{N_A}\}$. Just as in the *intra* case, the presence of a point of type A at those locations, under this reference random hypothesis, should not modify (in average) the density of type B stores: the local B spatial concentration $(N_B(A_i, r)) / (N_t(A_i, r) - N_A(A_i, r))$ should be close (in average) to the concentration over the whole town, $(N_B) / (N_t - N_A)$. We define the *inter* coefficient as

$$M_{AB} = \frac{N_t - N_A}{N_A N_B} \sum_{i=1}^{N_A} \frac{N_B(A_i, r)}{N_t(A_i, r) - N_A(A_i, r)} \tag{2}$$

where $N_A(A_i, r)$, $N_B(A_i, r)$ and $N_t(A_i, r)$ are respectively the A, B and total number of points in the r-neighbourhood of point A_i (not counting A_i), i.e. points at a distance smaller than r. It is straightforward to check that for all $r > 0$, we have $E[M_{AB}] = 1$.

We can also deduce a qualitative behaviour in the following sense: if the observed value of the *inter* coefficient is greater than 1, we may deduce that A stores have a tendency to attract B stores, whereas lower values mean a rejection tendency.

3 Analyzing Retail Stores Interactions

I now analyze in detail the interactions of stores of different trades, using the coefficients defined above.

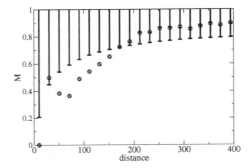

Fig. 2. Evolution of the *intra* coefficient for bakeries in the city of Lyon with respect to r, and (half) confidence interval with $\alpha = 0.05$. Data from CCI Lyon.

The figure 2 shows the practical importance of variance calculations for economic interpretations of the data. Although M_{AA} remains well below the reference value (i.e. 1), bakeries are significantly dispersed only until $150m$. For longer distances, their spatial locations approach a random pattern.

In the two following tables, I present other examples of interaction coefficients at $r = 100m$, together with the confidence intervals, for Paris, thanks to data kindly provided by Julien Fraîchard from INSEE. Table 1 shows the most aggregated activities.

Table 1. The most aggregated activities

activity	a	confidence interval at 95 %
textiles	5.27366	[0.979 , 1.021]
second-hand goods	3.47029	[0.9951 , 1.0049]
Jewellery	2.81346	[0.987 , 1.013]
shoes	2.60159	[0.9895 , 1.0105]
furniture, household articles	2.49702	[0.9846 , 1.0154]

Overall, the same activities are concentrated in Lyon and Paris. A simple economical rationale behind the concentrations or dispersions of retail activities is the following. Locating many stores at similar locations has two contradictory effects. First, it increases the attractiveness of the neighborhood by multiplying the offers. Second, it divides the generated demand among the stores. For some activities, the increase in demand is so high that it compensates the competition for customers. This is the case for when stores offer differentiated goods. Inversely, for stores offering more comparable products (such as bakeries), concentration does not increase the demand, and therefore would lead to a strong decrease in profit.

To illustrate the *inter* coefficient, I show in Table 2 the couples of activities that attract the most each other.

Table 2. The highest attractions between activities

activity 1	activity 2	a	confidence interval at 95 %
clothes	shoes	2.23216	[0.9978 , 1.0022]
Jewellery	Leather articles	2.12094	[0.984 , 1.016]
second-hand goods	household articles	2.10815	[0.9917 , 1.0083]
meat	fruits, vegetables	1.85213	[0.9906 , 1.0094]

4   Finding Retail Stores Communities

From the interaction coefficients measured above, one can define a network structure of retail stores. The nodes are the 55 retail activities (Table 3). The weighted[2] links are given by $a_{AB} \equiv \log(M_{AB})$, which reveal the spatial attraction or repulsion between activities A and B[3]. This retail network represents the first a social network with quantified "anti-links", i.e. repulsive links between nodes[4]. The anti-links add to the usual (positive) links and to the absence of any significant link, forming an essential part of the network. If only positive links are used, the analysis leads to different results, which are less satisfactory (see below).

To divide the store network into communities, I adapt the "Potts" algorithm[5] [19]. This algorithm interprets the nodes as magnetic spins and groups them in several homogeneous magnetic domains to minimize the system energy. Anti-links can then be interpreted as anti-ferromagnetic interactions between the spins. Therefore, this algorithm naturally groups the activities that attract each other, and places trades that repel into different groups. A natural definition [19,20] of the satisfaction ($-1 \leq s_i \leq 1$) of site i to belong to group σ_i is:

---

[2] Important differences introduced by including weighted links are stressed for example in [16].

[3] For a pair interaction to be significant, I demand that both a_{AB} and a_{BA} be different from zero, to avoid artificial correlations [17]. For Lyon's city, I end up with 300 significant interactions (roughly 10% of all possible interactions), of which half are repulsive.

[4] While store-store attraction is easy to justify (the "market share" strategy, where stores gather in commercial poles, to attract costumers), direct repulsion is generally limited to stores of the same trade which locate far from each other to capture neighbor costumers (the "market power" strategy). The repulsion quantified here is induced (indirectly) by the price of space (the sq. meter is too expensive downtown for car stores) or different location strategies. For introductory texts on retail organization ans its spatial analysis, see [18] and the Web book on regional science by E. M. Hoover and F. Giarratani, available at http://www.rri.wvu.edu/WebBook/Giarratani/contents.htm.

[5] Note that the presence of anti-links automatically ensures that the ground-state is not the homogeneous one, when all spins point into the same direction (i.e. all nodes belong to the same cluster). Then, there is no need then of a γ coefficient here.

$$s_i \equiv \frac{\sum_{j \neq i} a_{ij} \pi_{\sigma_i \sigma_j}}{\sum_{j \neq i} |a_{ij}|} \tag{3}$$

where $\pi_{\sigma_i \sigma_j} \equiv 1$ if $\sigma_i = \sigma_j$ and $\pi_{\sigma_i \sigma_j} \equiv -1$ if $\sigma_i \neq \sigma_j$.

To obtain the group structure, I run a standard simulated annealing algorithm [21] to maximize the overall site satisfaction:

$$K \equiv \sum_{i,j=1,55; i \neq j} a_{ij} \pi_{\sigma_i \sigma_j} \tag{4}$$

Pott's algorithm divides the retail store network into five homogeneous groups (Table I, note that the number of groups is not fixed in advance but a variable of the maximisation). This group division reaches a global satisfaction of 80% of the maximum K value and captures more than 90% of positive interactions inside groups. Except for one category ("Repair of shoes"), our groups are communities in the strong sense of Ref. [20]. This means that the grouping achieves a positive satisfaction for every element of the group. This is remarkable since hundreds of "frustrated" triplets exist[6]. Taking into account only the positive links and using the modularity algorithm [22] leads to two large communities, whose commercial interpretation is less clear.

Two arguments ascertain the commercial relevance of this classification. First, the grouping closely follows the usual categories defined in commercial classifications, as the U.S. Department of Labor Standard Industrial Classification System[7] (see Table 1). It is remarkable that, starting exclusively from location data, one can recover most of such a significant commercial structure. Such a significant classification has also been found for Brussels, Paris and Marseilles stores, suggesting the universality of the classification for European towns. There are only a few exceptions, mostly non-food proximity stores which belong to the "Food store" group. Second, the different groups are homogeneous in relation to correlation with population density. The majority of stores from groups 1 and 2 (18 out of 26) locate according to population density, while most of the remaining stores (22 out of 29) ignore this characteristic[8]. Exceptions can be explained by the small number of stores or the strong heterogeneities[9] of those activities.

---

[6] A frustrated (A, B, C) triplet is one for which A attracts B, B attracts C, but A repels C, which is the case for the triplet shown in Fig. 1.

[7] See for example the U.S. Department of Labor Internet page:
http://www.osha.gov/pls/imis/sic_manual.html.

[8] To calculate the correlation of store and population density for a given activity, I count both densities for each of the 50 Lyon's sectors. I then test with standard econometric tools the hypothesis that store and population densities are uncorrelated (zero slope of the least squares fit), with a confidence interval of 80%.

[9] Several retail categories defined by the Commerce Chamber are unfortunately heterogeneous: for example, "Bookstores and newspapers" refers to big stores selling books and CDs as well as to the proximity newspaper stand. Instead, bakeries are precisely classified in 4 different categories: it is a French commercial structure!

Table 3. Retail store groups obtained from Pott's algorithm. Our groups closely match the categories of the U.S. Department of Labor Standard Industrial Classification (SIC) System: group 1 corresponds to Personal Services, 2 to Food stores, 3 to Home Furniture, 4 to Apparel and Accessory Stores and 5 to Used Merchandise Stores. The columns correspond to: group number, activity name, satisfaction, correlation with population density (U stands for uncorrelated, P for Population correlated) and finally number of stores of that activity in Lyon. To save space, only activities with more than 50 stores are shown.

group	activity	s	pop corr	N_{stores}
1	bookstores and newspapers	1.00	U	250
1	Repair of electronic household goods	0.71	P	54
1	make up, beauty treatment	0.68	P	255
1	hairdressers	0.67	P	844
1	Power Laundries	0.66	P	210
1	Drug Stores	0.55	P	235
1	Bakery (from frozen bread)	0.54	P	93
2	Other repair of personal goods	1.00	U	111
2	Photographic Studios	1.00	P	94
2	delicatessen	0.91	U	246
2	grocery (surface $< 120m^2$)	0.77	P	294
2	cakes	0.77	P	99
2	Miscellaneous food stores	0.75	P	80
2	bread, cakes	0.70	U	56
2	tobacco products	0.70	P	162
2	hardware, paints (surface $< 400m^2$)	0.69	U	63
2	meat	0.64	P	244
2	flowers	0.58	P	200
2	retail bakeries (home made)	0.47	P	248
2	alcoholic and other beverages	0.17	U	67
3	Computer	1.00	P	251
3	medical and orthopaedic goods	1.00	U	63
3	Sale and repair of motor vehicles	1.00	P	285
3	sport, fishing, camping goods	1.00	U	119
3	Sale of motor vehicle accessories	0.67	U	54
3	furniture, household articles	0.62	U	172
3	household appliances	0.48	U	171
4	cosmetic and toilet articles	1.00	U	98
4	Jewellery	1.00	U	230
4	shoes	1.00	U	178
4	watches, clocks and jewellery	1.00	U	92
4	clothing	0.91	U	914
4	tableware	0.83	U	183
4	opticians	0.78	U	137
4	Other retail sale in specialized stores	0.77	U	367
4	Other personal services	0.41	U	92
4	Repair of boots, shoes	-0.18	U	77
5	second-hand goods	0.97	U	410
5	framing, upholstery	0.81	U	135

5   From Interactions to Location Niches

Thanks to the quantification of retail store interactions, I can construct a mathematical index to automatically detect promising locations for retail stores. Let's

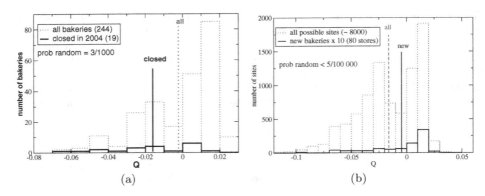

Fig. 3. The landscape defined by the quality index is closely correlated to the location decisions of bakeries. (a) The 19 bakeries that closed between 2003 and 2005 had an average quality of -2.2×10^{-3} to be compared to the average of all bakeries (4.6×10^{-3}), the difference being signifcative with probability 0.997. Taking into account the small number of closed bakeries and the importance of many other factors in the closing decision (family problems, bad management...), the sensitivity of the quality index is remarkable. (b) Concerning the 80 new bakeries in the 2005 database (20 truly new, the rest being an improvement of the database), their average quality is -6.8×10^{-4}, to be compared to the average quality of all possible sites in Lyon (-1.6×10^{-2}), a difference significant with probability higher than 0.9999.

take the example of bakeries. The basic idea is that a location that gathers many activities that are "friends" of bakeries (i.e. activities that attract bakeries) and few "ennemies", might well be a good location for a new bakery. The quality $Q_A(x,y)$ of an environment around (x,y) for an activity A as:

$$Q_A(x,y) \equiv \sum_{B=1,55} N_B(x,y) \tag{5}$$

where $N_B(x,y)$ represents the number of neighbor stores around x,y. To calculate the location quality for an existing store, one removes it from town and calculates Q at its location.

As often in social contexts, it is difficult to test empirically the relevance of our quality index. In principle, one should open several bakeries at different locations and test whether those located at the "best" places (as defined by Q) are on average more successful. Since it may be difficult to fund this kind of experiment, I use location data from two years, 2003 and 2005. It turns out (Fig. 3) that bakeries closed between these two years are located on significantly lower quality sites. Inversely, new bakeries (not present in the 2003 database) do locate preferently on better places than a random choice would dictate. This stresses the importance of location for bakeries, and the relevance of Q to quantify the interest of each possible site. Possibly, the correlation would be less satisfactory for retail activities whose locations are not so critical for commercial success.

6   Conclusions, Perspectives

Practical applications of Q are under development together with Lyon's Chamber of Commerce and Industry. A software called *LoCo* reads the location data of the town(s) under investigation and gives in a few seconds the top quality regions. this can help retailers to find good locations and/or city mayor's in improving commercial opportunities on specific town sectors. In a word, LoCo pumps the cleverness of social actors, inscribed in the "optimal" town configuration, and uses it to help finding good locations. Whether the actual store configuration is optimal or not is an open question. Clearly, no one expects all retailers to be able to choose the "best" location. However, one could argue that those that have selected bad locations perish, leading to a not too bad overall configuration. This analysis suggests a crude analogy with the Darwinian selection process, with variation and selection, which would be interesting to discuss further.

References

1. Hoover, E.M.: Location Theory and the Shoe and Leather Industries. Harvard University Press, Cambridge (1937)
2. Egami, T., Billinge, S.: Underneath the Bragg Peaks: Structural Analysis of Complex Materials. Pergamon Materials Series (2003)
3. Ward, J.S., Parker, G.R., Ferrandino, F.J.: Long-term spatial dynamics in an old-growth deciduous forest. Forest ecology and management 83, 189–202 (1996)
4. Hoover, E.M., Giarratani, F.: An Introduction to Regional Economics (1984)
5. Jensen, P.: Network-based predictions of retail store commercial categories and optimal locations. Physical Review E (Statistical, Nonlinear, and Soft Matter Physics) 74 (2006)
6. Ripley, B.D.: The second-order analysis of stationary point processes. Journal of Applied Probability 13, 255–266 (1976)
7. Besag, J.E.: Comments on ripley's paper. Journal of the Royal Statistical Society B 39, 193–195 (1977)
8. Ellison, G., Glaeser, E.L.: Geographic concentration in us manufacturing industries: A dartboard approach. Journal of Political Economy 105, 889–927 (1997)
9. Maurel, F., Sedillot, B.: A measure of the geographic concentration of french manufacturing industries. Regional Science and Urban Economics 29, 575–604 (1999)
10. Duranton, G., Overman, H.G.: Testing for localisation using micro-geographic data. The Review of Economic Studies 72, 1077 (2005)
11. Marcon, E., Puech, F.: Measures of the geographic concentration of industries: improving distance-based methods (2007)
12. Yule, G.U., Kendall, M.G.: An Introduction to the Theory of Statistics. Griffin, London (1950)
13. Unwin, D.J.: Gis, spatial analysis and spatial statistics. Progress in Human Geography 20, 540–551 (1996)
14. Openshaw, S.: The Modifiable Areal Unit Problem. Geo Books, Norwich (1984)
15. Briant, A., Combes, P.P., Lafourcade, M.: Do the size and shape of spatial units jeopardize economic geography estimations (2007)
16. Barthelemy, M., Barrat, A., Pastor-Satorras, R., Vespignani, A.: Rate equation approach for correlations in growing network models. Physica A 346 (2005)

17. Marcon, E., Puech, F.: Measures of the geographic concentration of industries: Improving distance-based methods (2007)
18. Berry, B.J., Parr, J.B., Epstein, B.J., Ghosh, A., Smith, R.H.: Market Centers and Retail Location: Theory and Application. Prentice-Hall, Englewood Cliffs (1988)
19. Reichardt, J., Bornholdt, S.: Detecting fuzzy communities in complex networks with a potts model. Phys. Rev. Lett. 93 (2004)
20. Radicchi, F., Castellano, C., Cecconi, F., Loreto, V., Parisi, D.: Defining and identifying communities in networks. Publ. Natl. Acad. Sci. USA 101, 2658–2663 (2004)
21. Kirkpatrick, S., Gelatt Jr., C.D., Vecchi, M.P.: Optimization by simulated annealing. Science 220, 671 (1983)
22. Newman, M.E.J., Girvan, M.: Community structure in social and biological networks. Proceedings of the National Academy Science USA 69, 7821–7826 (2004)

Change (Detection) You Can Believe in: Finding Distributional Shifts in Data Streams

Tamraparni Dasu[1], Shankar Krishnan[1], Dongyu Lin[2],
Suresh Venkatasubramanian[3], and Kevin Yi[4]

[1] AT&T Labs - Research
{tamr,krishnas}@research.att.com
[2] University of Pennsylvania
dongyu@wharton.upenn.edu
[3] University of Utah
suresh@cs.utah.edu
[4] Hong Kong University of Science and Technology
yike@cse.ust.hk

Abstract. Data streams are dynamic, with frequent distributional changes. In this paper, we propose a statistical approach to detecting distributional shifts in multi-dimensional data streams. We use relative entropy, also known as the Kullback-Leibler distance, to measure the statistical distance between two distributions. In the context of a multi-dimensional data stream, the distributions are generated by data from two sliding windows. We maintain a sample of the data from the stream inside the windows to build the distributions.

Our algorithm is streaming, nonparametric, and requires no distributional or model assumptions. It employs the statistical theory of hypothesis testing and bootstrapping to determine whether the distributions are statistically different. We provide a full suite of experiments on synthetic data to validate the method and demonstrate its effectiveness on data from real-life applications.

1 Introduction

The ability to discover trends, patterns and changes in the underlying processes that generate data is one of the most critical problems in large-scale data analysis and mining. In this paper, we focus on the problem of finding distributional shifts in multi-dimensional data streams. Data streams pose special challenges since there are constraints on access, storage and duration of access to the data. Streaming algorithms therefore need to be mindful of storage as well as computational speed. Since data streams are constantly changing, the algorithm needs to adapt and reflect these changes in a computationally efficient manner. Distributional shifts can be spurious, caused by glitches in the data, or genuine, reflecting changes in the underlying generative process. We will use the terms *distributional shifts* and *change detection* interchangeably in this paper.

N. Adams et al. (Eds.): IDA 2009, LNCS 5772, pp. 21–34, 2009.
© Springer-Verlag Berlin Heidelberg 2009

To be a viable algorithm, any change detection mechanism has to satisfy a number of criteria. Crucial features are:

- **Generality:** Applications for change detection come from a variety of sources, and the notion of *change* varies from setting to setting. Thus, a general approach to defining change is important.
- **Scalability:** Any approach must be scalable to very large (and streaming) datasets. An important aspect of scalability is dealing with multidimensional data. A change detection scheme must be able to handle multidimensional data directly in order to capture spatial relationships and correlations.
- **Statistical soundness:** By connecting a change detection mechanism to statistically rigorous approaches for significance testing, we quantify the ability to generalize while avoiding the arbitrariness of tuning parameters.

1.1   A Statistical Approach

A natural approach to detecting change in data is to model the data via a distribution. Nonparametric methods *make no distributional assumptions on the data.* Statistical tests that have been used in this setting include the Wilcoxon test, the Kolmogorov-Smirnov test, and their variants. Here, as before, the approach is to compute a *test statistic* (a scalar function of the data), and compare the computed values to determine whether a change has occurred.

The above tests attempt to capture a notion of *distance* between two distributions. A measure that is one of the most general ways of representing this distance is the *relative entropy* from information theory, also known as the *Kullback-Leibler* (or KL) distance. The KL-distance has many properties that make it ideal for estimating the distance between distributions [4, §12.8].

Intuitively, the KL-distance between distributions behaves like squared Euclidean distance in $\mathbb{R}^d$; here the "points" are distributions that lie on the simplex, rather than vectors in $\mathbb{R}^d$. Using the KL-distance allows us not only to measure the distance between distributions, but attribute a meaning to this value. Further, an information-theoretic distance can be defined independent of the inherent dimensionality of the data, and is even independent of the spatial nature of the data, when one invokes the *theory of types.* Thus, we can isolate the definition of change from the data representation, cleanly separating the computational aspects of the problem from the distance estimation itself.

There are advantages to the information-theoretic approach from a computational perspective as well. Tests like the Wilcoxon and Kolmogorov-Smirnov cannot be easily extended to data in more than a single dimension. This is principally because these tests rely on data being ordered (they are rank-based statistics), and thus in two or more dimensions, the lack of a unique ordering renders them ineffective.

2   Related Work

The Kullback-Leibler Distance. The Kullback-Leibler distance is one of the most fundamental measures in information theory. It also has a natural interpretation as a distance function between distributions, being a special case of an *Ali-Silvey* and *Bregman* distance. In both cases it emerges as the *unique* measure satisfying certain axioms over the unit simplex. Due to its relation to the log likelihood ratio and Neyman-Pearson classifiers, it has been used extensively in classification and model selection problems in machine learning [17,18]. Most recently, Johnson and Gruner [12] have suggested the idea of using the KL-distance to decode neural signals presented as time series. They also develop the idea of using bootstrap methods to evaluate the significance of their results.

Change Detection Schemes. A variety of change detection schemes have been studied in the past, examining static datasets with specific structure [3], time series data [13], and for detecting "burstiness" in data [15]. The definition of change has typically involved fitting a model to the data and determining when the test data deviates from the built model [11,9].

Ganti, Gehrke, Ramakrishnan and Loh [9] use a family of decision tree models to model the data, and define change in terms of the distance between model parameters that encode both topological and quantitative characteristics of the decision trees. They use bootstrapping to determine the statistical significance of their results.

The paper by Kifer, Ben-David and Gehrke [14] lays out a comprehensive non-parametric framework for change detection in streams. They exploit order statistics of the data, and define generalizations of the Wilcoxon and Kolmogorov-Smirnoff test in order to define distance between two distributions. Aggarwal [1] considers the change detection problem in higher dimensions based on kernel methods; however, his focus is on detecting the "trends" of the data movement, and has a much higher computational cost. Subramaniam *et al.* [21] propose a distributed, streaming, outlier and change detection algorithm using kernel methods in the context of sensor networks. Given a baseline data set and a set of newly observed data, Song, Wu, Jermaine and Ranka [20] define a test statistic called the *density test* based on kernel estimation to decide if the observed data is sampled from the baseline distribution.

In statistical literature, *Referential Distance* (RD) methods are used primarily for outlier detection. RD methods aim to detect outliers by computing a measure of how far a particular point is from the center of the data. The measure of *outlyingness* of a data point is based on the Mahalanobis distance. In the context of change detection, Referential Distance methods can be viewed as a dimensionality reduction approach. These methods are effective in very high dimensions, where, due to data sparsity, space partitioning techniques are known to be quite sensitive.

Computing Bootstraps. The bootstrap method was developed by Efron, and the book by Efron and Tibshirani [7] is a definitive reference on bootstrap techniques. KL-distance estimates have bias, and there has been some recent work [19] (for parametric families of distributions) on improving the bootstrap estimates of the KL-distance and removing bias.

Sampling from a Data Stream. Given a data stream of potentially unbounded size and a parameter k, the *reservoir sampling* algorithm of Vitter [22] maintains a uniform sample of size k from the stream seen so far. This algorithm works when data is only inserted into the stream. In streaming applications where the data expires after a certain age, it is nontrivial to extend the reservoir sampling algorithm. Babcock *et al.* [2] introduce the notion of *chain sampling* to solve the problem over a *sliding window*.

2.1   Our Contributions

In this paper, we present a general streaming algorithm for detecting distributional changes in multi-dimensional data streams within a statistical hypothesis testing framework. This work is an extension of an extended abstract [6] that described the basic framework of change detection using an information-theoretic approach. We extend it in several directions, some of which include:

- Our approach is tailored for streaming computation. Every step of our algorithm is amenable to streaming computation, and is space and time efficient. It automatically adapts to changes, captures global and local shifts.
- Our use of the Kullback-Leibler (KL) distance as well as the bootstrap computations are nonparametric, free of distributional and model assumptions, making our method general and widely applicable.
- We present a novel space partitioning scheme called the *kdq-tree* that we use to construct multi-dimensional histograms. Its size grows linearly with the dimensionality and the size of the data, at the same time generates cells with good aspect ratio. Both of these properties are important for accurate histograms.
- We overcome the issue of data sparsity in very high dimensions by transforming the original data to a lower dimensional space using referential distances and applying the KL change detection to the transformed data.
- We provide a strong statistical basis for deciding distributional shifts in the data stream using confidence intervals for sample proportions.
- Our method works naturally in a multivariate setting, unlike methods that rely on order statistics [14], and has been tested on a wide range of data streams including *i.i.d.*, time series data with seasonality and trends, and other types of distributional shifts.

3   Basic Overview of Our Approach

Let $x_1, x_2, \ldots$ be a stream of points in $\mathbb{R}^d$. A *window* $W_{i,n}$ denotes the sequence of points ending at x_i of size n: $W_{i,n} = (x_{i-n+1}, \ldots, x_i), i \geq n$. We will drop the subscript n when the context is clear. Distances are measured between distributions constructed from points in two windows W_t and $W_{t'}$. The choice of window size is not very crucial. Typically, we choose several windows whose size increases exponentially and we can run our algorithm on each of them independently. In the rest of this paper, we will describe our method for a fixed window size n.

Each window W_t defines an empirical distribution F_t. We maintain a sample of size k from W_t in a streaming fashion using the chain sampling algorithm [2] (see section 4.1). We compute the distance $d_t = d(F_t, F_{t'})$ from F_t to $F_{t'}$ using the samples, where t' is either $t - n$ (*adjacent window* model) or n (*fix-slide window* model) depending on the sliding window model we are using. Section 4.4 describes the computation of d_t in more detail. Formally, we assert the null hypothesis

$$H_0 : F_t = F_{t'}$$

and wish to determine the probability of observing the value d_t if H_0 is true.

To determine this, we generate a set of B *bootstrap estimates* $\hat{d}_i, i = 1 \ldots B$. These estimates form an empirical distribution from which we construct a critical region $[d_{H_0}(\alpha), \infty]$, where α denotes the desired *significance level* of the test. If d_t falls into this region, we consider that H_0 is invalidated. Since we test H_0 at every time step, in order to improve robustness, we only signal a change after we have accumulated a statistically significant run of invalidations. A simple, but naive, approach is to signal a change after we see γn distances larger than $d_{H_0}(\alpha)$ in a row, where γ is a small constant defined by the user. The idea is that a true change should be more persistent than a false alarm, which might be transient, and we term γ the *persistence factor*.

We can, however, use the theory of confidence intervals for sample proportions to determine whether the change is persistent. Under the null hypothesis, we expect an α proportion of distances to fall in the critical region. Due to sampling variability this proportion will change but stay statistically close to α. The statistical closeness is measured by a *confidence interval* containing the sample proportion of distances that fall in the critical region. Let $\varphi(w)$ be the observed proportion of distances that fall in the critical region $[d_0(\alpha), \infty]$ in the interval of length w. Then the lower bound, $\hat{\alpha}_{lo}$, is given by [8] $\hat{\alpha}_{lo} = (\varphi(w) + z_{1-\alpha/2}^2/2w - z_{1-\alpha/2}\sqrt{f(\varphi, w)})/(1 + z_{1-\alpha/2}^2/w)$ where $f(\varphi, w) = \varphi(w)(1 - \varphi(w))/w + z_{1-\alpha/2}^2/4w^2$ and z_q represents the q-quantile of the standard normal distribution with mean 0 and variance 1. We use the lower bound $\hat{\alpha}_{lo}$ of this confidence interval to determine a change (the upper bound is not relevant). If α is smaller than $\hat{\alpha}_{lo}$, we declare a statistically significant distributional shift. This approach removes the strict requirement of consecutive runs of null hypothesis invalidations to signal a change. If no change has been reported, we update the windows and repeat the procedure. The high-level algorithm is summarized in Algorithm 1.

Algorithm 1. Change detection algorithm (for a fixed window size)

$t \leftarrow 2n;\ t' \leftarrow n;$
Construct *chain sample* of size k from windows W_t and $W_{t'}$;
Compute $d_t = d(F_t, F_{t'})$;
Compute bootstrap estimate $\hat{d}_i, i = 1, \ldots, B$ and critical region $[d_{H_0}(\alpha), \infty]$;
$\varphi \leftarrow 0;$
while not at end of stream **do**
  if $d_t > d_{H_0}(\alpha)$ **then**
    $\varphi \leftarrow \varphi + 1;\ w \leftarrow w + 1;$
    Compute $\hat{\alpha}_{lo};$
    if $\alpha < \hat{\alpha}_{lo}$ **then**
      Signal **change**;
      $\varphi \leftarrow 0;\ w \leftarrow 0;$
      Start over;
    end if
  else
    $w \leftarrow w + 1;$
  end if
  Slide window W_t (and $W_{t'}$ if required);
  Update $d_t;$
end while

---

4 Details of the Algorithm

We start by defining the Kullback-Leibler distance, also called the *relative entropy* [4, Sec 2.3], which is used to compare distributions.

Definition 1. *The* relative entropy *or* Kullback-Leibler distance *between two probability mass functions* $p(x)$ *and* $q(x)$ *is defined as*[1]

$$D(p\|q) = \sum_{x \in \mathcal{X}} p(x) \log \frac{p(x)}{q(x)},$$

where the sum is taken (in the discrete setting) over the atoms of the space of events $\mathcal{X}$.

4.1 Constructing a Distribution from a Stream

To map our data streams to distributions, we use a spatial partitioning scheme (discussed in Section 4.3) to assigns points to cells. For a window W of size n and a sampling parameter k, we perform chain sampling [2] to maintain a sample of size k in W. When a data item x_i arrives at time i, it is chosen to be in the sample with probability $1/\min(i, n)$ (the original paper has an error in this expression). If the item is chosen, the algorithm also selects the timestamp of the element that will replace this item when it expires. This is done by choosing

---

[1] All logarithms are base 2 in this paper.

a random index j between $i + 1, i + 2, \ldots, i + n$. When the item x_j with the chosen timestamp arrives, it is stored, and the timestamp of the item that will replace x_j is selected. This builds a *chain* of items to use when the current sample expires. The above algorithm maintains a sample of size one inside the window. To produce a sample of size k, we perform k independent trials of the same algorithm. We have developed a template-based implementation of the chain sampling algorithm in C++ which makes it portable in a wide class of data streams.

Let $P_W(a) = \frac{N(a|W)}{k}$, where $N(a \mid W)$ is the number of points that fall into cell a. Then P_W forms an empirical distribution that the sequence of points maps to. To address the issue of zero counts in cells, we use a simple correction suggested by Krichevsky and Trofimov [16]. They replace the estimate $P_{\mathbf{w}}(a)$ by the estimate $P_{\mathbf{w}}(a) = \frac{N(a|\mathbf{w})+0.5}{k+|\mathcal{A}|/2}$, where $\mathcal{A}$ is the set of cells in space.

Given two windows W_1, W_2, the distance from W_1 to W_2 is

$$D(W_1 \| W_2) = \sum_{a \in \mathcal{A}} P_{\mathbf{W_1}}(a) \frac{P_{\mathbf{W_1}}(a)}{P_{\mathbf{W_2}}(a)}.$$

The justification for this approach comes from the *theory of types*, due to Csiszár and Körner [5]. It has been shown by Gutman [10] that empirical distributions constructed as above retain (asymptotically) the same classifier properties as true distributions; moreover, the relative ratios that we construct are the maximum likelihood estimators for the true distribution of the data.

4.2   Bootstrap Methods and Hypothesis Testing

The bootstrapping procedure works as follows: given the empirical distributions $\hat{P}$ derived from the counts P (see Section 4.1), we *sample* m sets $S_1, \ldots S_m$, each of size $2n$, with replacement. Treating the first n elements S_{i1} as coming from one distribution F, and the remaining n elements $S_{i2} = S_i - S_{i1}$ as coming from the other distribution G, we compute bootstrap estimates $\hat{d}_i = D(S_{i1} \| S_{i2})$.

Once we fix the desired significance level α, we choose the $(1 - \alpha)$-percentile of these bootstrap estimates as $d_{H_0}(\alpha)$. We call $[d_{H_0}(\alpha), \infty]$ the critical region, if $\hat{d} > d_{H_0}(\alpha)$, the measurement is statistically significant and invalidates H_0. Bootstrap procedures are a form of Monte Carlo sampling over an unknown distribution. Asymptotically, the bootstrap sample distribution approaches the true underlying distribution.

4.3   Data Structures

From now on, we will assume that the data points in the streams lie in a d-dimensional unit hypercube. The structure that we propose, called a *kdq-tree*, is a combination of two space partitioning schemes, the *k-d*-tree and the quad tree, and has the advantages of both structures.

We will describe the structure in two dimensions; generalization to high dimensions will be obvious. A *kdq*-tree is a binary tree, each of whose nodes is

Table 1. Default parameters values

Parameter	Symbol	Value
Min. cell side	δ	2^{-10}
Max. points in a cell	τ	100
Chain sample size	k	$0.4n$
Significance Level	α	1%
# bootstrap samples	B	800

Table 2. Running times with varying n and d

d	n	Construction (sec)	Update (msec)
4	10000	3.87	0.011
6	10000	4.61	0.019
8	10000	4.68	0.023
10	10000	5.23	0.027
10	20000	11.31	0.029
10	30000	19.83	0.029
10	40000	26.45	0.027

Each example is accompanied by a figure, (for example Figure 2) where the first panel is a *spatial representation* using a two-dimensional projection of the multidimensional data stream. Each period between change points is shown in different colors: blue red and green. The second panel shows a *temporal representation* of the stream by plotting a single arbitrarily chosen dimension against time. Changes detected by the KL and RD methods are shown as elongated tick marks on the X-axis, with red indicating changes detected by KL and black tick marks denoting changes detected by RD. Tick marks of different lengths correspond to changes reported at different significance levels α. Typically, every sharp distributional changes results in two changes being reported. The first change occurs when the windows start encountering the new distribution. As the new distribution fills up the windows, the mix of the old and new changes and when the new distribution is sufficiently dominant, another change is detected. If the change is sustained or gradual, changes will be reported continuously until the stream stabilizes.

Gradual change in 3-D data stream. The first experiment involved a 3-dimensional data stream where each component is a mixture of Gaussian and

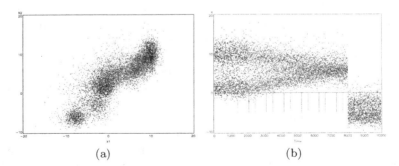

(a)   (b)

Fig. 1. (a) 2-D projection of a 3-D data stream exhibiting gradual change. The two initial clusters (blue) merge gradually over time (red) and separate suddenly (green). (b) Time series plot of one of the three variables to illustrate the stream behavior over time. Different colors (blue, green, red) are used to denote the stream between different change points. Both the methods report changes during the gradual transition of the stream (red).

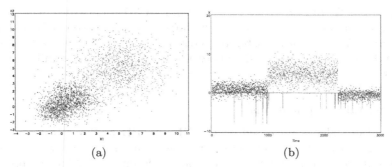

(a)   (b)

Fig. 2. (a) 2-D projection of a 100-D data stream which has level shifts and dispersion changes. (b) Time series plot of one of the variables. Tick marks of varying lengths indicate the different significance levels α=0.01, 0.05, 0.10. The KL method generates numerous alarms (red tick marks) while the RD method (black tick marks) generates alarms only near the change points.

Uniform distributions. The spatial representation is shown in Figure 1 (a) in blue, red and green. The temporal representation in Figure 1(b) shows that the data stream starts with two clusters (blue) of equal mass that start moving together at t=2000 and continue to get closer (red) until t=8000. At t=8000, they shift abruptly, as shown in green in Figure 1(a) and (b). The period of gradual change is punctuated by alarms denoted by long tick marks (KL=red, RD=black, α=0.01), which stop once the distribution becomes stable.

High dimensional (100-D) stream: level shift and dispersion change. The second experiment consisted of a 100-dimensional stream where each component was a mixture of standard distributions. The spatial representation in Figure 2(a) depicts a 2-D projection of the 100-D data stream. The stream has two change points, one at t=1000 and another at t=2250. The mean and the dispersion of the cluster change at each of the change points. We tested for change at three different levels of significance (α=0.01, 0.05, 0.01). KL generates numerous alarms even when the stream is stable indicating that KL is sensitive in higher dimensions. The RD method detects changes only near the change points.

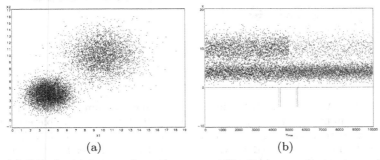

(a)   (b)

Fig. 3. (a) 5-D data stream where the mass shifts from one cluster to another. (b) Both the KL and RD method detect the change in density.

Shift in mass from one cluster to another in 5-D. The third experiment is based on a 5-dimensional stream, which starts with two clusters of equal mass. The mass shifts from one cluster to another at the change point $t=5000$, as shown in Figure 3. Both KL and RD are able to detect the mass shift, shown in Figure 3(b).

5.1   Real Life Applications

We applied our algorithm based on the KL and RD methods to two real life applications, each of which generates a multi-dimensional data stream. The applications are of critical importance to a large telecommunications corporation.

File Descriptor Streams. The calls made on a telecommunications network are logged and written to files in a highly specialized format. The files are gathered and processed for billing and network performance measurement purposes. A file monitoring process tracks the file stream by gathering at short intervals of times, file descriptors such as number of files received, their sizes and other characteristics. For the purposes of this paper, we track the sizes of three important file types across 3058 instances.

In Figure 4(b), the three file descriptor variables are shown as different colored dots. They are plotted on a staggered axis for better visual effect. The file descriptor data stream is well behaved as expected, since it is generated by a well designed, well established data gathering process, except for occasional hiccups caused by maintenance. Both the KL and RD methods detect the single change that was caused by a temporary software glitch shown by the red and black tick marks on the X-axis. The processing center was not aware of these distributional changes until we alerted them to it. They tracked down the glitch and improved their work flow, preventing significant revenue loss caused by unlogged calls.

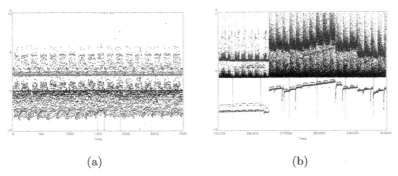

(a)   (b)

Fig. 4. Real Life Applications: (a) A file descriptor data stream generated during the processing of hundreds of thousands of files a day. The stream is stable except for a software glitch that caused a small change, detected by both KL and RD. (b) A server usage data stream characterized by volatility and change. Note: For the purpose of presentation in this plot, we shifted the means of the three variables in both plots (a) and (b).

Server Usage Streams. The second application concerns monitoring the performance and resource usage of a cluster of servers that supports an important e-commerce application. The application is critical enough that the servers need to be up all the time. For the purpose of this paper, we used three variables measured at 317,980 time points from this real-time data stream.

In Figure 4(a) we show a small section ($t = 250000$ to $t = 300000$) of the server usage data stream. It is extremely volatile and as a result, both the KL and RD methods detect changes (overlapping red and black tick marks) at regular intervals that correspond to the window size used in the change detection method. When a data stream is as volatile and dynamic as the server usage stream, change is constant. In such situations, engineers often focus on high-level aggregates like univariate moving average methods and threshold based techniques developed through consultation with subject matter experts.

6   Conclusion and Discussion

We have presented an efficient, nonparametric, fully streaming algorithm for detecting distributional changes in multi-dimensional data streams within a statistically rigorous hypothesis testing framework. We describe a novel data structure, the kdq-tree, to maintain general purpose multi-dimensional histograms that offer a granular summarization of the data stream. We draw upon the Kullback-Leibler distance to measure the distance between data stream distributions, either using the original data or its *referential distance*. We have tested it exhaustively and found it to be effective in both synthetic and real-life applications.

We are investigating potential extensions to this work. One aspect of our focus is incorporating dimensionality reduction techniques like LSH (Locality Sensitive Hashing) or MDS (Multi-Dimensional Scaling) to deal with high dimensional data sets. We have preliminary results using LSH that are encouraging.

References

1. Aggarwal, C.C.: A framework for diagnosing changes in evolving data streams. In: Proceedings of the ACM SIGMOD International Conference on Management of Data, pp. 575–586 (2003)
2. Babcock, B., Datar, M., Motwani, R.: Sampling from a moving window over streaming data. In: SODA, pp. 633–634 (2002)
3. Chawathe, S.S., Abiteboul, S., Widom, J.: Representing and querying changes in semistructured data. In: ICDE 1998, pp. 4–13 (1998)
4. Cover, T.M., Thomas, J.A.: Elements of Information Theory. John Wiley and Sons, Inc., Chichester (1991)
5. Csiszár, I., Körner, J.: Information Theory: Coding Theorems for Discrete Memoryless Systems. Academic Press, London (1981)
6. Dasu, T., Krishnan, S., Venkatasubramanian, S., Yi, K.: An information-theoretic approach to detecting changes in multi-dimensional data streams. In: Interface 2006 (2006)

7. Efron, B., Tibshirani, R.J.: An Introduction to the Bootstrap. Chapman and Hall, Boca Raton (1993)
8. Fleiss, J.L., Levin, B., Paik, M.: Statistical Methods for Rates and Proportions, 3rd edn. John Wiley and Sons, New York (2003)
9. Ganti, V., Gehrke, J., Ramakrishnan, R., Loh, W.-Y.: A framework for measuring differences in data characteristics. pp. 126–137 (1999)
10. Gutman, M.: Asymptotically optimal classification for multiple tests with empirically observed statistics. IEEE Trans. Inf. Theory 35, 401–408 (1989)
11. Hulten, G., Spencer, L., Domingos, P.: Mining time-changing data streams. In: KDD, pp. 97–106 (2001)
12. Johnson, D., Gruner, C.: Information-theoretic analysis of neural coding. Journal of Computational Neuroscience 10, 47–69 (2001)
13. Keogh, E., Lonardi, S., Chiu, B.Y.: Finding surprising patterns in a time series database in linear time and space. In: KDD, pp. 550–556 (2002)
14. Kifer, D., Ben-David, S., Gehrke, J.: Detecting changes in data streams. In: Proceedings of the 30th International Conference on Very Large Databases, pp. 180–191 (2004)
15. Kleinberg, J.: Bursty and hierarchical structure in streams. Data Mining and Knowledge Discovery 7(4), 373–397 (2003)
16. Krichevsky, R.E., Trofimov, V.K.: The performance of universal encoding. IEEE Trans. Inf. Theory 27, 199–207 (1981)
17. Pereira, F., Tishby, N., Lee, L.: Distributional clustering of English words. In: 31st Annual Meeting of the ACL, pp. 183–190 (1993)
18. Pietra, S.D., Pietra, V.D., Lafferty, J.: Inducing features of random fields. IEEE Trans. Pattern Analysis and Machine Intelligence 19, 380–393 (1995)
19. Shibata, R.: Bootstrap estimate of Kullback-Liebler information for model selection. Statistica Sinica 7, 375–394 (1997)
20. Song, X., Wu, M., Jermaine, C., Ranka, S.: Statistical change detection for multi-dimensional data. In: ACM SIGKDD 2007, pp. 667–676 (2007)
21. Subramaniam, S., Palpanas, T., Papadopoulos, D., Kalogeraki, V., Gunopulos, D.: Online outlier detection in sensor data using non-parametric models. In: VLDB 2006, pp. 187–198 (2006)
22. Vitter, J.S.: Random sampling with a reservoir. ACM Transactions on Mathematical Software 11, 37–57 (1985)

Exploiting Data Missingness in Bayesian Network Modeling

Sérgio Rodrigues de Morais[1] and Alex Aussem[2]

[1] University of Lyon, LIESP, INSA-Lyon, 69622 Villeurbanne, France
`sergio.rodrigues-de-morais@insa-lyon.fr`
[2] University of Lyon, LIESP, UCBL, 69622 Villeurbanne, France
`aussem@univ-lyon1.fr`

Abstract. This paper proposes a framework built on the use of Bayesian networks (BN) for representing statistical dependencies between the existing random variables and additional dummy boolean variables, which represent the presence/absence of the respective random variable value. We show how augmenting the BN with these additional variables helps pinpoint the mechanism through which missing data contributes to the classification task. The missing data mechanism is thus explicitly taken into account to predict the class variable using the data at hand. Extensive experiments on synthetic and real-world incomplete data sets reveals that the *missingness* information improves classification accuracy.

Keywords: Bayesian networks, missing data mechanism, pattern recognition, classification.

1   Introduction

The methods for coping with missing values can be grouped into three main categories [1]: inference restricted to complete data, imputation-based approaches, and likelihood-based approaches. Unfortunately, these methods are based on the assumption that the mechanism of missing data is ignorable. Under this assumption, the missing values can be inferred from the available data. This assumption is hard to test in practice (Statistical tests have been proposed, but these are restricted to a certain class of problems) and the decrease in accuracy may be severe when the assumption is violated. Some mechanisms leading to missing data actually possess information and the missingness of some variables can be a predictive information about other variables. Since the missingness mechanism contains information independent of the observed values, it requires an approach that can explicitly model the absence of data elements.

Encountering a situation where a portion of the missing data is inaccessible should not discourage the researcher from applying a statistically principled method. Rather, the attitude should be to account for as much of the mechanism as possible, knowing that these results will likely be better than those produced by naive methods. Moreover, the missing data mechanism is rarely completely inaccessible. Often, the mechanism is actually made up of both accessible and

N. Adams et al. (Eds.): IDA 2009, LNCS 5772, pp. 35–46, 2009.
© Springer-Verlag Berlin Heidelberg 2009

inaccessible factors. Thus, although a researcher may not be confident that the data present a purely accessible mechanism, covering as much of the mechanism as possible should be regarded as beneficial rather than detrimental.

In this study, we experiment a new graphical method of treating missing values, based on Bayesian networks (BN). We describe a novel approach that uses explicitly the information represented by the absence of data to help detect the missing mechanism and reduce the classification error. We create an additional dummy boolean variable to represent missingness for each existing variable that was found to be absent (missingness indicator approach). The advantages of using BNs include the following: (1) they can be used to predict a target variable in the face of uncertainty. (2) BNs can provide a valid output when any subset of the modeled variables is present. (3) The graphical structure of the BN representing the joint probability distribution of the variables can be used to help identify the missingness mechanism. Imputation and classification are handled the same way. To perform imputation, we treat each attribute that contains missing values as the class attribute, then fill each missing value for the selected class attribute with the class predicted from the model. The model include original random variable and artificially created variables for representing missingness. Our approach is based on the identification of relevant subsets of variables that jointly prove useful to construct an efficient classifier from data. We solve this feature subset selection (FSS) problem using Markov boundary (MB for short) learning techniques. A Markov boundary of a variable T is any minimal subset of $\mathbf{U}$ (the full set of variables) that renders the rest of $\mathbf{U}$ independent of T. Our idea is to train a classifier with these relevant variables as input to impute the missing entries of T. Once the missing data are imputed, visual inspection of the induced graph reveals useful information on the the missing data mechanism. Several experiments on synthetic and real-world incomplete data sets will be conducted to illustrate the usefulness of this approach.

2 Background

2.1 Deletion Process

According to [2], the assumptions about the missing data mechanisms may be classified into three categories: 1) missing completely at random (MCAR): the probability that an entry is missing is independent of both observed and unobserved values in the data set; 2) missing at random (MAR): the probability that an entry is missing is a function of the observed values in the data set; 3) informatively missing (IM) or Non-MAR (NMAR): the probability that an entry is missing depends on both observed and unobserved values in the data set. In order to specify the deletion processes, a dummy binary variable R_i may be associated with each random variable X_i. When R_i takes value '1', the entry $X_i = x_i$ is not observed and vice-versa. When the probability distribution of each R_i is independent of $X_1, \ldots, X_n$, the data may be seen as MCAR. When this probability distribution is a function of the observed values in the data set, data are MAR. Now, when this probability distribution is a function of the observed and unobserved entries, data are IM. For instance, when machine learning algorithms are applied to data collected during the course of clinical care, the absence of expected data elements is common and the mechanism through which a data element is missing often involves the clinical relevance of that data element in a specific patient [3,4]. Hence the need for methods that help to detect the censoring mechanism. Notice however that no method can tell for sure, under all scenarios, from the data alone whether the missing observations are IM (although it is possible to distinguish between MCAR and MAR).

2.2   Related Work

Recent studies have investigated the impact of imputation with Machine learning (ML) methods on the accuracy of the subsequently performed classification [5,6,4,7]. In contrast to statistical methods, ML algorithms generate a model from data that contain missing values, and then use the model to perform classification that imputes the missing values. These methods do not concentrate solely on identifying a replacement for a missing value, but on using available information to preserve relationships in the entire dataset. These studies have empirically been shown to perform better than ad-hoc methods. However they are not meant to detect the missing mechanism nor to use such mechanisms to improve accuracy of prediction. Indeed, unless auxiliary information is available, modeling an IM mechanism is usually not possible because the missing data mechanism in this case depends on the missing data themselves. Nonetheless, Jamshidian et al. [8] propose a simple postmodeling sensitivity analysis to distinguish between the missing data mechanisms of MCAR and IM. Their method is built upon the premise that if data are MCAR, the maximum likelihood estimates obtained based on various randomly selected subsets of the data, of the same size, would have the same asymptotic distribution, whereas this would not hold for data that are IM. Therefore, if a significant disagreement is observed between the two distributions, then this constitutes grounds to believe that the data are not MCAR and are possibly IM.

Recently, [3] experimented with a method of treating missing values in a clinical data set by explicitly modeling the absence of data. They showed that in most cases a Naive Bayesian network trained using the explicit missing value treatments performed better. However there method is unable to pinpoint explicitly the missing mechanism and their experiments focus on small clinical datasets and thus the results may not generalize to other settings. Note also that several approaches have been designed with a view to be 'robust' to the missing data mechanism [9,10]. No assumption about the unknown censoring mechanism is made, hence the "robustness". However, the utility of these methods is questionable when the percentage of missing data is high.

Our approach is different here in that we try to model the missing data mechanism explicitly and take it into account to impute the missing data using the data at hand. Moreover, we use highly scalable FSS methods based on recent Markov boundary learning algorithms to infer only the relevant variables because only a subset of the variables provided with explicit representations for absent values should contribute information. Once again, we do not claim that our method is able to detect the IM mechanism from data alone under all scenarios but its aim is to raise a flag if data are "possibly" IM, as done in [8].

2.3 Bayesian Networks

For the paper to be accessible to those outside the domain, we recall briefly the principles of Bayesian networks. Formally, a BN is a tuple $< \mathcal{G}, P >$, where $\mathcal{G} =< \mathcal{V}, \mathcal{E} >$ is a directed acyclic graph (DAG) with nodes representing the random variables $\mathcal{V}$ and P a joint probability distribution on $\mathcal{V}$. A BN structure $\mathcal{G}$ entails a set of conditional independence assumptions. They can all be identified by the *d-separation criterion* [11]. We use $X \perp_{\mathcal{G}} Y | \mathbf{Z}$ (resp. $X \perp_P Y | \mathbf{Z}$) to denote the assertion that X is d-separated from Y given $\mathbf{Z}$ in $\mathcal{G}$ (resp. in P). If $< \mathcal{G}, P >$ is a BN, $X \perp_P Y | \mathbf{Z}$ if $X \perp_{\mathcal{G}} Y | \mathbf{Z}$. The converse does not necessarily hold. We say that $< \mathcal{G}, P >$ satisfies the *faithfulness condition* if the d-separations in $\mathcal{G}$ identify *all and only* the conditional independencies in P, i.e., $X \perp_P Y | \mathbf{Z}$ iff $X \perp_{\mathcal{G}} Y | \mathbf{Z}$. A Markov blanket $\mathbf{M}_T$ of a variable T is any set of variables such that T is conditionally independent of all the remaining variables given $\mathbf{M}_T$.

Definition 1. $\mathbf{M}_T$ *is a Markov blanket of the* T *iff for all* $X \notin \mathbf{M}_T \cup \{T\}$, $X \perp_P T | \mathbf{M}_T$.

A Markov boundary, denoted by $\mathbf{MB}_T$, of T is any Markov blanket such that none of its proper subsets is a Markov blanket of T.

Theorem 1. *Suppose* $< \mathcal{G}, P >$ *satisfies the faithfulness condition. Then for all* T, *the set of parents, children of* T, *and parents of children of* T *is the unique Markov boundary of* T.

More thorough discussion Bayesian networks and Markov boundaries can be found in [12] for instance. From Theorem 1, a principled solution to the feature subset selection problem is to determine $\mathbf{MB}_T$. $\mathbf{MB}_T$ contains all and only the relevant variables that jointly prove useful to construct an efficient classifier from data.

2.4 Markov Boundary Learning Algorithms

In recent years, there have been a growing interest in inducing the MB automatically from data. Very powerful correct, scalable and data-efficient constraint-based (CB) algorithms have been proposed recently [13,14,15,16]. They search the MB of a variable without having to construct the whole BN first. Hence their ability to scale up to thousands of variables. CB methods systematically check the data for independence relationships to infer the structure. Typically, the algorithms run a χ^2 independence test in order to decide upon the acceptance or rejection of the null hypothesis of conditional independence (i.e., $X \perp_P Y | \mathbf{Z}$).

When no entry is missing in the database, the MB can be estimated efficiently with these methods. Unfortunately, when some entries are reported as unknown, the simplicity and efficiency of these methods are lost. The EM algorithm [2], the Markov chain Monte Carlo (MCMC) methods [17] including Gibbs sampling [18] are popular solutions to handle incomplete data sets with Bayesian networks, but all the above methods are based on the assumption that the mechanism of missing data is ignorable. Notice that all those methods do not make use of explicit representations of the missingness mechanisms with the aim of improving accuracy, that is, they are only based on observed values of the original random variables.

3   The Imputation Model

The problem of finding relevant predictive features is achieved in the context of determining the Markov boundary of the class variable that we want to predict. However, as some of these variables may have incomplete records as well, the idea is to induce a broader set of features that would not be strictly relevant for classification purposes, if the data set was complete, but that are still associated to the target. Therefore, the MB learning algorithm is called several times recursively to construct a local BN around the target. We call *MBLearning* the generic procedure applied for seeking the Markov boundary of a target from a data set. This procedure can be replaced by any of the current state-of-the-art Markov boundary searching algorithms, such as [13,14,15,16]. The local graph provides a broader picture of the features that carry some information about the target variable. If the data set was complete, these additional variables would deteriorate classification accuracy due to increasing design cost. This is not the case here as the variables in the true Markov boundary may be missing. A second important characteristic of the method presented in this section is that the scope of the search process is augmented by the addition of artificially created variables that explicitly represent missingness.

The iterative algorithm is called *Growing Markov Boundary* (*GMB* for short). *GMB* receives four parameters: the target variable (T), the maximal number of iterations (r), the maximal acceptable ratio for missing values (α) and the data set ($\mathcal{D}$). *GMB* proceeds as follows: first the scope of variables (**U**) is created in line 1 of the algorithm. **U** is composed by all the original random variables (X_i) and artificially created variables (R_i) representing missingness of their respective random variables. *MBLearning* is then first run on the target variable T (line 2). It is then run again repeatedly on the adjacent nodes and so on up to a radius of r around the target node (lines 5-13). A similar approach was proposed in [19], but it does not take into account missingness as a possible piece of information. After finishing the feature subset selection process, *GMB* creates at line 14 the local BN including the selection of existing and the dummy variables. The user-defined radius of the Bayesian network constructed by *GMB* (r) trades off accuracy and scalability. *MissRatio(X)* is the missing rate of X.

It is important to note that *MBLearning* builds the MB in the presence of missing data. As discussed above, the structural learning can be performed with EM or MCMC techniques. In this study, we adopt the simple 'available case analysis' (ACA), i.e., the χ^2 for the test $X \perp_P Y | \mathbf{Z}$ is calculated on the cases having

---

Algorithm 1. *GMB*

Require: T: target variable; r: maximal number of iterations for FSS; α: maximal acceptable ratio for missing values; $\mathcal{D}$: data set.
Ensure: BN: Bayesian network.

1: $\mathbf{U} = (X_{i..n} \cup R_{i..n})$
2: **Set1** $\leftarrow$ *MBLearning*$(T, \mathbf{U})$
3: $\mathbf{V} \leftarrow$ **Set1** $\cup T$
4: $I \leftarrow 1$
5: **while** $I < r$ **do**
6:     **Set2** $\leftarrow \emptyset$
7:     **for all** $(X_i \in$ **Set1**, such that *MissRatio*$(X_i) \geq \alpha)$ **do**
8:        **Set2** $\leftarrow$ **Set2** $\cup$ *MBLearning*$(X_i, \mathbf{U})$
9:     **end for**
10:    **Set1** $\leftarrow$ **Set2**
11:    $\mathbf{V} \leftarrow \mathbf{V} \cup$ **Set2**
12:    $I \leftarrow I + 1$
13: **end while**

14: **BN** $\leftarrow$ Build Bayesian Network for $\mathbf{V}$

---

non missing values for X, Y and $\forall j, Z_j \in \mathbf{Z}$. Of course, this will bias results if the remaining cases are not representative of the entire sample. Note however that ACA is not a valid procedure, even for MAR data, but its implementation is straightforward. Still, good performances are obtained in practice as we will see next. The MBOR algorithm [14] is used to implement *MBLearning*.

4 Experiments

We report now the results of our experiments on synthetic and real-world data. The focus of the experiments was to determine whether missingness contains information value or not. Our specific aims are : (1) to provide a graphical representation of the statistical relationships between the existing variables and the dummy missingness variables (2) to help identify the mechanism of the missing data, (3) to gain insight on the applicability and limitations of BN methods for handling missing data, and (4) to assess the merits of including missing data for imputation and classification problems. The improvement in performance was used as a measure of the information value provided by missingness. In the analysis, we compared the performance of prediction models using data sets with and without explicit representations of the absence of data elements.

4.1 Synthetic Data

Whittaker reports a data set [20] that involves six boolean risk factors $X_1, \ldots, X_6$ observed in a sample of 1841 employees of a Czech car factory. Ramoni and Sebastiani [9] considered these data and used a structure learning algorithm to output a structure that they used afterwards as a toy problem to learn the conditional probability tables from incomplete data sets. In these experiments, we use the same toy problem to illustrate our method. We assess how the use of explicit representation of missing data affects classification across a range of different amounts of missing values, sample size and missing data mechanisms. X_6 is our target variable. We caused about 5%, 15% and 25% of the values to be missing according to MCAR, MAR and IM mechanisms by modifying the probability tables of the toy BN reported in [9]. Fig. 1 illustrates each of the three missing mechanisms. The original BN consists of the 6 random variables $X_1, \ldots, X_6$ in plain line. Three sample sizes (i.e., 500, 1500 and 5000) are considered in the experiments.

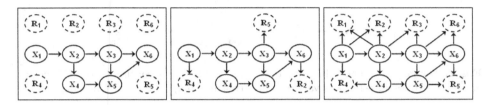

Fig. 1. Graphical representation of the MCAR, MAR and IM missing data mechanism on the BN reported by Ramoni and Sebastiani [9]. R_i takes value '1', when X_i is missing.

We run $GMB(X_6)$ for $r = 1, 2, 3$ and the local BN output by GMB was used as the classifier for X_6 using standard inference techniques. Fig. 5 summarizes the variability of the Kappa measure by 10-fold cross validation. The Kappa distribution over 50 data sets is illustrated in the form of boxplots. The Kappa measure assesses improvement over chance. The following ranges of agreement for the Kappa statistic suggested in the literature are: poor $K < 0.4$, good $0.4 < K < 0.75$ and excellent $K > 0.75$. As may be seen in Fig. 5, the prediction value derived from missing data appears to be useful for increasing the accuracy of the toy problem when the percentage of missing data is superior to 5%. The term 'MB' denotes the classifier using only the MB of the variable (without the use of the dummy variables R_j). The analysis presented here suggests that attention to missing data may improve the prediction accuracy. Further conclusions can be drawn from these results. In the MCAR case, the inclusion of the dummy variables cannot improve classification because they are independent of all the variables. The observed improvement for $r > 1$ is only due to the additional X_j variables that are found useful when others are missing. A radius $r > 1$ was not found to improve significantly the classification, compared to $r = 1$, when data are missing by MAR or IM. The usefulness of the dummy variables increases with the ratio of missing data when data are IM. This is very noticeable on the lower right plot (IM, 5000 instances and 25% missing data). Finally, the size of the data set has little influence on the results when data is MAR and IM.

4.2   Detecting the Missing Mechanism

In this section, we illustrate on a toy problem how augmenting the data with the missingness variables may, in some cases, help pinpoint the mechanism of missing data. A simple way to detect this mechanism is by visually inspecting the graph (DAG) induced by the BN structure learning algorithm. Under the faithfulness condition, the BN encodes the augmented probability distribution including the dummy boolean variables expressing missingness. It is important to note that the detection (when possible) of the IM mechanisms must necessarily be preceded by an imputation phase as explained below. Imputation of any variable X can easily be carried out by inferring the missing entries of X in the local network induced by $GMB(X)$.

Fig. 2. Toy network. The values of variable X are missing according to the IM mechanism. The missing rate, p, is varied from 0 (not missing) to $1/2$ (50% missing rate).

Consider the toy BN presented in Fig. 2 for the sake of illustration. As may be observed, the missing mechanism is IM because the absence of X depends on the value of X which is missing. We generated 1000 independent and identically distributed samples from this BN for various values of p. Notice that for $p = 1/2$, X is always missing when $X = 1$, therefore the Y does not carry any information anymore to impute X (the edge $Y \to X$ is lost) so the edge $X \to R_X$ can not be induced from the available data anymore. Similarly, when $p = 0$, X is not missing, R_X is a constant and the edge $(X \to R_X)$ does not exist anymore. In this experiment, GMB is run on X with parameters, r and α, set to $r = 1$, and $\alpha = 0.01$, for varying values of $p \in [0, 1/2]$. The goal is study to what extent the imputation of X thanks to Y can help detect the edge $X \to R_X$. Fig. 3 summarizes the accuracy of the detection process (i.e., the number of times the edge between X and R_X is detected divided the number of runs) as a function of the missing data ratio and the number of instances on the data set. As expected, the accuracy of the detection process increases with the number of instances. Interestingly, the non-ignorable missing mechanism is rapidly detected as p increases and falls off rapidly as p approaches $1/2$. Finally, provided that the information carried by Y is sufficient to impute X with reasonable assuredness and that the missing rate is sufficiently high, it is possible to detect that the missing data mechanism for X is non-ignorable.

Consider now a real-world data set, namely the Congressional Voting Records, available from the Machine Learning Repository. We would like to shed some light into the unknown missing mechanism in a real scenario. The data set describes votes for each of the 435 member of the US House of Representative on the 16 key issues during the 1984. Hence, the data set consists of 435 cases on 17 binary attributes. X_1 (Class Name - democrat or republican),

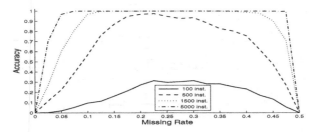

Fig. 3. Average accuracy, over 1000 runs, in detecting the IM mechanism of toy problem shown in Fig. 2, for 500, 1500 and 5000 instances, as a function of the missing rate p.

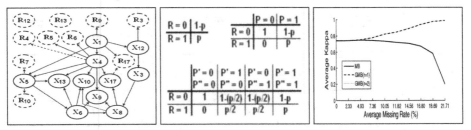

Fig. 4. Left: Bayesian network representing the joint probability distribution of the variables on the *Congressional Voting Records Data Set* X_4 is our class variable. Center: conditional probability tables used for the R_j variables according to the number of parents. Right: average prediction accuracy as a function of the missing rate.

X_2 (handicapped-infants), X_3 (water-project-cost-sharing), X_4 (adoption-of-the-budget-resolution), X_5 (physician-fee-freeze), X_6 (el-salvador-aid), X_7 (religious-groups-in-schools), X_8 (anti-satellite-test-ban), X_9 (aid-to-nicaraguan-contras), X_{10} (mx-missile), X_{11} (immigration), X_{12} (synfuels-corporation-cutback), X_{13} (education-spending), X_{14} (superfund-right-to-sue), X_{15} (crime), X_{16} (duty-free-exports), X_{17} (export-administration-act-south-africa). We consider X_4 as our target variable. There are 289 values reported as unknown. Although these missing entries amount to 4% of the data set, the number of incomplete cases is 203, more than 45% of the total. An important feature of this data is that unknown entries, and hence what member of the US House of Representative did not vote on, can be predictive.

To learn the structure, we run $GMB(X_j)$ for $r = 1$ on each existing variable X_j in the network and we used the BN output by GMB as the classifier using standard inference techniques to fill in the missing entries for X_j. When all missing values were imputed, the overall structure was constructed and oriented by standard CB techniques. The resulting DAG depicted in Figure 4 provides a complete picture of the missing mechanisms. As may be seen, X_{12} and X_{13} are MCAR as their respectively dummy variables are singletons; X_3, X_5, X_6, X_7, X_9 and X_{10} are MAR; X_4 and X_{17} seem to be IM because R_4 and R_{17}

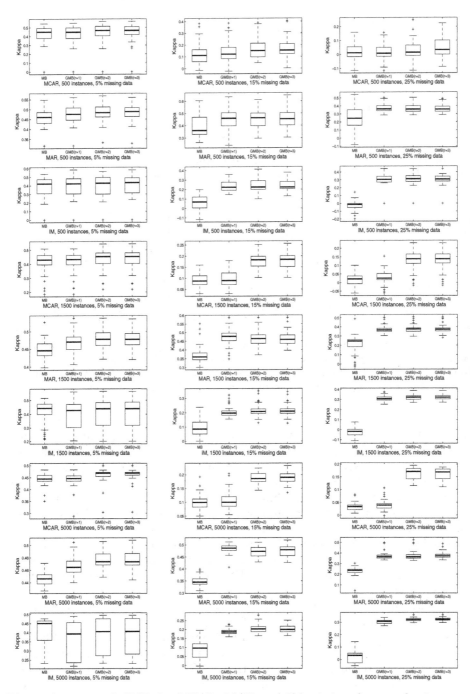

Fig. 5. Accuracy of GMD for MCAR, MAR and IM missing data mechanisms on synthetic data, for 5%, 15% and 25% missing values, and 500, 1500 and 5000 instances

are directly linked to their corresponding variable. Variables 'adoption-of-the-budget-resolution' and 'export-administration-act-south-africa' are expected to be non-ignorable. Moreover, even if 'party affiliation' is not directly associated to 'aid-to-nicaraguan-contras' and 'export-administration-act-south-africa', their missingness do carry some information. Nonetheless, explicit inclusion of missingness information was not found to have any significant effect on the accuracy of classification. It not surprising as the missing entries amounts to only 4% in total. This is in agreement with the previous conclusions on synthetic data. In order to analyse the impact of missing data on the Congressional Voting Records Data Set, the probability tables of the missingness variables R_j were modified as illustrated on Figure 4 (middle) in order to increase the ratio missing data. The network was used for generating several data sets of 435 samples (as in the original data set), but with increasing amounts of missing data. Three probability tables are considered according to the number of parents. The parameter p was varied from 0 to 0.9 in order to generate data sets with different amounts of missing data. The average prediction accuracy, over 200 experiments, is reported on the right plot as a function of the missing rate. Again, we observe significant improvements for missing rates above 5% with $GMB(X_4)$, for both $r = 1$ and $r = 2$, compared to using solely the Markov boundary of X_4.

5 Conclusion

In this article, we addressed data missingness issue when building a classifier. Although absence of data is usually considered a hindrance to accurate prediction, our conclusion is that the absence of some data elements in the data sets can be informative when the percentage of missing data is greater than 5%. Our approach provides a graphical representation of the statistical relationships between the existing variables and the dummy missingness variables that may help identify the mechanism of the missing data. In certain cases, our method was able to detect non-ignorable non-response from data alone after the missing data were imputed. Therefore, it would be interesting for work to be performed to ascertain the probability distributions for which the IM detection is possible.

References

1. Little, R., Rubin, D.: Statistical analysis with missing data. Wiley Interscience, Hoboken (2002)
2. Dempster, A.P., Laird, N.M., Rubin, D.B.: Maximum likelihood from incomplete data via the EM algorithm. J. Roy. Statist. Soc. Ser. B 39(1), 1–38 (1977)
3. Lin, J., Haug, P.: Exploiting missing clinical data in Bayesian network modeling for predicting medical problems. Journal of Biomedical Informatics 41, 1–14 (2008)
4. Siddique, J., Belin, T.: Using an approximate Bayesian bootstrap to multiply impute nonignorable missing data. Computational Statistics & Date Analysis 53, 405–415 (2008)
5. Saar-Tsechansky, M., Provost, F.: Handling missing values when applying classification models. Journal of Machine Learning Research 8, 1625–1657 (2007)

6. Farhangfara, A., Kurganb, L., Dyc, J.: Impact of imputation of missing values on classification error for discrete data. Pattern Recognition 41, 3692–3705 (2008)
7. Corani, G., Zaffalon, M.: Learning reliable classifiers from small or incomplete data sets: The naive credal classifier 2. Journal of Machine Learning Research 9, 581–621 (2008)
8. Jamshidian, M., Mata, M.: Postmodeling sensitivity analysis to detect the effect of missing data mechanisms. Multivariate Behavioral Research 43, 432–452 (2008)
9. Ramoni, M., Sebastiani, P.: Robust learning with missing data. Machine Learning 45(2), 147–170 (2001)
10. Aussem, A., Rodrigues de Morais, S.: A conservative feature selection algorithm with missing data. In: IEEE International Conference on Data Mining ICDM 2008, Pisa, Italy, pp. 725–730 (2008)
11. Pearl, J.: Probabilistic Reasoning in Intelligent Systems: Networks of Plausible Inference. Morgan Kaufmann, San Francisco (1988)
12. Neapolitan, R.E.: Learning Bayesian Networks. Prentice-Hall, Englewood Cliffs (2004)
13. Peña, J., Nilsson, R., Björkegren, J., Tegnér, J.: Towards scalable and data efficient learning of Markov boundaries. International Journal of Approximate Reasoning 45(2), 211–232 (2007)
14. Rodrigues de Morais, S., Aussem, A.: A novel scalable and data efficient feature subset selection algorithm. In: Daelemans, W., Goethals, B., Morik, K. (eds.) ECML PKDD 2008, Part II. LNCS (LNAI), vol. 5212, pp. 298–312. Springer, Heidelberg (2008)
15. Tsamardinos, I., Brown, L., Aliferis, C.: The max-min hill-climbing Bayesian network structure learning algorithm. Machine Learning 65(1), 31–78 (2006)
16. Yaramakala, S., Margaritis, D.: Speculative Markov blanket discovery for optimal feature selection. In: IEEE International Conference on Data Mining, pp. 809–812 (2005)
17. Myers, J., Laskey, K., Levitt, T.: Learning Bayesian networks from incomplete data with stochastic search algorithms. In: Proceedings of the 15th Conference on Uncertainty in Artificial Intelligence, UAI 1995 (1995)
18. Geman, S., Geman, D.: Stochastic relaxation, gibbs distributions, and the Bayesian restoration of images. IEEE Trans. Pattern Anal. Machine Intell. 6(6), 721–741 (1984)
19. Peña, J.M., Björkegren, J., Tegnér, J.: Growing Bayesian network models of gene networks from seed genes. Bioinformatics 40, 224–229 (2005)
20. Whittaker, J.: Graphical Models in Applied Multivariate Analysis. Wiley, New York (1990)

DEMScale: Large Scale MDS Accounting for a Ridge Operator and Demographic Variables

Stephen L. France[1] and J. Douglas Carroll[2]

[1] Lubar School of Business, UW – Milwaukee, 3202 N. Maryland Avenue., Milwaukee, Wisconsin, 53201-0742
france@uwm.edu
[2] Rutgers University, Graduate School of Management, Newark, New Jersey, 07102-3027
dcarroll@rci.rutgers.edu

Abstract. In this paper, a method called DEMScale is introduced for large scale MDS. DEMScale can be used to reduce MDS problems into manageable sub-problems, which are then scaled separately. The MDS items can be split into sub-problems using demographic variables in order to choose the sections of the data with optimal and sub-optimal mappings. The lower dimensional solutions from the scaled sub-problems are recombined by taking sample points from each sub-problem, scaling the sample points, and using an affine mapping with a ridge operator to map the non-sample points. DEMScale builds upon the methods of distributional scaling and FastMDS, which are used to split and recombine MDS mappings. The use of a ridge regression parameter enables DEMScale to achieve stronger solution stability than the basic distributional scaling and FastMDS techniques. The DEMScale method is general, and is independent of the MDS technique and optimization method used.

Keywords: MDS, Visualization, Ridge Operator, PCA.

1 Introduction

Multidimensional scaling (MDS) can be described as a set of techniques for interpreting similarity or dissimilarity data. Typically, an MDS procedure takes proximity data as input and then embeds the data in a metric (usually Euclidean) space. The proximity data may be gathered directly or calculated from a higher dimensional configuration using an appropriate metric. The original MDS techniques such as classical multidimensional scaling (CMDS) [33][34] and Kruskal's distance based metric and nonmetric scaling [17][18] were initially designed to deal with small scale data sets. Over the last 40 years researchers in psychometrics, statistics, and data mining have developed techniques to scale up MDS in order to improve MDS as a tool for large scale exploratory data analysis and visualization.

CMDS [33]-[34] is based on a singular value decomposition of a derived "scalar products" matrix. Kruskal's MDS procedure [17][18] is based upon fitting the best ordinary least squares (OLS) approximation to the distances in the low dimensional space (d_{ij}). The original STRESS function is given in (1). The STRESS function is typically optimized using a gradient descent algorithm or by function majorization [12].

N. Adams et al. (Eds.): IDA 2009, LNCS 5772, pp. 47–58, 2009.
© Springer-Verlag Berlin Heidelberg 2009

We abbreviate this variant of MDS as DMMDS (distance based metric MDS). Variants on STRESS include SSTRESS [31], which uses squared distances, and the Sammon mapping criterion [27], which is designed to account for nonlinearity in data structures.

$$STRESS = \sqrt{\sum_i \sum_j \left(d_{ij} - \hat{d}_{ij}\right)^2 \Big/ \sum_i \sum_j d_{ij}^2} \qquad (1)$$

MDS is often used as a 'visualization' procedure, with the derived lower dimensional solution being plotted visually. MDS procedures have been combined with visual mapping tools such as GGobi [29]. Modern visualization techniques need to be able to cope with the large scale data sets prevalent in the modern world. Assuming symmetric input, a total of $n \times (n - 1) / 2$ distinct distances/dissimilarities are required as input for MDS procedures. Even for moderate sized data sets, these data require a large amount of memory. This memory requirement restricts the use of MDS for large data sets. Table 1 lists some of the main papers that present methods for scaling up MDS to larger data sets. Our definition of MDS is rather narrow, including only techniques directly related to CMDS or DMMDS; we do not include neural network based learning techniques, such as self organizing maps. We split the innovations into several categories. The categories are the initialization of a starting solution (IN), fast heuristic optimization techniques (OP), and techniques for reducing the memory required by the MDS algorithm for storing distance information (RE).

Table 1. Literature summary for large scale MDS

Authors	MDS	IN	OP	RE	Description
Basalaj (1999) [1]	DS			Y	Clustering groups + add points
Brandes and Pich (2007) [4]	CS			Y	MetricMap - Nystrom
Brodbeck and Girarfin (1998) [5]	DS		Y	Y	Stoc. scaling + interpolation
Chalmers (1996) [6]	DS		Y	Y	Stochastic scaling
De Silva and Tenenbaum (2004) [7]	CS			Y	LandmarkMDS - Nystrom
Faloutsos and Lin (1995) [8]	CS			Y	Fastmap - Nystrom
Groenen and Heiser (1996) [10]	DS		Y		Tunnelling heuristic
Groenen, Heiser, and Meulman (1999) [11]	DS		Y		Distance Smoothing
Groenen, Mathar, and Heiser (1995) [12]	DS		Y		Function majorization
Heiser and de Leeuw (1986) [14]	CS		Y		SMACOF – Iterative updates
Jourdan and Melancon (2004) [16]	DS			Y	Stoc. scaling + bins
Morrison and Chalmers (2002) [20]	DS			Y	Stoc. scaling + bins
Morrison, Ross, and Chalmers (2003) [21]	DS			Y	Stoc. scaling + bins
Naud (2004, 2006) [22, 23]	DS			Y	distributional scaling
Platt (2006) [25]	CS			Y	Nystrom algorithms
Trosset and Groenen (2005) [35]	CS	Y	Y		Fast initial Construction
Wang, Wang, Shasta, and Zhang (2005) [36]	CS			Y	MetricMap - Nystrom
Williams and Muntzer (2004) [37]	DS		Y	Y	Progressive stoc. scaling

Basalaj [1] develops an incremental scaling technique that uses a minimum spanning tree to order objects in terms of structural importance, performs full scaling on a subset of the most important points, and then uses single scaling (minimizing STRESS while only allowing the configuration of one point to change) on each of the remaining points in turn. Naud [22][23] develops a variant of Basalaj's technique, choosing points close to the center of k-Means clustering solutions rather than using a minimum spanning tree, and incrementally adding multiple points rather than adding single points.

Quist and Yona [26] introduce a technique called distributional scaling. They split the solution points into k clusters, as per Naud, but they then scale the points in each of these clusters, creating a separate lower dimensional embedding for each cluster. They then use an affine transformation to map the points together, based upon a solution created from sample points selected from each cluster. FastMDS [38] splits the source matrix into separate sub-matrices randomly and performs classical MDS on each of these submatrices. Sampling points are taken from each of the submatrices and then classical MDS is performed on the sample points. FastMDS configures the overall solution using an affine mapping, in a similar way to [26].

There are several techniques that perform CMDS on a subset of the points to be scaled and then fill in the remaining points using a mapping or triangulation relative to the original data. These techniques include FastMap [8], MetricMap [36], and Landmark MDS [7]. FastMap, MetricMap, and Landmark MDS are all variants on the Nyström approximation of the eigenvectors and eigenvalues of a matrix [25].

Chalmers [6] develops an algorithm that for each solution point stochastically selects a subset of points for calculating the STRESS measure. Two lists of points are created at each iteration. If a randomly generated point has a smaller distance than the maximum on the first list then the point is added to the first list, otherwise it is added to the second list. The first neighborhood list becomes more local as the algorithm progresses while the second list remains random. [20][21] develop a hybrid approach, combining [6] with an interpolation method described in [5].

2 Methodology

In this paper we generalize, empirically test, and expand on the techniques developed for distributional scaling and FastMDS. Both distributional scaling and FastMDS work in a similar manner, but for different varieties of MDS. The advantages of these techniques over the others described are that optimal mappings (assuming a global optimum has been found for distance based techniques) can be guaranteed for the subsets of the solution that are scaled separately and that the procedure used to split and recombine solutions is independent of the type of MDS used. DEMScale can be used to reduce MDS problems into manageable sub-problems, which are then scaled separately. The lower dimensional solutions are recombined by taking sample points from each sub-problem, scaling the sample points, and using an affine mapping with a ridge operator to map the non-sample points.

Using DEMScale, one can take advantage of the optimal subset property by splitting solutions using demographic variables, thus controlling the subsets of the data across which there is optimal solution recovery. Consider the following situation.

Patients in a large scale epidemiological study are scaled based upon answers to a large scale questionnaire and measurements from health indicators. One researcher is interested in exploring differences between males and females and between different socioeconomic groups within prescribed geographic areas. A second researcher is interested in differences between smokers and non-smokers for subjects in different socio-economic groups. For researcher 1, the MDS problem could be split via geographical location, giving optimal mappings for all subjects in a single geographical location. For researcher 2, the MDS problem could be split by socio-economic group, giving optimal mappings for all subjects in certain socioeconomic groups. DEMScale is independent of the variety of MDS used and the algorithm is as general as possible. At a minimum, the maximum size of the subgroups of items to be scaled must be specified. If the data items are not split using demographic variables, then the data are split randomly in step 3.

3 DEMScale Algorithm

1. Set the maximum number of input items to be scaled at any one time as *Maxopt*. Set *Divide* to be the reciprocal of the proportion of items in each group that are to be scaled in the combined solution.
2. Split the data set into groups $G_1, \cdots, G_n$, based upon chosen demographic variables or by any other grouping method (e.g., a clustering solution). If the demographic variables are ordered (with the variable ordered last the most important not to split on), then iterate through the list, splitting using each demographic variable until all groups have less than *Maxopt* variables. If not using any grouping method at this stage then there is one group, G_1, containing all of the items in the dataset.
3. If $|G_i| > Maxopt$ for Group i then split the Group into $floor(|G_i| / Maxopt) + 1$ groups randomly.
4. Calculate a distance matrix for each group so there are matrices $\mathbf{D}_1, \ldots, \mathbf{D}_n$.
5. Calculate a lower dimensional solution using MDS for each group to retrieve lower dimensional solutions $\mathbf{L}_1, \ldots, \mathbf{L}_n$. This step is independent of the type of MDS solution used.
6. Take $s = floor(|G_i| / Divide)\}$ items randomly from each $\mathbf{L}_i$, combine these items, and calculate the distance matrix $\mathbf{D}_{n+1}$ from the combined items.
7. Use MDS to calculate a lower dimensional embedding $\mathbf{M}$ from $\mathbf{D}_{n+1}$.
8. Use (3) to calculate affine least squares mappings between each subset of s points of $\mathbf{L}_i$ and each $\mathbf{M}_i$ for $i = 1$ to n, where $\mathbf{M}_i$ is the subset of $\mathbf{M}$ corresponding to the points sampled from $\mathbf{L}_i$.
9. For each $\mathbf{L}_i$, use $\mathbf{A}$ to calculate the final mapping $\mathbf{L}_i\mathbf{A}$ of all of the points in $\mathbf{L}_i$.

For each subset of items, the affine mapping is used to calculate the position of each of the points not included in the sample points in the main solution. The least squares function to be minimized for the estimation of the affine mapping is given in (2).

$$Min \ (\mathbf{M}_i - \mathbf{S}_i\mathbf{A})'(\mathbf{M}_i - \mathbf{S}_i\mathbf{A}) + \lambda\mathbf{A}'\mathbf{A} \tag{2}$$

$\mathbf{M}_i$ is an $s \times d$ subset of $\mathbf{M}$, corresponding to the points sampled from $\mathbf{L}_i$, s is the number of sample points taken, and d is the number of dimensions of the mapping.

$\mathbf{S}_i$ is an $s \times (d + 1)$ matrix containing a column of 1's and the columns of the lower dimensional solution for the s points sampled from $\mathbf{L}_i$.

$\mathbf{A}$ is a $(d + 1) \times d$ matrix of the estimated affine mapping from the points in the sub-solution to the final mapping.

The mapping is solved using (3).

$$\mathbf{A} = (\mathbf{S}_i'\mathbf{S}_i + \lambda \mathbf{I}_{d+1})^{-1}\mathbf{S}_i'\mathbf{M} \tag{3}$$

If $\lambda = 0$ for the affine mapping, then DEMScale reduces to distributional scaling for DMMDS and to FastMDS for CMDS. The rationale behind ridge regression is that standard ordinary least squares (OLS) regression is liable to over fit the training data, leading to sub-optimal results when applied to test data. By the Gauss-Markov theorem, OLS regression provides the best unbiased estimator for the regression problem, but a biased estimator such as ridge regression [15][19] may give a more reliable predictor when the estimators are used on data drawn from the same population as the test data used to fit the regression equation. In the case of DEMScale, we hypothesize that the use of ridge regression will lessen the chance of over fitting on the sample points and lead to more stable mappings.

4 Computational Complexity

The worst case computational complexity of DEMScale can be calculated based upon the MDS technique used, the method of optimization or calculation of the solution, and the maximum number of items in each subgroup. One can calculate the computational complexity relative to the original problem. Take the worst case running time for the original MDS problem to be (4), where $f(N)$ is a function of the number of points N in the solution. One can make the assumption that $f(N)$ is monotone increasing with N, as in the context of MDS it would be nonsensical for solution time to decrease with increasing problem size.

$$CC(\mathit{full}) = O(f(N)) \tag{4}$$

Consider the situation where $\mathit{Maxopt} = M$ and $M \leq N$. The N items are split into subgroups of points, where the maximum size of the subgroups of points is M. Consider the points evenly split between groups (with remainder points split randomly amongst groups) then an inequality for the minimum number of subgroups n is given in (5), where $|G_i|$ is the number of items in group i.

$$N/M \leq n = \lceil N/M \rceil \leq (N/M) + 1, \text{ where } |G_i| \leq M \tag{5}$$

The MDS procedure must be run for each subgroup. A formula for the worst case complexity for the n DEMScale sub-problems is given in (6).

$$CC(DS) \leq O\big(((N/M) + 1) \times (f(M))\big) \tag{6}$$

Theorem: Given the function $k_1 \cdot N^p < f(N) < k_2 \cdot N^q$ for all N, arbitrary constants k_1 and k_2, and some $p \geq 1$ and $q \geq p$, then the N items to be scaled can be split into any number of groups with any permutation of items in groups so that $|G_i| \leq M$ and equation (6) still holds.

Partition the problem into two distinct cases:

Case 1: The number of groups n is minimal and $n = \lceil N / M \rceil$. This case is covered by (6) as $n < ((N / M) + 1)$ and by assumptions of monotonicity $f(|G_i|) \leq f(M)$ for all subgroups i. This gives (7).

$$O\left(f\left(|G_1|\right)+f\left(|G_2|\right)+\cdots+f\left(|G_n|\right)\right) \leq O\left(\left((N/M)+1\right)\times\left(f\left(M\right)\right)\right) \tag{7}$$

Case 2: The number of groups $n > \lceil N / M \rceil$. One needs to show that for any item splitting scheme $G2$ with $n_2 > \lceil N / M \rceil$ subgroups there is an item splitting scheme $G1$ with $n_1 = \lceil N / M \rceil$ subgroups, so that given $= k_1 \cdot N^p < f(N) < k_2 \cdot N^q$ and $p \geq 1$ and $q \geq p$ then the inequality holds.

$$O\left(f\left(|G2_1|\right)+\cdots+f\left(|G2_{n_2}|\right)\right) \leq O\left(f\left(|G1_1|\right)+\cdots+f\left(|G1_{n_1}|\right)\right) \tag{8}$$

Without any loss of generality one can create a permutation of the groups so the groups are ordered in decreasing size.

$$|G2_{p1}| \geq |G2_{p2}| \geq \cdots \geq |G2_{pn_1}| \cdots \geq |G2_{pn_2}| \tag{9}$$

Now there is some $G1$ so that all the N items are divided amongst $n_1 = \lceil N / M \rceil$ subgroups so that $|G1_i| \leq M$ for all i and that there are either M or $M - 1$ items in each group. Take each of the items in groups $G2_{pn_1+1}$ to $G2_{pn_2}$ and add them iteratively to one of the groups $G2_1$ to $G2_{n_1}$ with the least number of items. At each stage the change in the value of the computational requirements is equal to (10).

$$\left(f\left(|G2_i|+1\right)-f\left(|G2_i|\right)\right)-\left(f\left(|G2_j|-1\right)-f\left(|G2_j|\right)\right) \tag{10}$$

As $|G2_i| \geq |G2_j|$, then for any polynomial function $f(N) = k_1 \cdot N^p$ where $p \geq 1$, (10) is positive. As each change is positive and the permutation scheme with $n_1 = \lceil N / M \rceil$ groups (call this $G1$) has a greater value than the starting permutation scheme then (11) holds.

$$O\left(f\left(|G2_1|\right)+f\left(|G2_2|\right)+\cdots+f\left(|G2_{n_2}|\right)\right) \leq O\left(\left((N/M)+1\right)\times\left(f\left(M\right)\right)\right), \tag{11}$$

for arbitrary $n_2 > \lceil N / M \rceil$ and for $f(N) = k_1 \cdot N^p$ where $p \geq 1$.

The theorem follows by using the squeeze theorem with $f(N) = k_1 \cdot N^p$ and $f(N) = k_2 \cdot N^q$, as both these polynomials are continuous for all N.

For example, consider an MDS algorithm with computational complexity $O(N^4)$. An example would be Isomap [32], which has a dynamic programming step with $O(N^4)$ that dominates the other steps). If the maximum subgroup size M is set as $N^{1/2}$ then the computational complexity for the subgroup calculations is given as (12).

$$O\left(\left(N^{1-1/2}+1\right)\times N^{4/2}\right) \approx O\left(N^{5/2}\right) \tag{12}$$

If the number of sample points taken from each of the runs is less than $N^{1/2}$ then the complexity of this step is $O(N^2)$. For the example given, there are approximately $N^{1/2}$ (the number of subgroups) affine mappings. The complexity of the affine mapping is dominated by the calculation of $\mathbf{L}_i' \, \mathbf{L}_i$. (as other matrix calculations involve $s \times (d + 1)$ or $(d + 1) \times (d + 1)$ items). The matrix $\mathbf{L}_i$ has $N^{1/2} \times (d + 1)$ items and $N^{1/2}$ affine mappings have a complexity of $N^{1/2} \times [\, (d + 1)^2 \cdot N^{1/2} \,]= k \cdot N$, where $k = (d + 1)^{1/2}$. Thus the overall complexity is given in (13) and the complexity of performing MDS on the subgroups dominates the computational complexity of the other stages.

$$CC = k_1 \cdot N^{5/2} + k_2 \cdot N^2 + k_3 N \approx O\left(N^{5/2}\right) \tag{13}$$

5 Experimentation and Results

The experimentation on DEMScale has two major purposes. The first is to check the accuracy of the DEMScale mappings relative to mappings created using the entire set of data at one time. The second is to find the optimal ridge regression parameter, λ, and to show that the use of ridge regression helps stabilize the affine mapping and thus produce more reliable solutions. In order to test the stability of the mapping, one must compare the solutions constructed with affine mappings on the same sub-solutions, but using differently randomly selected sets of sample points. The average similarity between solutions can then be taken as a measure of stability.

In order to ensure that the results are not skewed by the peculiarities of the MDS technique used, DEMScale was tested with both CMDS and DMMDS. We implemented DMMDS using three different MDS functions (STRESS, SSTRESS, and SAMMON) and optimized these functions using a second order gradient descent algorithm. Newton gradient based optimization sets convergence criterion as a gradient step per variable, so $S \cong 1 / (N \times d)$, where N is the number of points and d is the dimensionality of the derived solution. Thus, the computational complexity can be considered per iteration when comparing between running MDS on all items and running MDS on multiple subsets of items. DEMScale was tested on multiple datasets, which are summarized in Table 2.

Table 2. Experimental datasets

Data	Dims	Type	Data set description
fbis	2463	Sparse Count	TREC (1999)
isolet	617	Real	Fanty and Cole (1990) [9]
CRM	142	Real	Neslin, Gupta, Kamakura, Lu, Mason (2006) [24]
Madelon	31472	Real	TREC (1999)

As different MDS techniques are being tested, comparison metrics independent of the MDS technique are required. In order to compare the accuracy of the solution recovery, the inter-item distances from the derived solution must be compared with the input proximities. In order to test the stability of the affine mapping, the inter-item distances must be compared between solutions mapped from the same sub-solutions, but with different items sampled for the affine mapping. To compare configurations, we use the congruence coefficient [3], which gives the uncentered correlation between distances in the solutions being compared. We tested the algorithm using a multi-factorial design. The factors used were as follows:

MDS Algorithms: CMDS and DMMDS were tested. Three different functions were implemented for DMMDS. These were STRESS, SSTRESS, and SAMMON.

Maxopt: *Maxopt* of values $M = N / 4$, $M = N / 8$, and $M = N / 16$, where N is the size of the data set.

Splitting: If items are split using demographic variables then it is unlikely that item subgroups will have equal numbers of variables. Therefore a random splitting technique was used; $n - 1$ split points in the data were randomly generated from the uniform distribution, in order to create $n = \lceil N / M \rceil$ subgroups. If subgroups had more than M items then the subgroups were split, as per step 3 in the DEMScale algorithm. Very small groups of items may lead to singularity in the affine mapping matrix. We thus set the minimum group size as *Maxopt* / 4.

Sample points: Three different values for the number of sample points taken from each subgroup were used. These were $S = M / 4$, $S = M / 8$, and $S = M / 16$.

Data sets: 2000 items were selected from each of the datasets given in Table 2, giving a maximum proximity matrix size of $(500 \times 499) / 2$ for $M = N / 4$. For each dataset a proximity matrix was calculated from the data using an appropriate metric. N.B. Larger datasets of up to 20,000 items were tested and visualized, but these datasets required a larger number of splits and smaller number of sample points than used in the experiment.

For each value of S, five runs were made, with S sample points selected randomly for each run. The average values of the congruence (c) were calculated across the five runs as measures of solution quality. The average values of c were calculated across the $(5 \times 4) / 2 = 10$ distinct pairs of lower dimensional solutions as a measure of solution stability and consistency. In total there were $4 \times 4 \times 3 \times 3 = 144$ different experimental conditions. Five runs were made per experimental condition, totaling 720 runs. All lower dimensional solutions were embedded in two dimensions due to this dimensionality being the most commonly used for visualization applications.

The solution quality data were analyzed using the MANOVA technique. The value of the average congruence between the input configuration and the lower dimensional solution (i.e., solution quality) was taken as the first dependent variable. The R^2 value ((SST - SSE) / SSE) of the affine mapping for overall solution construction was taken as the second dependent variable. Using Wilks's Lamba, all factors and interactions between factors were significant, with $p < 0.001$ for all values. Tests of between subject effects were significant for all combinations of dependent variables and factors, again with $p < 0.001$ for all values. The between subject effects were all significant except for *MDSType* × *RRParam* for congruence and for *MaxOpt* × *RRParm* for R^2.

Post hoc tests were carried out on the marginal means of *MDSType*, *MDSOpt*, and *Divide*. The Scheffé test [28] was used, as it is a fairly conservative and robust post-hoc test. At the 95% confidence level, there are differences between the means of each pairwise combination of MDS techniques. The order (strongest performance first) is STRESS>SAMMON>SSTRESS>CMDS. The order of the performance in terms of the R^2 values of the affine mapping is STRESS> SSTRESS>SAMMON>CMDS. The STRESS function uses a least squares criterion, so the results may have something to do with the fact that the OLS affine mapping works well in combination with the solutions produced with the STRESS metric.

The results of the post-hoc tests for the *MDSOpt* and *Divide* independent variables are straightforward and easily explained. As *MDSOpt* increases, so does the quality of the mapping. Due to a reduction in computation complexity, smaller group sizes lead quicker run times and lower memory requirements. As *Divide* becomes smaller the quality of the mapping improves. Smaller values of *Divide* result in more points in the between groups MDS solution and thus more accurate affine mappings. There is a strong interaction affect, resulting in a large drop off in performance for small values of *MaxOpt* and large values of *Divide*.

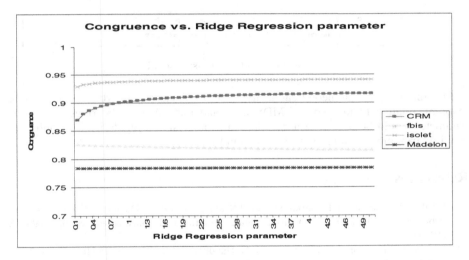

Fig. 1. Congruence vs. Ridge Regression parameter for all files

Fig. 1 shows the performance of the algorithm relative to the value of the ridge regression parameter. The results described by this graph are somewhat inconclusive, but show that for some data sets, the use of ridge regression can positively affect the overall quality of solution mapping. For three of the four files there is a large jump in performance when a ridge regression parameter is used. In fact, some of the runs with no ridge parameter produced no results due to the singularity of the matrices. This may be due to correlated dimensions in the L_i. For DMMDS, irrespective of the data set, it would be useful to use a ridge parameter in order to avoid possible matrix singularity and the failure of the affine mapping. As the dimensions in the CMDS solutions are uncorrelated, the matrix singularity is unlikely to occur.

The stability of the solutions was calculated by finding the average congruence between the four runs for each experimental condition. The solution stability is plotted against the ridge parameter in Fig. 2. One can see an increase in stability for a small ridge parameter. The optimal value of the ridge parameter is independent of the data set tested on.

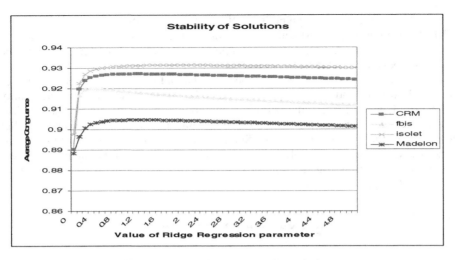

Fig. 2. Stability of ridge regression solutions

For future work, a theoretical analysis of the matrix properties of the affine mapping relative to the type of MDS used and the value of the ridge regression parameter would be useful. An excellent such analysis on regularization applied to multi set canonical correlation is described in [31].

References

1. Basilaj, W.: Incremental Multidimensional Scaling Method for Database Visualization. In: Erbacher, R.F., Pang, A. (eds.) Visual Data Exploration and Analysis, vol. VI, pp. 149–158 (1999)
2. Benzecri, J.P.: Correspondence Analysis Handbook. Marcel Dekker, Inc., New York (1992)
3. Borg, I., Leutner, D.: Measuring the Similarity between MDS Configurations. Multivariate Behavioral Research 20, 325–334 (1985)
4. Brandes, U., Pich, C.: Eigensolver methods for progressive multidimensional scaling of large data. In: Kaufmann, M., Wagner, D. (eds.) GD 2006. LNCS, vol. 4372, pp. 42–53. Springer, Heidelberg (2007)
5. Brodbeck, D.L., Girardin, D.L.: Combining Topological Clustering and Multidimensional Scaling for Visualising Large Data Sets. Unpublished Paper (Accepted for, but Not Published in, Proc. IEEE Information Visualization 1998) (1998)
6. Chalmers, M.: A Linear Iteration Time Layout Algorithm for Visualizing High-Dimensional Data. In: Yagel, R., Nielson, G.M. (eds.) Proceedings of the 7th Conference on Visualization, pp. 127–132. IEEE Computer Society Press, Los Alamitos (1996)
7. de Silva, V., Tenenbaum, J.B.: Sparse Multidimensional Scaling using Landmark Points. Technical Report, Stanford University (2004)
8. Faloutos, C., Lin, K.: FastMap: A Fast Algorithm for Indexing, Data-Mining and Visualization of Traditional and Multimedia Datasets. In: Cary, M., Schneider, D. (eds.) 1995 ACM SIGMOD International Conference on Management of Data, pp. 163–174. ACM, New York (1995)

9. Fanty, M., Cole, R.: Spoken Letter recognition. In: Lippman, R.P., Moody, J., Touretzky, D.S. (eds.) Advances in Neural Information Processing Systems 3, pp. 220–226. Morgan Kaufmann, San Mateo (1990)

10. Groenen, P.J.K., Heiser, W.J.: The Tunneling Method for Global Optimization in Multidimensional Scaling. Psychometrika 61, 529–550 (1996)

11. Groenen, P.J.K., Heiser, W.J., Meulman, J.J.: Global Optimization in Least-Squares Multidimensional Scaling. Journal of Classification 16, 225–254 (1999)

12. Groenen, P.J.K., Mathar, R., Heiser, W.J.: The Majorization Approach to Multidimensional Scaling for Minkowski Distances. Journal of Classification 12, 3–19 (1995)

13. Guyon, I., Li, J., Mader, T., Pletscher, P.A., Schneider, G., Uhr, M.: Competitive Baseline Methods Set New Standards for the NIPS 2003 Feature Selection Benchmark. Pattern Recognition Letters 28, 1438–1444 (2007)

14. Heiser, W.J., de Leew, J.: SMACOF-I. Technical Report UG-86-02, Department of Data Theory, University of Leiden (1986)

15. Hoerl, A.E., Kennard, R.W.: Ridge Regression: Biased Estimation for Nonorthogonal Problems. Techometrics 42, 80–86 (2000)

16. Jourdan, F., Melancon, G.: Multiscale Hybrid MDS. In: Proceedings of the Information Visualisation Eighth International Conference, pp. 388–393. IEEE Computer Society, Washington (2004)

17. Kruskal, J.B.: Multidimensional Scaling for Optimizing a Goodness of Fit Metric to a Nonmetric Hypothesis. Psychometrika 29, 1–27 (1964)

18. Kruskal, J.B.: Nonmetric Multidimensional Scaling: A Numerical Method. Psychometrika 29, 115–129 (1964)

19. Marquardt, D.W., Snee, R.D.: Ridge Regression in Practice. The American Statistician 29, 3–20 (1975)

20. Morrison, A., Ross, G., Chalmers, M.: Fast Multidimensional Scaling through Sampling, Springs, and Interpolation. Information Visualization 2, 68–77 (2003)

21. Morrison, A., Chalmers, M.: A Hybrid Layout Algorithm for Sub-Quadratic Multidimensional Scaling. In: Wong, P., Andrews, K. (eds.) Proceedings of the IEEE Symposium on Information Visualization, pp. 152–158. IEEE, Los Alamitos (2002)

22. Naud, A.: An Accurate MDS-Based Algorithm for the Visualization of Large Multidimensional Datasets. In: Rutkowski, L., Tadeusiewicz, R., Zadeh, L.A., Żurada, J.M. (eds.) ICAISC 2006. LNCS (LNAI), vol. 4029, pp. 643–652. Springer, Heidelberg (2006)

23. Naud, A.: Visualization of High-Dimensional Data Using an Association of Multidimensional Scaling to Clustering. In: 2004 IEEE Conference on Cybernetics and Intelligent Systems, pp. 252–255 (2004)

24. Neslin Scott, A., Sunil, G., Kamakura, W.A., Lu, J., Mason, C.H.: Defection Detection: Measuring and Understanding the Predictive Accuracy of Customer Churn Models. Journal of Marketing Research 43, 204–211 (2006)

25. Platt, J.C.: FastMap, MetricMap, and Landmark MDS are all Nyström Algorithms. Microsoft Research Working Paper (2006)

26. Quist, M., Yona, G.: Distributional Scaling. An Algorithm for Structure Preserving Embedding of Metric and Nonmetric Spaces. Journal of Machine Learning Research 5, 399–430 (2004)

27. Sammon, J.W.: A Nonlinear Mapping for Data Structure Analysis. IEEE Transactions on Computers 18, 401–409 (1969)

28. Scheffé, H.: The Analysis of Variance. John Wiley & Sons, New York (1959)

29. Swayne, D.F., Lang, D.T., Buja, A., Cook, D.: GGobi: Evolving from XGobi into an Extensible Framework for Interactive Data Visualization. Computational Statistics & Data Analysis 43, 423–444 (2003)
30. Takane, Y., Yanai, H., Hwang, H.: Regularized Multiple-Set Canonical Correlation Analysis. Psychometrika 73, 753–775 (2008)
31. Takane, Y., Young, F.W., de Leew, J.: Nonmetric Individual Differences Multidimensional Scaling: An Alternating Least Squares Method with Optimal Scaling Features. Psychometrika 42, 7–67 (1977)
32. Tenenbaum, J.B., de Silva, V., Langford, J.C.: A Global Geometric Framework for Nonlinear Dimensionality Reduction. Science 290, 2319–2323 (2000)
33. Torgerson, W.S.: Theory and Methods of Scaling. Wiley, New York (1958)
34. Torgerson, W.S.: Multidimensional Scaling, I: Theory and Method. Psychometrika 17, 401–419 (1952)
35. Trosset, M.W., Groenen, P.J.F.: Multidimensional Scaling Algorithms for Large Data Sets. Computing Science and Statistics 37 (2005)
36. Wang, J.T.L., Wang, X., Shasta, D., Zhang, K.: MetricMap: An Embedding Technique for Processing Distance-Based Queries in Metric Spaces. IEEE Transactions on Systems, Man, and Cybernetics 35, 973–987 (2005)
37. Williams, M., Munzner, T.: Steerable, Progressive Multidimensional Scaling. In: Ward, M., Munzer, T. (eds.) IEEE Symposium on Information Visualisation 2004, pp. 57–64. IEEE Computer Society, Washington (2004)
38. Yang, T., Liu, J., McMillan, L., Wang, W.: A Fast Approximation to Multidimensional Scaling. In: Proceedings of IEEE Workshop on Computation Intensive Methods for Computer Vision, pp. 1–8 (2006)

How to Control Clustering Results?
Flexible Clustering Aggregation

Martin Hahmann, Peter B. Volk, Frank Rosenthal, Dirk Habich,
and Wolfgang Lehner

Dresden University of Technology, Database Technology Group
dbinfo@mail.inf.tu-dresden.de

Abstract. One of the most important and challenging questions in the
area of clustering is how to choose the best-fitting algorithm and parameterization to obtain an optimal clustering for the considered data.
The clustering aggregation concept tries to bypass this problem by generating a set of separate, heterogeneous partitionings of the same data
set, from which an aggregate clustering is derived. As of now, almost
every existing aggregation approach combines given crisp clusterings on
the basis of pair-wise similarities. In this paper, we regard an input set
of soft clusterings and show that it contains additional information that
is efficiently useable for the aggregation. Our approach introduces an
expansion of mentioned pair-wise similarities, allowing control and adjustment of the aggregation process and its result. Our experiments show
that our flexible approach offers adaptive results, improved identification
of structures and high useability.

1 Introduction

Data clustering is an important data-mining technique, commonly used in various domains [1,2,3]. This technique can be utilized for an initial exploration
of scientific data and often builds the basis for subsequent analysis techniques.
Generally, clustering is defined as the problem of partitioning a set of objects
into groups, so-called clusters, so that objects in the same group are similar,
while objects in different groups are dissimilar [3]. Following this definition, a
well-defined measure for similarity between objects is required [1,3].

In this area, the selection of the best-fitting clustering algorithm, including
parameterization, from the multitude of options is a non-trivial task, especially
for users who have little experience in this area. However, this selection issue
is vital for the clustering result quality [3] and therefore, users usually conduct
the following steps in an iterative way until a satisfying result is achieved: *algorithm selection, parameter selection, clustering* and *evaluation*. This iterative
approach is tedious work and requires profound clustering knowledge. Therefore,
an alternative approach to make clustering more applicable for a wide range of
non-clustering experts in several domains is desirable.

On a conceptual level this issue can be tackled by applying the clustering
aggregation technique. Fundamentally, clustering aggregation combines an ensemble of several partitionings of a data set, generated using different algorithms

N. Adams et al. (Eds.): IDA 2009, LNCS 5772, pp. 59–70, 2009.
© Springer-Verlag Berlin Heidelberg 2009

and/or parameters, into a final clustering result and hence avoids the fixation on only one clustering method. As demonstrated in various papers, the quality and robustness of the aggregated clustering result increase in comparison with the input clusterings [4,5,6,7]. The proposed aggregation techniques can be classified into three basic classes: (i) pair-wise assignment class [7,8,9], (ii) hypergraph-based class [7], and (iii) cluster correspondence class [6,10,11].

Most aggregation techniques are members of the first class, relying on pair-wise assignments and on an associated majority decision. Typically, they use a set of crisp clustering results as basic input. A crisp result, e.g., one determined by k-means [2,3] or DBSCAN[1], assigns each object exclusively to the cluster having the highest similarity with the object. Therefore, a pair of objects can be located: (i) in the same or (ii) in different clusters. For the final assignment case of a pair in the aggregate, the assignment occurring most in the input set of clusterings is selected. For details about the second class refer to Strehl et al. [7] and to Boulis et al. [6] and Topchy et al. [12] for examples of class three. In comparison, class one utilizes the most information about the data and is thus presumed to be the optimal approach for aggregation at the moment.

The aggregation concept increases the clustering quality as well as the robustness [4,5,6,7] and frees the user from selecting the optimal algorithm and parameterization. But the aggregation process itself is not controllable in an efficient way. If an user obtains an unsatisfying aggregate, the only adjustment option consists of the modification of input clusterings. To allow control and result flexibility, we present an enhanced aggregation concept. The first enhancement concerns the aggregation input. Instead of using crisp results, we utilize soft clustering results—assigning each object its similarity to all determined clusters—obtained via algorithms like *FCM* [13] or refinement techniques like a-posteriori [4]. This fine-grained information is efficiently useable to expand pair-wise assignments making them more accurate. With this second enhancement, we are able to (i) revise the aggregation itself and (ii) introduce user-friendly control options.

To summarize, we propose our novel flexible aggregation concept in this paper. The contributions—also reflecting the structure of the paper—are as follows: We start with a detailed description of already available clustering aggregation concepts in Section 2 and highlight several drawbacks. In Section 3, we expand the pair-wise assignments for soft clusterings and introduce a novel significance score. Based on these expansions, we propose our flexible aggregation method in Section 4. In Section 5, we conduct an exhaustive evaluation and present future research aspects, before we conclude the paper in Section 6.

2   Preliminaries

For the explanations made in this paper, we assume the following **setting:** let $\mathcal{D}$ be a dataset $\{x_1, \cdots, x_n\}$ consisting of n points—also called objects—and $\mathcal{C}$ be a cluster ensemble $\{C_1, \cdots, C_e\}$, created with different algorithms and parameterizations. Each $C_l \in \mathcal{C}(1 \leq l \leq e)$ has k_l clusters $c_1, \cdots, c_{k_l}$, satisfying $\bigcup_{i=1}^{k_l} c_i = \mathcal{D}$. Based on this, the common **goal** is the construction of an aggregate clustering $\hat{C}$ by combining all members of $\mathcal{C}$.

Utilizing a given set of crisp clusterings, each point $x_i \in D$ has a unique label denoting its cluster assignment. Regarding the pair-wise similarities of two points in C_l, two pair-wise assignment cases (*pa*-cases) are definable: (i) a_+ for objects with equal cluster labels that are located in the same cluster of C_l and (ii) a_- for object pairs featuring different labels, indicating membership in separate clusters of C_l. To construct $\hat{C}$, the *pa*-case of every pair of points from $\mathcal{D}$ is determined for each clustering of $\mathcal{C}$. After that, the *pa*-case that is dominant throughout $\mathcal{C}$ is selected to hold for the respective pair in $\hat{C}$ [5,7,8,9]. **Example**: two objects $x_1; x_2$ that belong to the same cluster in 7 out of 10 clusterings of $\mathcal{C}$ also belong to the same cluster in $\hat{C}$.

To use soft clusterings as input for the clustering aggregation, we need to update our setting. Each point $x_i \in \mathcal{D}$ is now assigned to all clusters of C_l to a certain degree. Thus, the assignment information of x_i in C_l is denoted as a vector $\vec{v_i}$ with the components $v_{ip}(1 \leq p \leq k_l)$ describing the relation between x_i and the p-th cluster of C_l. Clustering aggregation based on soft assignments is challenging because it requires the determination of *pa*-cases using vectors. We are able to simply adopt the previous approach by stating that x_i and x_j are members of the same cluster if their assignment vectors $\vec{v_i}$ and $\vec{v_j}$ are equal by components. This condition is very strict and would most likely lead to nearly no a_+ assignments. Therefore, this constraint is softened and the a_+ case now holds for objects with similar assignment vectors.

This principle is employed by available aggregation concepts for soft input sets [4,14]. Both approaches use well-known distance measures—e.g. the euclidean distance in [14]—to calculate the similarity between vectors and to derive the *pa*-cases. If the calculated vector similarity exceeds a certain threshold, the resp. points are considered as a_+ or else as a_-. Per definition, the approaches of [4,14] do not deal with aggregation control. Their major problem, described subsequently, concerns the handling of soft assignments, using only common distance measures. For evaluation in this context, we assume the following experimental **setup**: a clustering C_l with $k_l = 2$, a set of 121 vector pairs $\vec{v_i}; \vec{v_j}$, satisfying $\sum_{p/q=1}^{2} v_{ip/jq} = 1, i \neq j, 0 \leq v_{ip/jq} \leq 1$, where $v_{ip/jq}$ are multiples of 0.1.

We start by applying the L_2 norm resp. *euclidean* distance to our setup. In Fig. 1(a), the obtained results are shown; (i) via x- and y-coordinates a vector pairing is specified, while (ii) the corresponding z-value represents the L_2 distance for this pair. **Example**: the pair $\vec{v_i}^\top = (1,0)$ and $\vec{v_j}^\top(0,1)$ (western corner of Fig. 1(a)) has a distance of $\sqrt{2}$. Basically, L_2 is a non-directional distance function only considering the norm, which is a major drawback when measuring similarity of vectors in this case. Thus, pairs $\vec{v_i}; \vec{v_j}$ can have equal L_2 distances regardless of x_i and x_j actually being in the same cluster or not. **Example**: the pair $\vec{v_i}^\top = (0.1, 0.9)$ and $\vec{v_j}^\top(0.3, 0.7)$ is located in cluster 2, i.e. a_+ holds; pair $\vec{v_k}^\top = (0.6, 0.4)$; $\vec{v_l}^\top(0.4, 0.6)$ is separated in clusters 1 and 2, i.e. a_-. Although *pa*-cases are actually different, both pairs have the same L_2 distance of $\sqrt{0.08}$. It is obvious that this can lead to incorrect decisions in the construction of $\hat{C}$, especially if thresholds or clustering algorithms are employed. Consequently, vector direction is vital for an accurate interpretation of *pa*-cases.

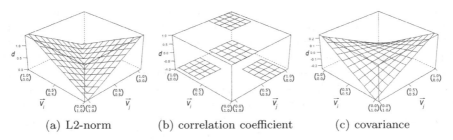

(a) L2-norm   (b) correlation coefficient   (c) covariance

Fig. 1. Different distance measures applied to 2-dimensional vectors

Next, we examine distance metrics considering the direction resp. composition of vectors. First, we look at the *Pearson correlation coefficient* (ϱ) assuming a_+ for positive and a_- for negative linear dependency between $\vec{v_i}$ and $\vec{v_j}$. In Fig.1(b), we can see two pairs of separated planes as results of our experiment. When examining vector pairs and their corresponding ϱ, we can confirm our assumption about the relation between the value of $\varrho(\vec{v_i}, \vec{v_j})$ and *pa*-cases. The correlation coefficient has two advantages: (i) direction awareness and (ii) a direct link between the *pa*-case and the algebraic sign of the ϱ-value.

Regarding Fig.1(b), we notice gaps between the planes. These originate from vector pairs where at least one member has zero variance ($\sigma^2 = 0$). The *Pearson correlation coefficient* is defined as the ratio of the covariance of two vectors and the product of their standard deviations. Therefore, $\sigma^2 = 0$ leads to a division by zero, making ϱ undefined. To get rid of this problem, we exclude the mentioned division from ϱ, reducing it to the *covariance*.The results for this last experiment are shown in Fig. 1(c). We observe a behavior similar to ϱ, but in contrast there are no undefined areas and continuous values. The last two experiments have shown a special behavior of ϱ and covariance for vectors with $\sigma^2 = 0$. While ϱ is not defined for these cases, the covariance yields zero.

Vectors $\vec{v_i}$ with $\sigma^2 = 0$ are an interesting phenomenon in our soft clustering scenario. They satisfy $\forall v_{ip}|v_{ip} = \frac{1}{k_l}$, stating that the respective object x_i has equal relations with all clusters of C_l. Thus, it is impossible to determine an explicit cluster affiliation for this object. We refer to such cases as *fully balanced* assignments. Since we cannot decide in which cluster an object x_i with a fully balanced assignment is situated in, it is also impossible to determine a *pa*-case for a pair $x_i; x_j$ if at least one member has a fully balanced assignment. Until now, all clustering aggregation approaches assume only two possible *pa*-cases, a_+ and a_-. With the emergence of fully balanced assignments, a novel additional *pa*-case can be defined covering object pairs with undecidable assignments.

3   Expanding Pair-Wise Assignments

The previous section has shown that the determination of pair-wise assignments is a non-trivial task in a scenario utilizing soft cluster mappings. Existing approaches use common distance functions to solve this problem. But our experiments brought up two major flaws of this concept: **(i)** not all distance functions can be effectively applied and **(ii)** fully balanced assignments are ignored. In this section we expand the concept of pair-wise assignments to fix the aforementioned problems and introduce a novel significance score for *pa*-cases.

3.1   A Novel Pair-Wise Assignment

In our preliminaries, we described *fully balanced* assignments as a special kind of assignments that make the identification of an explicit cluster relation impossible. Until now, the concept of pair-wise assignments has been restricted to two possible cases that both need definite cluster affiliations. Therefore, proper handling of fully balanced cases requires a novel third assignment case. This case covers undecidable pair-wise assignments and will be denoted as $a_?$. To correctly determine a *pa*-case for any pair $x_i; x_j$ in a clustering C_l, we need to know if a $\vec{v_i}$ is fully balanced. This is the case if each component of $\vec{v_i}$ equals $\frac{1}{k_l}$. An additional form of undecidable assignments, which we denote as *balanced*, occurs with vectors having more than one maximum component v_{ip}. Assume e.g. an object x_i with $\vec{v_i}^\top = (0.4, 0.4, 0.2)$ for a clustering C_l with $k_l = 3$ clusters. Although we can state that x_i is not a member of cluster 3, it is impossible to specify whether the object effectively belongs to cluster 1 or 2. In contrast, a vector $\vec{v_i}^\top = (0.6, 0.2, 0.2)$ containing multiple equal but not maximal components v_{ip} is not critical. As long as the maximum v_{ip} is singular, we can derive a clear cluster affiliation. Based on this observation, we define a balance-detection function $b(\vec{v_i})$ testing if an object x_i has a fully balanced or a balanced assignment. If $\vec{v_i}$ contains multiple maxima, hence showing no clear cluster affiliation, the function $b(\vec{v_i})$ results in *true*; otherwise $b(\vec{v_i})$ yields *false*.

Next, we need to decide whether x_i and x_j belong to the same partition of C_l or not. Therefore, we regard the strongest cluster affiliation of x_i i.e. the maximum v_{ip}. If the maximum components v_{ip} and v_{jq} of two vectors $\vec{v_i}; \vec{v_j}$, are located in the same dimension of their respective vectors, x_i and x_j belong to the same cluster. In contrast, objects with maximum components in different dimensions of $\vec{v_i}$ are located in different clusters. Based on this, we define a co-occurrence function $c(\vec{v_i}, \vec{v_j})$, stating whether $x_i; x_j$ are part of the same cluster:

$$c(\vec{v_i}, \vec{v_j}) \begin{cases} 1 & \text{if } \{p|v_{ip} = max(\vec{v_i})\} \cap \{q|v_{jq} = max(\vec{v_j})\} \neq \emptyset \\ -1 & \text{otherwise} \end{cases} \quad (1)$$

$$case(x_i, x_j) = \begin{cases} 1, & \text{if } c(\vec{v_i}, \vec{v_j}) = 1 \text{ and } \neg(b(\vec{v_i}) \vee b(\vec{v_j})) \\ -1, & \text{if } c(\vec{v_i}, \vec{v_j}) = -1 \\ 0, & \text{otherwise} \end{cases} \quad (2)$$

where $i \neq j$, $1 \leq (p, q) \leq k_l$ and $max(\vec{v_i})$ returns the maximum component of $\vec{v_i}$. Now, we can create a function $case()$ (eq.2) that determines the *pa*-case of any object pair in a given clustering C_l. Our function $case(x_i, x_j)$ returns 1 if a_+ holds for x_i and x_j. This is the case if no object has a balanced or fully balanced $\vec{v_i}$ and if both objects are clearly related with the same cluster of C_l. The result -1 denotes the *pa*-case a_-. There, it is not relevant if balanced objects are part of the pair in question. Assume for C_l with $k_l = 3$ a balanced $\vec{v_i} = (0.4, 0.4, 0.2)$ and $\vec{v_j} = (0.1, 0.1, 0.8)$. Since the maximum components are in different dimensions, a_- holds. Although we cannot decide to which cluster x_i belongs, it is definitely not the cluster x_j belongs to. For undecidable cases like pairs containing fully balanced objects or pairs with balanced assignments that co-occur ($c(\vec{v_i}, \vec{v_j}) = 1$), $case()$ yields 0, indicating $a_?$. Our function $case()$ solves the problems described at the beginning of this section and allows the correct determination of one of our three pa-cases for any arbitrary object pair.

3.2   Introducing Significance

By definiton, our novel $a_?$ case is limited to specific vector compositions, whereas the remaining two pa-cases apply for nearly all possible object pairs resp. a wide range of $\vec{v_i}$'s. Therefore, it is obvious to bring up the question of significance. In other words, is a decision for a certain pair of objects made with more or less confidence than for other pairs?

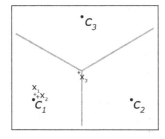

Fig. 2. $\vec{v_i}$ with different significance

Consider the example shown in Fig. 2, of a clustering C_l with $k_l = 3$ clusters and their respective centroids c_1, c_2 and c_3. The grey lines show the borders of the area of influence each cluster has. An object located on those lines or at intersection points has an equal degree of similarity with adjacent clusters and has thus a balanced resp. fully balanced assignment. The two depicted objects x_1 and x_2 have a very strong relation with c_1 and only negligible links with the remaining clusters of C_l. For this example, our function $case(x_1, x_2)$ results in 1, hence a_+ would be stated for x_1 and x_2. Now regard object x_3: it still has the strongest degree of similarity with c_1 but it also has a nearly equal similarity with c_2 and c_3, bringing x_3 very close to a fully balanced assignment. Nevertheless, $case(x_1, x_3)$ determines that x_1 and x_3 both belong to cluster c_1, which is correct in this example. Regarding both resulting pa-cases, we would intuitively say that the one made for x_1, x_2 has more confidence.

When the significance of a pa-case is evaluated in a subjective way, two properties have to be respected: **(i)** $\vec{v_i}$ and $\vec{v_j}$ should show an explicit cluster relationship i.e. show high dissimilarity to the fully balanced assignment, like $x_1; x_2$ in Fig. 2; **(ii)** $\vec{v_i}$ and $\vec{v_j}$ should have a high component-wise similarity, which is also the case for $x_1; x_2$ in Fig. 2. Examples of pairs maximizing both factors are the corners of the planes, shown in Fig. 1. It is plausible to assume that, starting from these locations, the significance should decrease when approaching the middle of the plane or one of its bisectors (one of the grey lines in Fig. 2 resp.), where balanced or fully balanced assignments are located. As we can see in Fig. 1(c), the covariance partly shows this desired behavior of high values at the corners and low resp. zero values in the middle of the plane. Using this observation, we define a significance measure $s(\vec{v_i}, \vec{v_j})$—similar to the covariance—that returns a significance score for a pa-case determined for a pair $x_i; x_j$ in a clustering C_l.

$$s(\vec{v_i}, \vec{v_j}) = \sum_{p,q=1}^{k_l} \left(|v_{ip} - \frac{1}{k_l}| \cdot |v_{jq} - \frac{1}{k_l}| \right) (i \neq j, p = q, 1 \leq (p,q) \leq k_l)) \quad (3)$$

A high value from $s(\vec{v_i}, \vec{v_j})$ indicates a high significance of the determined pa-case. Now, we are able to determine a pa-case and an additional significance value for any pair of objects. To simplify matters we combine $s(\vec{v_i}, \vec{v_j})$ and $case()$ into one single function: $case^+(x_i, x_j) = case(x_i, x_j) \cdot s(\vec{v_i}, \vec{v_j})$. With this, the result interpretation changes slightly. Now, the determined pa-case is denoted by the algebraic sign of the result, while its absolute value represents the confidence of the decision. For undecidable pa-cases the function simply yields 0.

Regarding significance, at this point, we can only evaluate it in relation to other significance values, stating e.g that a_+ for $x_1; x_2$ has a higher significance than for $x_1; x_3$. To make assumptions about the significance on an absolute scale, we need to normalize our results, so that $case^+(x_i, x_j)$ yields 1 if a_+ holds and -1 if a_- holds with maximum significance. Therefore, we require the results of $case^+$ for the mentioned cases. We will illustrate this normalization with some examples, beginning with the most significant a_+ case. An example for this case, in a C_l with $k_l = 3$, would be given for $x_i; x_j$ with $\vec{v_i}^\top = \vec{v_j}^\top = (1, 0, 0)$. As simplification we assume for these examples that $v_{ip} = 1$ and $v_{ip} = 0$ can occur. Actually the strict definition for soft cluster assignments demands $\forall v_{ip} | 0 < v_{ip} < 1$. In this example, the most significant a_+ leads to $case^+ = \frac{2}{3}$. Using this setting, the most significant a_- occurs e.g. for $\vec{v_i}^\top = (1, 0, 0)$ and $\vec{v_i}^\top = (0, 0, 1)$ and results in $case^+ = -\frac{5}{9}$. We can see that the absolute values differ for both maximum significance cases. The reason for this behavior is $s(\vec{v_i}, \vec{v_j})$. It measures the distance from the fully balanced assignment in each dimension. We already know that $0 < v_{ip} < 1$, by $\frac{1}{k_l}$ this range is divided into two intervals. These have equal sizes for $k_l = 2$ but become disproportionate as k_l increases resp. $\frac{1}{k_l}$ decreases. This means that $v_{ip} > \frac{1}{k_l}$ can have a higher maximum distance to $\frac{1}{k_l}$ than $v_{ip} < \frac{1}{k_l}$. Based on this we define a norm considering k_l and integrate it into our $case^+$ method, thus creating our final function $case^{\|+\|}$:

$$case^{\|+\|}(x_i, x_j) = \frac{case^+(x_i, x_j)}{\|k_l\|}; \quad \|k_l\| = \begin{cases} 1 - \frac{1}{k_l} & \text{if } case(x_i, x_j) = 1 \\ -\frac{4}{k_l^2} + \frac{3}{k_l} & \text{if } case(x_i, x_j) = -1 \\ 1 & \text{if } case(x_i, x_j) = 0 \end{cases}$$
$$(4)$$

4 Flexible Clustering Aggregation

In this section, we describe how the expansions introduced in the previous section are integrated into the clustering aggregation to make it flexible and enable result adjustments. For the basic aggregation procedure, we adopt the idea described by Gionis et al. in [5]. Using our function $case^{\|+\|}$, we determine the pa-case for every object pair in all clusterings of $\mathcal{C}$. When deciding on the assignment case for $x_i; x_j$ in the aggregated result $\hat{C}$, we enact a majority decision and choose the

pa-case occuring the most for $x_i; x_j$. If no majority can be identified, e.g. if all three pa-cases have equal occurences, we decide for $a_?$ for the corresponding pair in the aggregate, since the final/global assignment is effectively undecidable. With this method, we can construct an aggregate but we are still lacking flexibility resp. control.

To achieve this control, we utilize the significance information provided by $case^{\|+\|}$ and filter all pa-cases according to their significance. This can be done with a filtering function that returns 0 if $case^{\|+\|}(x_i, x_j)| \leq t$ and $case^{\|+\|}(x_i, x_j)$ otherwise. The threshold t specifies the minimum amount of significance a pairwise assignment needs to have to be considered as decidable. Therefore, all assignments not exceeding t are classified as $a_?$ with zero significance. With this we are able to create an area of undecidability that allows us to mark not only balanced/fully balanced assignments as $a_?$, but also those assignments in their close proximity. Lets regard our example in Fig. 2 again: undecidable assignments are located on the grey lines and for pair $x_1; x_3$, a_+ holds with low significance. If we apply filtering, the grey lines of Fig. 2 expand and form an area of undecidability that can be described as a union of circles centered at intersection points and broadened lines/stripes. With increasing t the circles radii and width of stripes also increase. If the t-defined area is big enough to enclose x_3, its assignment becomes undecidable. Under these conditions, the pair $x_1; x_3$ is classified as $a_?$. Basically, via filtering we guarantee a minimal confidence for all decidable pa-cases.

In Fig. 3(a), the results of our function $case^{\|+\|}$ for the experimental setting from Section 2 are shown. We can observe our desired behavior of absolute and maximal significance scores at the plane corners. Take for example the western corner at $\vec{v_i}^\top = (1,0)$ and $\vec{v_j}^\top = (1,0)$, the pa-case for this pair is a_+ with maximum significance, so $case^{\|+\|}$ yields 1 at this point. The significance drops linearly towards and equals zero at the planes middle and its bisectors. The middle of the plane is specified by $\vec{v_i}^\top = (0.5, 0.5)$ and $\vec{v_j}^\top = (0.5, 0.5)$. This pair is composed of two objects with fully balanced assignments, making it undecidable i.e. $case^{\|+\|}$ yields zero. When we apply *filter* with threshold $t = 0.3$, the results change to Fig. 3(b). A flat area has formed around the center of the plane and

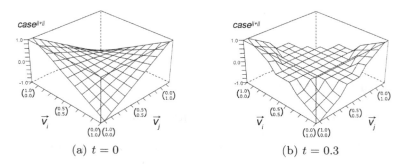

(a) $t = 0$ (b) $t = 0.3$

Fig. 3. Results of $case^{\|+\|}$ with and without filtering

its bisectors. None of the object pairs in this area satisfies the filtering criterion, and hence, is classified as $a_?$.

With the methods proposed so far, we are able to determine one of our three pa-cases on the aggregate level and can control the amount of $a_?$ via t. With this, we define stable cores in our aggregate–a_+,a_- robust against t–and around them, areas of undecidable $a_?$'s. These areas are the key to result flexibility and we introduce some examples of $a_?$-handling in the next section.

5 Evaluation

For the experiments in this section, we used a synthetic dataset consisting of 1500 objects. These objects form 7 clusters, where 2 clusters are very close but not linked and two cluster pairs are connected via bridges of different length and width. The dataset structure is depicted in Fig. 4. We used k-means [2] to generate our input clusterings. Due to the characteristics of our dataset and k-means, it is very unlikely that we obtain a good clustering using iterations with only single algorithm runs.

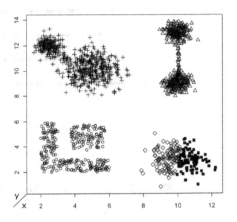

Fig. 4. $\hat{C}$ using scalar aggregation

By applying existing aggregation approaches, we can already improve results, even for disadvantageous algorithm-dataset combinations. Therefore, we generate a $\mathcal{C}$ with 10 input clusterings using $k = \{2, 3, \cdots, 10, 15\}$ and different initializations that is aggregated using the technique described in [5]. Fig. 4 shows the obtained result, consisting of five clusters, where three clusters might be divided further while the remaining two clusters could be merged. We see that with clustering aggregation, a useful partitioning can be obtained even if singular algorithm execution yields suboptimal results. But this aggregation result is still not optimal and if the user wants to adjust it, he/she has to repeat the cycle: (i) modify parameters/algorithms of $\mathcal{C}$; (ii) recreate $\mathcal{C}$; (iii) execute aggregation; (iv) evaluate $\hat{C}$ until the desired adjustments occur.

For our flexible clustering aggregation, we use the same setup as before but change the algorithm to FCM [13], a soft version of k-means. Concerning the handling of $a_?$, we have to regard two alternatives, since we cannot determine if undecidable pairs are in the same cluster or not. Therefore, we define two strategies: one mapping $a_?$ to a_+ and another one that maps it to a_-. We let t run from zero to one in steps of 0.1, obtaining 11 aggregation results, one for each t. Fig. 5(a), shows the distribution of the determined pa-cases for all C_l and $\hat{C}$, monitored over all runs. Each block of the matrix displays the ratio of the pa-cases with reference to all object pairs of a C_l (specified by row) subjected to

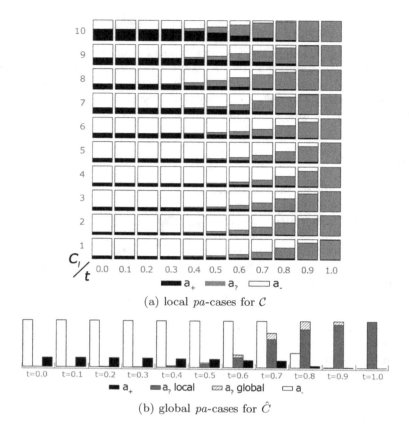

(a) local *pa*-cases for $\mathcal{C}$

(b) global *pa*-cases for $\hat{C}$

Fig. 5. Evaluation results

filtering using t (specified by column). We observe that the number of $a_?$ rises with increasing t, whereas different C_l show different levels of robustness towards t. The distributions of the global *pa*-cases leading to $\hat{C}$ are shown in Fig. 5(b). We notice again that with increasing t the number of $a_?$ rises. In this diagram, $a_?$ *local* indicates $a_?$ as dominant, while $a_?$ *global* implies multiple dominant *pa*-cases and thus undecidability on the aggregate level. A major part of our future work will be the utilization of this significance information for the construction of $\mathcal{C}$ resp. evaluation of its clusterings.

We now adjust the aggregate by modifying t and $a_?$-handling, while $\mathcal{C}$ remains untouched. We choose $a_? \rightarrow a_+$ and increase t. With $t = 0.1$, we obtain the result shown in Fig. 6(a), where the two clusters in the lower right have been fused due to the points along the border between both former clusters. Having nearly equal affiliations to both clusters, they lead to *pa*-cases with low significance. Therefore, $a_?$ starts to occur near the border when $t = 0.1$ is applied. Since we map $a_?$ to a_+, both clusters are connected. If t increases further, more clusters connect leading to a unification of all datapoints at $t = 0.4$. This *merge* strategy

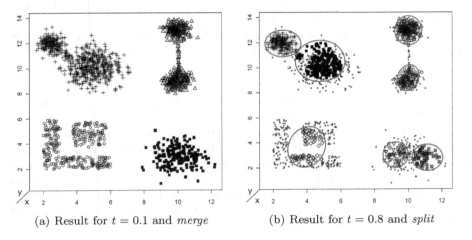

(a) Result for $t = 0.1$ and *merge*   (b) Result for $t = 0.8$ and *split*

Fig. 6. Aggregation results

is very delicate since one single pair classified as a_+ is enough to merge whole clusters, that would otherwise be very dissimilar.

Next, we use $a_? \rightarrow a_-$, which yields the result shown in Fig. 4 at $t = 0$ and observe no changes in $\hat{C}$ until $t = 0.4$. At this point, an additional cluster forms and contains all objects the algorithm was unable to assign to a cluster because they are labeled $a_?$ and hence a_- in all of C. Those objects are put into a *noise* cluster for convenience and presentation. Actually, each object is a singleton cluster for itself, since no affiliations to other objects or existing clusters can be determined, which is a novel trait that cannot occur in existing aggregation approaches. When we increase t, this *noise* grows, especially in areas equally influenced by multiple clusters. Fig. 6(b) shows the aggregation result for $t = 0.8$ with *noise* marked as *. We notice that the clusters in the upper quadrants were split by the *noise*. As $t \rightarrow 1$, all objects of the dataset become members of the *noise* cluster. During our experiments, we discovered that in contrast to our *merge* approach, this *split* strategy leads to slighter changes of the clustering aggregate. Part of our future work will deal with the construction of additional $a_?$-handling strategies as well as finding a method allowing independent selection of the handling strategy for each individual undecidable pa-case.

We showed that reasonable adjustments of $\hat{C}$ are possible using filtering and our proposed strategies *merge* and *split*. These adjustments can be easily made, since the required parameters can be abstracted to simple options. The handling strategies for $a_?$ effectively compare to "more clusters" for *split* and "fewer clusters" for *merge*, while t describes "how strong" each strategy is enforced. In summary, this section illustrated that clustering aggregation is beneficial even for algorithms not fitting the data. Futhermore, it described limitations of existing approaches and how to overcome them, using our flexible aggregation. Unfortunately, control and clustering flexibility come at the cost of runtime. Like all aggregation approaches utilizing pair-wise assignments, e.g. [5], our approach has a complexity of $\mathcal{O}(n^2)$, with n being the number of data objects. Additionally, our approach use vector calculations that add to the runtime. Runtime optimization is a general field of research in the clustering area and a major part of our future research in particular but not focus of this paper.

6 Conclusion

In this paper, we proposed our flexible clustering aggregation approach that allows the construction of a clustering aggregate from a set of separate soft clustering results. We described the challenges of pair-wise assignments and decidability in this scenario and introduced (i) novel tools like the $a_?$ pair-wise assignments to master these challenges, (ii) a significance measure for pa-cases as well as (iii) a controllable aggregation process. All this enables our approach to produce adjustable results. We also simplified and abstracted our proposed parameters, thus allowing user-friendly and -guided identification of structures hidden from existing aggregation methods.

References

1. Ester, M., Kriegel, H.P., Sander, J., Xu, X.: A density-based algorithm for discovering clusters in large spatial databases with noise. In: Proc. of KDD (1996)
2. Forgy, E.W.: Cluster analysis of multivariate data: Efficiency versus interpretability of classification. Biometrics 21 (1965)
3. Jain, A.K., Murty, M.N., Flynn, P.J.: Data clustering: a review. ACM Comput. Surv. 31 (1999)
4. Zeng, Y., Tang, J., Garcia-Frias, J., Gao, G.R.: An adaptive meta-clustering approach: Combining the information from different clustering results. In: Proc. of CSB (2002)
5. Gionis, A., Mannila, H., Tsaparas, P.: Clustering aggregation. In: Proc. of ICDE (2005)
6. Boulis, C., Ostendorf, M.: Combining multiple clustering systems. In: Boulicaut, J.-F., Esposito, F., Giannotti, F., Pedreschi, D. (eds.) PKDD 2004. LNCS (LNAI), vol. 3202, pp. 63–74. Springer, Heidelberg (2004)
7. Strehl, A., Ghosh, J.: Cluster ensembles — a knowledge reuse framework for combining multiple partitions. Journal of Machine Learning Research 3 (2002)
8. Filkov, V., Skiena, S.S.: Heterogeneous data integration with the consensus clustering formalism. In: Rahm, E. (ed.) DILS 2004. LNCS (LNBI), vol. 2994, pp. 110–123. Springer, Heidelberg (2004)
9. Fred, A.L.N., Jain, A.K.: Robust data clustering. In: Proc. of CVPR (2003)
10. Dimitriadou, E., Weingessel, A., Hornik, K.: Voting-merging: An ensemble method for clustering. In: Dorffner, G., Bischof, H., Hornik, K. (eds.) ICANN 2001. LNCS, vol. 2130, p. 217. Springer, Heidelberg (2001)
11. Long, B., Zhang, Z.M., Yu, P.S.: Combining multiple clusterings by soft correspondence. In: Proc. of ICDM (2005)
12. Topchy, A.P., Jain, A.K., Punch, W.F.: Combining multiple weak clusterings. In: Proc. of ICDM (2003)
13. Bezdek, J.C.: Pattern Recognition with Fuzzy Objective Function Algorithms. Plenum, New York (1981)
14. Habich, D., Wächter, T., Lehner, W., Pilarsky, C.: Two-phase clustering strategy for gene expression data sets. In: Proc. of SAC (2006)

Compensation of Translational Displacement in Time Series Clustering Using Cross Correlation

Frank Höppner[1] and Frank Klawonn[2,3]

[1] Department of Economics
University of Applied Sciences Braunschweig/Wolfenbüttel
Robert Koch Platz 10-14, D-38440 Wolfsburg, Germany
[2] Department of Computer Science
University of Applied Sciences Braunschweig/Wolfenbuettel
Salzdahlumer Str. 46/48, D-38302 Wolfenbuettel, Germany
[3] Helmholtz Centre for Infection Research
Department for Cell Biology
Inhoffenstr. 7, D-38124 Braunschweig, Germany

Abstract. Although k-means clustering is often applied to time series clustering, the underlying Euclidean distance measure is very restrictive in comparison to the human perception of time series. A time series and its translated copy appear dissimilar under the Euclidean distance (because the comparison is made pointwise), whereas a human would perceive both series as similar. As the human perception is tolerant to translational effects, using the cross correlation distance would be a better choice than Euclidean distance. We show how to modify a k-means variant such that it operates correctly with the cross correlation distance. The resulting algorithm may also be used for meaningful clustering of time series subsequences, which delivers meaningless results in case of Euclidean or Pearson distance.

1 Introduction

Finding typical patterns within a set of short time series by cluster analysis is a common approach in data analysis. Short time series might arise explicitly, for instance, as growth curves of bacteria or populations under varying conditions, but short time series can also be extracted from one long time series in the form of a sliding window. However, clustering such short time series derived from sliding windows can easily lead to meaningless results [13], a problem that will be addressed in more detail later on in this paper.

Among the applied clustering methods, k-means with Euclidean distance is most frequently used [12]. Typically, normalization of the time series is carried out, so that scaling of the measurements and the basic level of the time series do not have any influence. The common choice is z-score normalization. In this case, the Euclidean distance corresponds to the Pearson correlation coefficient up to a constant factor – thus, cluster analysis of time series with the Euclidean distance with z-score normalized data as the distance measure is (almost) the same as clustering with the Pearson correlation as the distance measure [2].

N. Adams et al. (Eds.): IDA 2009, LNCS 5772, pp. 71–82, 2009.
© Springer-Verlag Berlin Heidelberg 2009

Although z-score normalization and the Pearson correlation coefficient rule out differences based on scaling of the measured variable, they do not ensure a suitable time series alignment in case the observed "patterns" do not start at the same time (e.g., depending on the start of recording the data). In order to group time series with a similar patterns that are shifted in time, cross correlation appears to be a better choice. This paper discusses how to incorporate the cross correlation distance into k-means clustering. We want to emphasize, that our main purpose is *not* to advocate the cross correlation distance as the best measure to compare time series[1], but to show, once the decision to use cross correlation has been made, *how* to modify k-means appropriately to guarantee correct objective function-optimization.

For reasons of robustness [15], rather than simple k-means clustering we will use fuzzy clustering which is nothing else than a reformulation of k-means clustering with continuous membership degrees. Nevertheless, the proposed approach does not depend on fuzzy clustering and works in the same way with crisp clustering.

2   Brief Review of Fuzzy c-Means and Noise Clustering

Given a dataset $\{x_1, \ldots, x_n\} \subset \mathbf{R}^p$, k-means as well as fuzzy c-means (FCM) clustering [3] aims at minimizing the following objective function

$$f = \sum_{i=1}^{c} \sum_{j=1}^{n} u_{ij}^m d_{ij} \tag{1}$$

under the constraints

$$\sum_{i=1}^{c} u_{ij} = 1 \qquad \text{for all } j = 1, \ldots, n \tag{2}$$

where $u_{i,j}$ is the membership degree of data object x_j to cluster i and $d_{ij} = \|x_j - v_i\|^2$ denotes the squared Euclidean distance between data vector x_j and prototype $v_i \in \mathbf{R}^p$ representing cluster i. While k-means assumes a crisp assignment $u_{i,j} \in \{0, 1\}$, its *continuous* counterpart allows $u_{ij} \in [0, 1]$. c is the chosen number of clusters and m is the so-called fuzzifier, controlling (in case of fuzzy c-means) how much clusters may overlap.

Clustering is thus considered as a nonlinear optimization problem which is usually solved by an alternating scheme. The prototypes are chosen randomly in the beginning or by some suitable initialization strategy. Fixing the cluster prototypes, the optimal choice for the membership degrees is given by[2]

$$u_{ij} = \frac{1}{\sum_{k=1}^{c} \left(\frac{d_{ij}}{d_{kj}}\right)^{\frac{1}{m-1}}} \tag{3}$$

---

[1] If the series are dilated, measures such as dynamic time warping may be an option.
[2] If $d_{ij} = 0$ for one or more clusters, we deviate from (3) and assign x_j with membership degree 1 to the or one of the clusters with $d_{ij} = 0$ and choose $u_{ij} = 0$ for the other clusters i.

which is used as an update equation for the membership degree. Fixing the membership degrees, the best choice for the prototypes is

$$v_i = \frac{\sum_{j=1}^{n} u_{ij}^{m} x_j}{\sum_{j=1}^{n} u_{ij}^{m}}. \tag{4}$$

The alternating scheme is repeated until the algorithm converges, i.e., no more (or almost no) changes happen.

A very simple extension of k-means and fuzzy c-means clustering to cope with outliers is noise clustering [7]. In the set of prototypes, an additional noise cluster is included. All data have a fixed (usually large) distance d_{noise} to the noise cluster. As soon as the distance of some data x to the nearest cluster p comes close to d_{noise}, the noise cluster gains a considerable fraction of the total membership degree, thereby reducing the influence of x with respect to p. Noise clustering simply requires to exchange (3) by

$$u_{ij} = \frac{1}{\left(\frac{d_{ij}}{d_{\text{noise}}}\right)^{\frac{1}{m-1}} + \sum_{k=1}^{c} \left(\frac{d_{ij}}{d_{kj}}\right)^{\frac{1}{m-1}}} \tag{5}$$

and represents and effective mean to reduce the influence of noise and extract cluster prototypes more clearly.

In the objective function (1), the number of clusters c must be known or specified in advance. This is, of course, an unrealistic assumption. There are various approaches to determine the number of clusters automatically (for an overview, we refer to [4,10]). It is also possible to find clusters step by step, extending the idea of noise clustering [8]. A detailed discussion of methods for determining the number of clusters is out of the scope of this paper.

3   Measuring Time Series Similarity

Suppose we have two time series r and s, consisting of T samples each $r = (r_1, r_2, \ldots, r_T) \in \mathbf{R}^T$. The *squared Euclidean distance* between two time series r and s is given by:

$$d_E(r, s) = \sum_{t=1}^{T} (r_t - s_t)^2 \tag{6}$$

For time series analysis, it is often recommended to normalize the time series either globally or locally to tolerate vastly differing ranges [12]. Another prominent measure for time series comparison is the Pearson correlation coefficient, which measures the correlation ϱ between two random variables X and Y:

$$\varrho_{X,Y} = \frac{E[(X - \mu_X)(Y - \mu_Y)]}{\sigma_X \sigma_Y} \tag{7}$$

where μ_X denotes the mean and σ_X the standard deviation of X. We obtain a value of ± 1 if X and Y are perfectly (anti-) correlated and a value of ≈ 0

if they are uncorrelated. In order to use the Pearson correlation coefficient as a distance measure for time series it is desirable to generate low distance values for positively correlated (and thus similar) series. The *Pearson distance* is therefore defined as

$$d_P(r, s) = 1 - \varrho_{r,s} = 1 - \frac{\frac{1}{T}\sum_{t=1}^{T}(r_t - \mu_r)(s_t - \mu_s)}{\sigma_r \sigma_s} \tag{8}$$

such that $0 \leq d_P(r, s) \leq 2$. One can show that k-means clustering (via Euclidean distance) on z-score normalized time series is (almost) equivalent to k-means clustering using the Pearson correlation distance[3] [2].

Sometimes time series do not perfectly align in time and the best correspondence is obtained when shifting both series against each other. Two identical time series, one of them shifted by δ in time, may appear uncorrelated under the Pearson coefficient. The normalized cross correlation takes this shift δ into account and measures the Pearson correlation between r and a series s shifted by δ:

$$\varrho_{r,s}(\delta) = \frac{\frac{1}{T}\sum_{t=-T}^{T}(r_t - \mu_r)(s_{t+\delta} - \mu_s)}{\sigma_r \sigma_s} \tag{9}$$

with $s_t = r_t = 0$ for $t < 1$ and $t > T$. Cross correlation can help to overcome the missing alignment of the series by choosing δ such that the Pearson correlation becomes maximal. We define the cross correlation distance d_X as the best Pearson coefficient we may achieve for an optimal lag of δ where $-T \leq \delta \leq T$:

$$d_X(r, s) = 1 - \max\{\varrho_{r,s}(\delta) \mid -\Delta \leq \delta \leq \Delta\} \tag{10}$$

4 Compensating Translational Displacement

There are many more distance measures for comparing time series (cf. [12]), but often the proposed distance measures are simply plugged into an existing clustering algorithm without taking influence on the internal steps of the respective algorithm. But at least for objective function-based clustering (such as k-means or fuzzy c-means) replacing the distance measure alone is not sufficient: to operate correctly, the prototype update step has to be adapted to the chosen distance measure. The objective of this paper is therefore not to propose a new distance measure (as Pearson correlation and cross correlation are already well-established), but to modify the prototype update step in fuzzy c-means clustering such that prototypes are optimized w.r.t. cross correlation rather than Euclidean distance.

Let us denote the data series by x_j, $1 \leq j \leq n$, and the prototype series by p_i, $1 \leq i \leq c$. The cross correlation clustering (CCC) algorithm aims at minimizing the objective function

$$f = \sum_{i=1}^{c}\sum_{j=1}^{n}u_{i,j}^m d_X(p_i, x_j)$$

---

[3] It is *almost* equivalent because a normalization of the obtained prototypes is missing, but the absence of this re-scaling is usually neglectable in terms of results.

subject to (2). We introduce for any pair of time series (p_i, x_j) the lag parameter $\delta_{i,j}$ and reformulate f as

$$f = \sum_{i=1}^{c} \sum_{j=1}^{n} u_{i,j}^{m} (1 - \varrho_{p_i, x_j}(\delta_{i,j}))$$

The optimization will be carried out by alternating optimization in three steps: (a) optimize w.r.t. $\delta_{i,j}$ assuming prototypes and memberships being constant (Sect. 4.1), (b) optimize w.r.t. prototypes assuming lags and memberships being constant (Sect. 4.2), and (c) optimize w.r.t. memberships assuming lags and prototypes to be constant. The last step (c) is independent of the distance and therefore identical to update equations (3) or (5).

4.1 Efficient Calculation of the Optimal Lag

Calculating the cross correlation for a given value of δ is $O(T)$ in case of discrete time series of length T. Exploring the full range of possible δ-values is linear in T, too, so the overall complexity of distance estimation becomes $O(T^2)$. This time can be reduced to $O(T \log T)$ using the Fast Fourier Transform. (Unnormalized) cross correlation can be interpreted as a convolution $r \star s$ of two time series

$$(r \star s)(\delta) = \sum_{t=1}^{T} r_t s_{\delta-t}$$

Note that with convolution we have a factor of $s_{\delta-t}$ whereas cross correlation uses $s_{\delta+t}$. By reversing series s in time ($\bar{s}_t = s_{T-t}$) we overcome this difference.

However, we need to convolve two *normalized series* r and s: Suppose the first half of s correlates perfectly with the second half of r, but the remainder of the series are random noise. If we would normalize the respective halves of r and s individually, and calculate the correlation we would obtain a coefficient of 1.0. But if we would simply normalize r and s once beforehand, a lag of $\delta = T/2$ would deliver a different coefficient because the mean and variance considers also those parts of the time series that are otherwise masked out by the shift. Therefore, the normalization has to be carried out individually for each possible lag δ.

For notational convenience, we consider all series having indices ranging from $-T+1$ to $2T$ (filled up with zeroes) and denote an offset of δ by $s^{+\delta}$ as shown in this example for $T = 4$:

$$
\begin{aligned}
r = &\quad (3\,1\,3\,5) \qquad \in \mathbf{R}^4 \\
s = &\quad (1\,3\,5\,2) \\
s^{+0} = &\, (0\,0\,0\,0\,1\,3\,5\,2\,0\,0\,0\,0) \\
s^{+1} = &\, (0\,0\,0\,0\,0\,1\,3\,5\,2\,0\,0\,0) \\
s^{-2} = &\, (0\,0\,1\,3\,5\,2\,0\,0\,0\,0\,0\,0)
\end{aligned}
$$

We define $\chi \in \mathbf{R}^T$ as $\chi_t = 1$ for any $1 \le t \le T$ and $\chi_t = 0$ otherwise. For instance

$$
\begin{aligned}
\chi = &\, (0\,0\,0\,0\,1\,1\,1\,1\,0\,0\,0\,0) \\
\chi^{-2} = &\, (0\,0\,1\,1\,1\,1\,0\,0\,0\,0\,0\,0)
\end{aligned}
$$

The calculation of mean and standard deviation and subsequent normalization of two series, say r and s^δ, has to be carried out for the all indices t with $\chi_t \cdot \chi_t^\delta = 1$.

Fortunately, this normalization can also be carried out efficiently: By means of a series of partial sums $\hat{r}_0 = r_0$, $\hat{r}_{t+1} = \hat{r}_t + r_{t+1}$ and $\hat{\hat{r}}_0 = r_0^2$, $\hat{\hat{r}}_{t+1} = \hat{\hat{r}}_t + r_{t+1}^2$ we obtain the mean for the respective subseries $r_{i..j}$ from $(\hat{r}_j - \hat{r}_{i-1})/(j - i + 1)$ and the variance from $(\hat{\hat{r}}_j - \hat{\hat{r}}_{i-1})/(j - i + 1) - (\hat{r}_j - \hat{r}_{i-1})^2/(j - i + 1)^2$ due to $\mathrm{Var}[X] = E[X^2] - (E[X])^2$. Therefore the determination of the optimal lag δ remains $O(T \log T)$ even in the case of normalized correlation coefficients.

Revisiting the last example, we note that series r and s correlate perfectly for different values of δ, for instance, $\delta = +1$ and $\delta = -2$. In general we can expect high values of normalized correlation for $\delta \approx \pm T$ because then only a few values have to correlate incidentally. Therefore we introduce a bias in d_X towards preferable *long* matches. The distance d_X is multiplied by an additional *overlap factor* of $\frac{T - |\delta_{i,j}|}{T}$ where $T - |\delta_{i,j}|$ is the number of valid index positions shared by both series. For small lags we thus obtain an almost unaltered correlation coefficient whereas for large values of δ a possibly high coefficient is penalized due to its limited relevance. In our example, the unique best lag for s and r is now $\delta = 1$.

4.2   Determination of the Prototypes

As already mentioned, k-means clustering of z-score normalized series is (almost) identical to clustering via Pearson correlation, that is, the prototypes are obtained by the weighted mean of the series that are associated to the prototype (followed by a normalization step in case of Pearson correlation) [2]. Once the optimal lags for cross correlation have been determined, we consider them in the second step of alternating optimization as being constant and the cross correlation distance reduces to Pearson distance for prototypes p_i and the shifted data series $x_j^{+\delta_{i,j}}$.

As we have seen in the previous section, any two series p_i and $x_j^{+\delta_{i,j}}$ share different ranges of indizes and the prototype calculation has to be carried out pointwise (that is, index by index). Then minimization w.r.t. the prototypes leads to the following update equations (which are a generalization of the weighted mean to shifted series):

$$v_{i,t} = \sum_{j=1}^{n} u_{i,j}^m \cdot x_{j,t}^{+\delta_{i,j}}$$

$$w_{i,t} = \sum_{j=1}^{n} u_{i,j}^m \cdot \chi_{j,t}^{+\delta_{i,j}}$$

$$p'_{i,t} = \frac{v_{i,t}}{w_{i,t}}$$

This calculation is not restricted to the indices $t = 0 \ldots T$, but carried out for the full range of $t = -T + 1 \ldots 2T$. If there are many data series x_j whose second

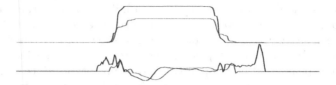

Fig. 1. Example for $w_{i,t}$ (top) and $p'_{i,t}$ (bottom) for $t = -T + 1 \ldots 2T$. To locate the best offset t_0 we identify the area of width T with the highest sum of weights $(\mathrm{argmax}_{t_0=-T+1..T} \sum_{t=1}^{T} w_{i,t_0+t})$.

half matches the first half of a prototype p, a higher *overlap factor* may possibly be achieved next time if the prototype would be shifted appropriately. The optimization w.r.t. the overlap factor is accomplished by analyzing the intermediate result $w_{i,t}$: These individual weights per index position indicate how often an optimal match to a data series has involved this index position. Figure 1 shows an example: the graph at the bottom shows the resulting pointwise mean (full width of $3T$) and the graph at the bottom the accumulated weight per index position. The optimal (T-dimensional) prototype subvector of p'_i is then found at

$$t_0 = \mathrm{argmax}_{t_0=-T+1..T} \sum_{t=1}^{T} w_{i,t_0+t}$$

Using a series of partial sums, this optimal position can be found in $O(T)$.

4.3   Interpretation of the Noise Distance

The third step of alternating optimization involves the membership update. For the case of correlation coefficients, the noise clustering approach is particularly appealing. As fuzzy c-means is a partitional clustering algorithm, all data series have to be assigned to the clusters, including those that do not correlate to any of the prototypes (or correlate negatively). By selecting a threshold correlation coefficient of, say, $d_{\mathrm{noise}} = 0.5$ the noise cluster attracts the membership of poorly correlating series, thereby avoiding a contamination of the clusters with poor matches and preventing a blurring of the cluster prototypes. Since the correlation coefficient has a fixed range of $[0, 2]$, the noise distance is easily interpretable.

5   Experimental Evaluation

We demonstrate the proposed method for clustering time series as well as clustering of time series subsequences (STS).

5.1   Clustering (Whole) Time Series

As a first test case, we consider the *hill & valley* dataset from the UCI repository [1]: Each record represents 100 points on a two-dimensional graph. When plotted in order the points will create either a hill (a bump in the terrain) or a valley (a dip in the terrain). We have z-score normalized the data and applied k-means clustering and cross correlation clustering, the results are shown in Fig. 2. As the peaks occur at different places, calculating the mean will eliminate all the individiual peaks and ends up with a noisy prototype. Both k-means prototypes converge roughly to the mean of the full data set, the two classes are not recovered by the prototypes. In contrast, the translational displacement is identified by the CCC algorithm. The hills and valleys are perfectly gathered in the respective clusters.

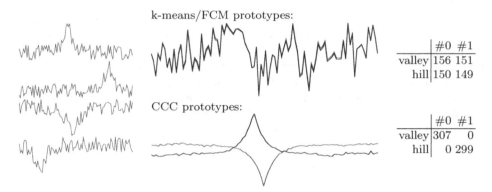

k-means/FCM prototypes:

	#0	#1
valley	156	151
hill	150	149

CCC prototypes:

	#0	#1
valley	307	0
hill	0	299

Fig. 2. Results on the hill & valley dataset. On the left, four example series are shown. The final prototypes and confusion matrix is given for k-means/FCM (top) and CCC (bottom).

The *synthetic control chart* dataset (taken from [14]) is also frequently used for the evaluation of time series clustering methods: The dataset contains 600 examples of synthetically generated control charts from six different classes: normal, cyclic, increasing trend, decreasing trend, upward shift, and downward shift (cf. left column of Fig. 3). Although started with 6 clusters, k-means/FCM ends up with three different prototypes only (always two prototypes are almost identical). The increasing and decreasing trend is well recovered, but these clusters also cover many examples from the up-shift and down-shift class. As the abrupt step in these series occurs at different points in time, k-means/FCM cannot detect them as individual clusters. The same argument applies to the cyclic series which have different phases. Again, CCC performs much better, all classes but one are recovered by the respective clusters. The first class (normal) consists of noise only, therefore it does not correlate to any existing cluster nor with other noisy series. Thus, the examples from this case are distributed among the other clusters by chance. The now superfluous sixth cluster is used to split the cyclic cluster into two prototypes. (The "optimal" number of clusters is thus 5 for CCC).

The results for the *cylinder-bell-funnel* (CBF) dataset, taken from [14], are shown in Fig. 4. It consists of 20 instances of three translated and dilated basic

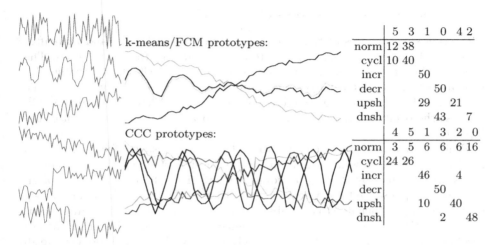

k-means/FCM prototypes:

	5	3	1	0	4	2
norm	12	38				
cycl	10	40				
incr			50			
decr				50		
upsh			29		21	
dnsh				43	7	

CCC prototypes:

	4	5	1	3	2	0
norm	3	5	6	6	6	16
cycl	24	26				
incr			46	4		
decr			50			
upsh			10	40		
dnsh					2	48

Fig. 3. Results on the synthetic control chart dataset. On the left, one example from each class is shown. The final prototypes and confusion matrix is given for k-means/FCM (top) and CCC (bottom).

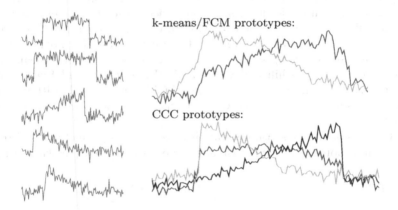

k-means/FCM prototypes:

CCC prototypes:

Fig. 4. Results on the cylinder-bell-funnel dataset. On the left, two examples from the cylinder class, one from the bell and two from the funnel class are shown. The final prototypes are presented for k-means/FCM (top) and CCC (bottom).

shapes (five example series in left column of Fig. 4). In the shown example run of k-means/FCM two of the three clusters occassionally collapsed into one proto-type, but in general k-means/FCM performs quite well on this dataset. Although the effects of dilation cannot be compensated by the CCC algorithm, the overall quality of the clusters is superior. A comparison between k-means/FCM and CCC with respect to the steepness of the flanks in the bell and funnel patterns reveals that the CCC patterns are much closer to the original patterns while the k-means/FCM clusters are somewhat blurred.

5.2   Clustering Subsequences of Time Series

Clustering subsequences of time series has been proposed in [6] and thereafter been used by many authors as a tool to extract patterns from time series. The resulting clusters, however, were of poor quality in practice [9], being very similar to translated and dilated trigonometric functions. A deeper analysis led to the conclusion that subsequence time series clustering is completely meaningless [13], because the resulting trigonometric patterns appeared to be independent of the input data.

While there are different attempts to explain this undesired effect [13,11,5], an intuitive explanation why subsequence clustering fails is easily given: the input series are obtained by shifting a sliding window of fixed length over the original series. Suppose we have a noisy series with a single bump (cf. hill and valley dataset in Fig. 2) then due to the subsequence generation this bump will re-occur at any location in the input series. The detection of a single cluster or pattern would be the desired result, but k-means clustering with $k = 1$ would average all the series and as the bump never repeats itself at the same spot, the bump gets completely blurred. So the problem is caused by the translation of the original pattern – and the CCC algorithm seems well prepared to compensate this displacement. Therefore, it can be considered as a promising candidate to overcome the problem of meaningless clusters in time series subsequence clustering.

For the hill and valley, CBF and ECG datasets we have concatenated all the series to a single, long time series and created a new dataset by moving a sliding window along the resulting series. Figure 5 shows the result of k-means/FCM and CCC in case of the hill and valley and CBF dataset, and Fig. 6 for the ECG200 dataset. In all cases, the k-means/FCM clustering algorithm delivers trigonometric shapes, whereas the CCC clusters correspond well to the underlying patterns.

The CCC algorithm has also been applied to real data, namely wind strength data measured hourly on a small island in the northern sea (shown in Fig. 6).

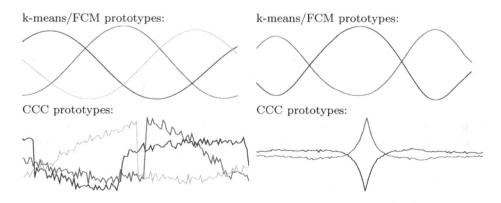

Fig. 5. Clustering of time series subsequences. Left: CBF dataset, right: hill & valley dataset.

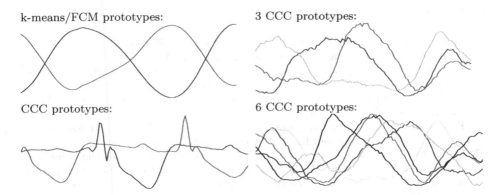

k-means/FCM prototypes: 3 CCC prototypes:

CCC prototypes: 6 CCC prototypes:

Fig. 6. Clustering of time series subsequences. Left: ECG dataset, right: windstrength dataset.

The sliding window was 5 days long. With real data we do not expect such distinct shapes as in the artificial datasets CBF or hill & valley, because the noise ratio is much higher and different patterns are not separated from each other by a period of time where the series remains constant (which makes it easy to distinguish the *pattern* from the *non-pattern* parts). Furthermore weather phenomena are cyclic in nature (consider the alternation of land and see breeze) and we expect the discovered patterns to exhibit such a cyclic nature. Nevertheless, the resulting prototypes clearly deviate from the sinusoidal prototypes obtained from standard k-means/FCM clustering. For instance, the rise and fall of wind strength have very different slopes in the various prototypes, some patterns contain long periods of still air, etc. The resulting prototypes clearly differ from those obtained from the other datasets.

6   Conclusions

The distance function used by a clustering algorithm should always be adapted carefully with respect to the problem at hand. If time series data has to be clustered and no dilational effects are expected, cross correlation appears to be a promising candidate. For such a situation, we have shown how the objective function-based clustering algorithms k-means/fuzzy c-means have to be modified in order to operate with this distance. The prototype update step of k-means is revised to handle the alignment of time series appropriately. The negative influence of outliers or data series that poorly correlate to any of the clusters is reduced by means of a noise cluster.

The resulting cross-correlation clustering (CCC) algorithm solves the problem of clustering unaligned time series. It can be applied to short time series (whole series clustering), but also to time series subsequence (STS) clustering. This is particularly interesting because most standard clustering algorithms fail with STS clustering.

References

1. Asuncion, A., Newman, D.: UCI machine learning repository (2007), http://www.ics.uci.edu/~mlearn/MLRepository.html
2. Berthold, M.R., Höppner, F.: On clustering time series using euclidean distance and pearson correlation. Technical report, University of Konstanz (2008)
3. Bezdek, J.: Pattern Recognition with Fuzzy Objective Function Algorithms. Plenum Press, New York (1981)
4. Bezdek, J., Keller, J., Krishnapuram, R., Pal, N.: Fuzzy Models and Algorithms for Pattern Recognition and Image Processing. Kluwer, Boston (1999)
5. Chen, J.R.: Useful clustering outcomes from meaningful time series clustering. In: AusDM 2007: Proceedings of the sixth Australasian conference on Data mining and analytics, Darlinghurst, Australia, pp. 101–109. Australian Computer Society, Inc. (2007)
6. Das, G., Lin, K.-I., Mannila, H., Renganathan, G., Smyth, P.: Rule discovery from time series. In: Proc. of the 4th ACM SIGKDD Int. Conf. on Knowl. Discovery and Data Mining, pp. 16–22. AAAI Press, Menlo Park (1998)
7. Davé, R.: Characterization and detection of noise in clustering. Pattern Recognition Letters 12, 657–664 (1991)
8. Georgieva, O., Klawonn, F.: Dynamic data assigning assessment clustering of streaming data. Applied Soft Computing 8, 1305–1313 (2008)
9. Höppner, F.: Time series abstraction methods – a survey. In: Proceedings GI Jahrestagung Informatik, Workshop on Knowl. Discovery in Databases, Dortmund, Germany, September 2002. Lecture Notes in Informatics, pp. 777–786 (2002)
10. Höppner, F., Klawonn, F., Kruse, R., Runkler, T.: Fuzzy cluster analysis. Wiley, Chichester (1999)
11. Idé, T.: Why does subsequence time-series clustering produce sine waves? In: Fürnkranz, J., Scheffer, T., Spiliopoulou, M. (eds.) PKDD 2006. LNCS (LNAI), vol. 4213, pp. 211–222. Springer, Heidelberg (2006)
12. Keogh, E., Kasetty, S.: On the need for time series data mining benchmarks: A survey and empirical demonstration. Data Mining and Knowledge Discovery 7(4), 349–371 (2003)
13. Keogh, E., Lin, J., Truppel, W.: Clustering of time series subsequences is meaningless: implications for previous and future research. In: Proc. IEEE Int. Conf. on Data Mining (ICDM), pp. 115–122 (2003)
14. Keogh, E., Xi, X., Wei, L., Ratanamahatana, C.A.: The UCR time series classification/clustering homepage (2006), www.cs.ucr.edu/~eamonn/time_series_data/
15. Klawonn, F.: Fuzzy clustering: Insights and a new approach. Mathware and Soft Computing 11, 125–142 (2004)

Context-Based Distance Learning for Categorical Data Clustering

Dino Ienco, Ruggero G. Pensa, and Rosa Meo

Dept. of Computer Science, University of Torino, Italy
{ienco,pensa,meo}@di.unito.it

Abstract. Clustering data described by categorical attributes is a challenging task in data mining applications. Unlike numerical attributes, it is difficult to define a distance between pairs of values of the same categorical attribute, since they are not ordered. In this paper, we propose a method to learn a context-based distance for categorical attributes. The key intuition of this work is that the distance between two values of a categorical attribute A_i can be determined by the way in which the values of the other attributes A_j are distributed in the dataset objects: if they are similarly distributed in the groups of objects in correspondence of the distinct values of A_i a low value of distance is obtained. We propose also a solution to the critical point of the choice of the attributes A_j. We validate our approach on various real world and synthetic datasets, by embedding our distance learning method in both a partitional and a hierarchical clustering algorithm. Experimental results show that our method is competitive w.r.t. categorical data clustering approaches in the state of the art.

1 Introduction

Clustering is a popular data mining technique that enables to partition data into groups (clusters) in such a way that objects inside a group are similar, and objects belonging to different groups are dissimilar [1]. Clearly, the notion of similarity is central in such a process. When objects are described by numerical (real, integer) features, there is a wide range of possible choices. Objects can be considered as vectors in a n-dimensional space, where n is the number of features. Then, many distance metrics can be used in n-dimensional spaces. Among them, probably the most popular metric is the Euclidean distance (or 2-norm distance), which is a special case of Minkowski distance (also called p-norm distance). Given two objects, these measures only depend on the difference between the values of the feature vectors.

In data mining applications, however, data are often described by categorical attributes that take values in a (usually finite) set of unordered nominal values. This makes impossible even to rank or compute differences between two values of the feature vectors. For categorical data the simplest comparison measure is *overlap* [2]. The proximity between two multivariate categorical entities is proportional to the number of attributes in which they match. Other metrics,

N. Adams et al. (Eds.): IDA 2009, LNCS 5772, pp. 83–94, 2009.
© Springer-Verlag Berlin Heidelberg 2009

such as the Jaccard coefficient, are derived from *overlap* and have been adopted in several (partitional and hierarchical) clustering algorithms [3,4,5].

Clearly, these distance metrics do not distinguish between the different values taken by the attribute, since they only measure the equality between pair of values. This is a strong limitation for a clustering algorithm, since it prevents to capture similarities that are clearly identified by human experts. For instance, given an attribute like *city*, which takes values in the set {*Paris, Rome, Florence*} it is obvious that Florence is more similar to Rome than to Paris, from a geographic point of view. However, in some other contexts, Paris might be more similar to Rome, since both of them are capitals, and they may share similar behaviors.

In literature some measures that take into consideration the context of the features, have also been employed but refer to continuous data, e.g., Mahalanobis distance.

In this paper we present a new methodology to compute a context-based distance between values of a categorical variable and apply this technique to clustering categorical data with both partitional and hierarchical techniques. For the introduction of our technique, consider the dataset described in figure 1(a), with only two categorical attributes: *City(Milan, Turin, Florence)* and *Sex(M,F)*. The contingency table in Figure 1(b) shows how these values are distributed. We observe that *City=Florence* occurs only with *Sex=Female*, and the *City=Turin* occurs only with *Sex=Male*. The value *City=Milan* occurs both with *Sex=Male* and *Sex=Female*. From this distribution of data, we infer that, in this particular context, *Florence* is more similar to *Milan* than to *Turin* because the probability to observe a person of a given sex is closer.

Sex	City
Male	Turin
Female	Milan
Male	Turin
Male	Milan
Female	Florence

(a)

	Turin	Milan	Florence
Female	0	1	1
Male	2	1	0

(b)

Fig. 1. A toy dataset (a) and its related contingency table (b)

From this example we can deduce that the distribution of the co-occurrence table may help to define a distance between values of a categorical attribute. To this purpose, we propose a two-step method:

1. for each categorical attribute X, first identify a suitable context constituted by a set of attributes $Y \neq X$, such that each attribute belonging to the context is correlated to the attribute X.
2. compute a distance matrix between any pair of values (x_i, x_j) of X: we take into account the distribution of x_i and x_j in objects having the same values for the context attributes.

The key contribution of our work are the following:

- we introduce a new method to compute the distance between any pair of values of a specific categorical attribute;
- we define a distance-learning approach which is independent of the employed distance-based clustering;
- we show the impact of our approach within two different distance-based clustering algorithms.

We will also show that our approach is scalable w.r.t. to the numbers of instances in the dataset, and can manage thousands of categorical attributes.

2 Related Work

Clustering is an important task in data mining, in information retrieval and in a wide range of analytical and scientific applications [1]. The goal of clustering is to find a partition of the instances according to a predefined distance measure or an objective function to optimize. The problem is particularly difficult when categorical attributes are involved in the clustering process. In literature, many approaches to categorical data clustering have been proposed. Most of them try to optimize a global objective function without using any notion of distance between the values of the same attribute. Furthermore they suffer in terms of efficiency and time complexity with large data sets.

One of the first work in the field of categorical clustering is K-MODES [3]. It tries to extend K-Means algorithm for categorical data. A cluster is represented as a data point which is composed by the most frequent value in each attribute domain. Therefore, in K-MODES the similarity of an unlabeled data point and a cluster representative can be simply calculated by the overlap distance [2].

Another approach to categorical clustering is ROCK [4]. It employs links to measure similarity/proximity between pairs of data points. An instance belongs to the neighborhood of another instance if the Jaccard similarity between them exceeds a user-defined threshold. It heuristically optimizes a cluster quality function with respect to the number of links in an agglomerative hierarchical way. The base algorithm has cubic complexity in the size of the data set, which makes it unsuitable for large datasets.

LIMBO [5], is a scalable hierarchical categorical clustering algorithm built on the Information Bottleneck framework. As a hierarchical algorithm, LIMBO is not as fast as partitional methods. The algorithm builds Distributional Cluster Features (DCF) trees to summarize the data in k clusters, where each node contains statistics on a subset of instances. Starting from DCF and the number of clusters k a scan over the whole data set is performed to assign each instance to the cluster with the closest DCF.

CLICKS [6] is a clustering algorithm based on graph/hypergraph partitioning. In general the cost of clustering with graph structures is acceptable, provided that the underlying data is low dimensional. CLICKS finds clusters in categorical datasets based on a search method for k-partite maximal cliques. The vertices of the graph are the value of the different attributes and there is an edge between two vertexes if the two attribute-values occur in the same instance. All maximal k-partite cliques in the graph are enumerated and the support of the candidate cliques within the original dataset is verified to form the final clusters.

Alternative approaches include combinatorial algorithms [7] and entropy-based methods [8,9]

A first attempt of computing a distance for categorical attributes is [10] where the authors propose a probabilistic framework which considers the distribution of all the attributes in the dataset. However they only compare their approach with K-MODES and on small and low dimensional datasets.

3   The DILCA Method

In this section we present DILCA (DIstance Learning in Categorical Attributes) for computing distances between any pair of values of a categorical attribute.

Let us consider a set $F = \{X_1, X_2, \ldots, X_m\}$ of m categorical attributes. We refer to the cardinality of an attribute (or feature) X as $|X|$. Let $D = \{d_1, d_2, \ldots, d_n\}$ be a dataset of instances defined over F. We denote by x_i a specific value of an attribute X.

From the example in Section 1 it turns out that the distribution of values of an attribute can be informative about the way in which another attribute is distributed in the dataset objects. Thanks to this method we can infer a context-based distance between any pair of values of the same attribute. In real applications there are several attributes: for this reason our approach is based on two steps:

1. selection of a relevant subset of the whole attributes set that we use as the context for a given attribute;
2. computation of the distance measure between pair of values of the same attribute using the context defined in the previous step.

Context selection. We investigate the problem of selecting a good (informative) set of features w.r.t. a given one. This is a classic problem in data mining named feature selection. Feature selection is a preprocessing step of data mining. Its goal is to select a subset of relevant and not redundant features and discard all the other ones w.r.t. a given class attribute (supervised feature selection [11]). In this branch of research many approaches for measuring the correlation/association between two variables have been proposed. An interesting metrics is the *Symmetric Uncertainty*, introduced in [12]. This measure is a correlation-based measure inspired by information theory. Symmetric Uncertainty is derived from entropy: it is a measure of the uncertainty of a random variable. The entropy of a random variable X is defined as:

$$H(X) = -\sum_i P(x_i) \log_2(P(x_i))$$

where $P(x_i)$ is the probability of the value x_i of X. The entropy of X after having observed the values of another variable Y is defined as:

$$H(X|Y) = -\sum_j P(y_j) \sum_i P(x_i|y_i) \log_2(P(x_i|y_i))$$

where $P(x_i|y_i)$ is the probability that $X = x_i$ after we have observed that $Y = y_i$. The information about X provided by Y is given by the *information gain* [13] which is defined as follows:

$$IG(X|Y) = H(X) - H(X|Y)$$

When $IG(X|Y) > IG(Z|Y)$ then the feature X is more correlated to Y than Z. Moreover, the Information gain is symmetrical for two random variables X and Y [12].

The Symmetrical Uncertainty is then defined as follows:

$$SU(X,Y) = 2 \cdot \frac{IG(X|Y)}{H(X) + H(Y)}$$

This measure varies between 0 and 1 (1 indicates that knowledge of the value of either X or Y completely predicts the value of the other variable; 0 indicates that X and Y are independent). The advantage of Symmetrical Uncertainty (SU) w.r.t. Information Gain is that this measure is not biased by the number of values of an attribute.

During the first step, we select a set of context attributes for a given target attribute X. This context, named *context(X)*, is such that the attributes Y belonging to this set have a high value of $SU(X,Y)$. Determining an adequate number of attributes for *context(X)* is not trivial. We propose to use an heuristic to set this number. The heuristic is based on the mean value of SU for a specific target attribute X. Given a target attribute X we want to compute $SU(X,Y)$, for each attribute $Y \neq X$. We denote this Symmetric Uncertainty $SU_X(Y) = SU(X,Y)$. The mean of this quantity is:

$$E[SU_X] = \frac{\sum_{Y \in F \setminus X} SU_X(Y)}{|F| - 1}$$

To determine the context of an attribute X we use the features that satisfy the following inequality:

$$context(X) = \{Y \neq X \ s.t. \ SU_X(Y) \geq \sigma E[SU_X]\}$$

where $\sigma \in [0,1]$ is a trade-off parameter that controls the influence of the mean value. According to this heuristic, at least one attribute is assigned to *context(X)*. This is simple to demonstrate: if $SU_X(Y)$ is the same for all Y then $SU_X(Y) = E[SU_X]$ for all Y; in this case all Y would be selected in *context(X)*. If there exists at least one attribute Y such that $SU_X(Y) \geq E[SU_X]$, then those attributes Y would be selected.

Distance computation. The second step of our approach consists in computing the distance between each pair of values of the considered feature. To compute this distance between x_i and x_j where $x_i \in X$, $x_j \in X$ we use the following formula:

$$d(x_i, x_j) = \sqrt{\sum_{Y \in context(X)} \sum_{y_k \in Y} (P(x_i|y_k) - P(x_j|y_k))^2} \tag{1}$$

For each context attribute Y we compute the conditional probability for both values x_i and x_j given the values $y_k \in Y$ and then we apply the Euclidean distance. Our intra-attribute distance measure is an application of the Euclidean distance. As such, our definition of distance is a metric.

We introduce now our algorithm which, for each attribute X, computes the similarity matrices between any pair of values of X.

Algorithm 1 shows the procedure adopted to compute the correlation matrix between each pair of features based on Symmetric Uncertainty. This algorithm takes as parameter the entire data set D. Function $feature(D)$ returns the set of all the features contained in D. Then the algorithm computes the co-occurrence table (CO_{XY}) between each pair of attributes using function $ComputeCoOccurrenceTable(D, X, Y)$. It computes the joint probability of (X, Y). These tables are used to compute the Symmetric Uncertainty between attributes to be stored in $matrixSU$.

Algorithm 2 computes the distance matrix between the values of the target attribute X. At the first line it selects the vector storing the correlation between X and all other features Y. During the second step it computes the mean of the vector and then it selects the features that will be included in the context of X. When the features are chosen, the distance matrix for the values of attribute X is built, using (1).

Algorithm 1. computeCorrelationMatrix(D)

1: **for all** $X, Y \in feature(D)|X \neq Y$ **do**
2:     CO_{XY} = ComputeCoOccurrenceTable(D,X,Y)
3:     $matrixSU[X][Y] = SU(CO_{XY})$
4: **end for**
5: **return** $matrixSU$

Complexity. Before computing the distance matrix, we must compute $l = m*(m-1)/2$ matrices $CO_{X,Y}$. These l matrices store the co-occurrence of the values between any pair of attributes. To build these matrices we need to perform a complete scan of the entire data set. We use the co-occurrence matrices to compute $matrixSU$. $matrixSU$ is $m \times m$, where m is the number of attributes in the dataset. Using $matrixSU$ we can compute $E[SU_X]$ and then select the right context making use of σ. To compute the distance matrix for each attribute we can use l co-occurrence matrices without the necessity of further scans of the

---

Algorithm 2. DILCA$(matrixSU,X,\sigma)$

1: $VectorSU_X = MatrixSU[X]$
2: $E = \text{computeMean}(VectorSU_X)$
3: $context(X) = \emptyset$
4: **for all** $y \in VectorSU_X$ **do**
5:   **if** $VectorSU_X[y] \geq \sigma E$ **then**
6:     $\text{insert}(Y,context(X))$
7:   **end if**
8: **end for**
9: **for all** $x_i, x_j \in X | x_i \neq x_j$ **do**
10:   $DistanceMatrix[x_i][x_j] = \sqrt{\sum_{Y \in context(X)} \sum_{y_k \in Y} (P(x_i|y_k) - P(x_j|y_k))^2}$
11: **end for**
12: **return** $DistanceMatrix_X$

---

dataset. From this, we derive that our algorithm only needs to scan the entire dataset once. In conclusion, our approach is $O(nm^2)$, with n the number of instances and m the number of involved attributes.

4   Experiments and Results

In this section we present a comprehensive evaluation of our approach. Since our method enables to use distance based approaches for clustering, we coupled it with two standard methods: a partitional one, and a hierarchical one. We compared both of them with state-of-the-art techniques for categorical data clustering.

Evaluation Measures for Clustering. Determine the clustering quality provided by an algorithm is often a hard and subjective task. Therefore, we use two objective criteria to evaluate the results: Accuracy and Normalized Mutual Information.

The first considers the original class label as a mean to evaluate clustering results. Assume that the instances in D have been already classified in p classes $\{p_1, p_2, ..., p_P\}$. Consider a clustering algorithm that partitions D into c clusters $\{cl_1, cl_2, ..., cl_C\}$. We refer to a one-to-one mapping, f, from classes to clusters, such that each class p_i is mapped to the cluster $cl_j = f(p_i)$. The *classification error* of the mapping is defined as:

$$E = \sum_{i=1}^{P} |p_i \cap \overline{f(p_i)}|$$

where $|p_i \cap \overline{f(p_i)}|$ measures the number of objects in class p_i that received the wrong label. The optimal mapping between clusters and classes is the one that minimizes the classification error. We use E_{min} to denote the classification error of the optimal mapping. Then to obtain the Accuracy we compute the following formula:

$$Acc = 1 - \frac{E_{min}}{|D|}$$

The second metrics provides an information that is independent of the number of clusters [14]. This measure takes its maximum value when the clustering partition matches completely the original partition. We can consider NMI as an indicator of the purity of the clustering results. NMI is computed as the average mutual information between any pair of clusters and classes:

$$\mathbf{NMI} = \frac{\sum_{i=1}^{C} \sum_{j=1}^{P} x_{ij} \log \frac{n * n_{ij}}{n_i n_j}}{\sqrt{\sum_{i=1}^{C} n_i \log \frac{n_i}{n} \sum_{j=1}^{P} n_j \log \frac{n_j}{n}}}$$

where n_{ij} is the cardinality of the set of objects that occur both in cluster i and in class j; n_i is the number of objects in cluster i; n_j is the number of objects in class j; n is the total number of objects. C and P are respectively the number of clusters and the number of classes.

Datasets for Categorical Clustering Evaluation. For the evaluation of our distance learning approach on categorical data, we used two collections of datasets. The first collection consists in real world data sets downloaded from the UCI Machine Learning Repository [15]. The second collection contains synthetic datasets produced by a data generator [16] using Gaussian distributed random attributes. The main characteristics of these datasets are summarized in Table 1. Notice that *Breast-w* and *Sonar* contain numerical variables. Indeed, they have been discretized using the supervised method proposed in [17].

Table 1. Datasets characteristics

Dataset	Type	Instances	Features	Values	Classes
Votes	Real	435	16	32	2
Mushroom	Real	8124	22	117	2
Breast-w	Real	699	9	29	2
Sonar	Real	208	60	81	2
SynA	Synth	1000	50	1000	5
SynB	Synth	3000	20	600	4

4.1 Experimental Settings and Results

Here, we report on the performance results of K-MODES$_{DILCA}$ (DILCA coupled with a simple K-MODES algorithm) and HCL$_{DILCA}$ (DILCA coupled with Ward hierarchical clustering (HCL). We compared them with ROCK [4] and LIMBO [5]. For all the algorithms we set the number of clusters equal to the number of classes. We implemented K-MODES$_{DILCA}$ within the WEKA platform [17], a Java open source library that provides machine learning and data mining algorithms. HCL$_{DILCA}$ was implemented over the Java Murtagh's implementation of HCL[1].

---

[1] http://astro.u-strasbg.fr/~{}fmurtagh/mda-sw/

We run the experiments on a PC with a 1.86GHz Intel Pentium M processor, 1024MB of RAM running Linux. For each particular algorithm we used the following setting:

- For K-MODES$_{DILCA}$ we varied parameter σ between 0 to 1 (with steps of 0.1) and we report the value which gave the best results. Since the initial partition is random, we run many times the algorithm and then we report the average result in terms of accuracy and Normalized Mutual Information.
- For HCL$_{DILCA}$ we varied parameter σ between 0 to 1 with step of 0.1 and we report the value which gave the best results. Since the hierarchical algorithm returns a dendrogram which, at each level, contains a different number of clusters, we considered the level corresponding to the number of clusters equal to the number of classes.
- For ROCK we set the threshold parameter between 0.2 to 1 with steps of 0.05. Also for this algorithm we retained the best obtained result.
- For LIMBO we set ϕ parameter between 0 to 1 with steps of 0.25 and we report the best obtained result. This parameter influences the information loss during the merging phase.

In Table 2 we report the result of the comparative evaluation with other clustering algorithms. For each dataset we report the average Accuracy in percentage and the average Normalized Mutual Information achieved by each algorithm. In almost all the experiments, our approach achieves the best results in at least one category of clustering, and in some cases (Sonar and Votes), the performance parameters are sensibly better than in ROCK and LIMBO. The only exception is SynA, where NMI computed for ROCK is slightly higher than NMI achieved by HCL$_{DILCA}$. However, the Accuracy measured for ROCK is lower than the one achieved by HCL$_{DILCA}$. Moreover, we observed that ROCK is very sensitive to the parameter value, and in many cases this algorithm produces one giant cluster that includes instances from more classes.

Table 2. Experiments on real and synthetic data

Dataset	K-MODES$_{DILCA}$		HCL$_{DILCA}$		ROCK		LIMBO	
	Acc.	NMI	Acc.	NMI	Acc.	NMI	Acc.	NMI
Sonar	**71.63%**	**0.9912**	55.29 %	0.0191	56.25 %	0.0093	66.35%	0.0843
Votes	87.59%	0.4892	**89.89%**	**0.5195**	83.90%	0.3446	87.12%	0.4358
Breast-w	**95.99%**	**0.7435**	94.13%	0.6650	76.68%	0.2570	55.94%	0.0009
Mushroom	**89.02%**	0.5518	**89.02%**	**0.5938**	50.57%	0.05681	88.95%	0.5522
SynA	89.50%	0.7864	**94.30%**	0.8641	80.3%	**0.8965**	87.6%	0.7540
SynB	100%	1.0000	100%	1.0000	100%	1.0000	26.77%	0.0017

Impact of σ. To evaluate the impact of σ parameter we use *Mushroom* and *Votes* datasets. For each dataset we plot the behavior of K-MODES$_{DILCA}$. We let vary the parameter σ from 0 to 1 with steps of 0.1. When the parameter is equal to 0 all the features are included in the context. We observed that the influence of different settings of σ w.r.t. accuracy is small (curves are omitted here). In both datasets, the variation in accuracy is very low (less than 0.50%). Although there is no general law about how to choose this parameter, we estimate that its impact is less important than standard clustering parameters (such as, the number of clusters, or other algorithm-specific parameters).

4.2   Scalability of DILCA

We introduce now a study on the scalability of our distance learning approach, by analyzing the overall computational time of K-MODES$_{DILCA}$ and HCL$_{DILCA}$, and the portion of time needed to perform distance computation (DILCA). We evaluate the scalability varying the two dimensions of the dataset that may have an impact on time performances.

The first dimension is the number of instances. For this purpose, we generated 30,000 synthetic instances described by 100 attributes, then we built 30 datasets containing from 1000 to 30,000 instances. We report the results in Figure 2(a). The second dimension is the number of attributes: we generated a synthetic dataset consisting in 5,000 attributes and 1,000 instances. Then we built 10 datasets containing from 100 to 5000 features. The results are depicted in Figure 2(b). We perform also some experiments to evaluate the impact of parameter σ on running time using the complete dataset (see Figure 2(c)).

As the value of σ enables to select a different number of attributes in the context of the target attribute, this parameter could have an impact on the time performances of our approach. In Figure 2(c) we observe how changes of σ influence the execution time of our algorithms. For the part that computes the distance we can see that the higher the value of σ the lower the time used to build the intra-attribute distance matrix.

We also compared HCL$_{DILCA}$ with LIMBO (which is also hierarchical) on the scalability w.r.t. the number of features. We used another synthetic dataset with 1,000 instances and 1,000 attributes from which we built 10 datasets containing a variable number of features: from 100 to 1,000 features. We report the results in figure 2(e). In Figure 2(f), we show a comparison between the two algorithms on the dataset composed by an increasing number of instances. We observe that HCL$_{DILCA}$ is faster than LIMBO w.r.t. the size and dimensionality of the dataset.

Finally, we also investigated on the time spent by the HCL$_{DILCA}$ to perform clustering. In Figure 2(d) we report the time spent by the three parts of HCL coupled with DILCA. The three curves represent respectively the time spent by DILCA to compute distances, the time spent to compute the point-to-point distance matrix given as input to the hierarchical algorithm, and the effective time spent by Ward algorithm to build the dendrogram. In any case the most consistent portion of the overall computation time is employed to calculate the point-to-point distance measure between each pair of instances.

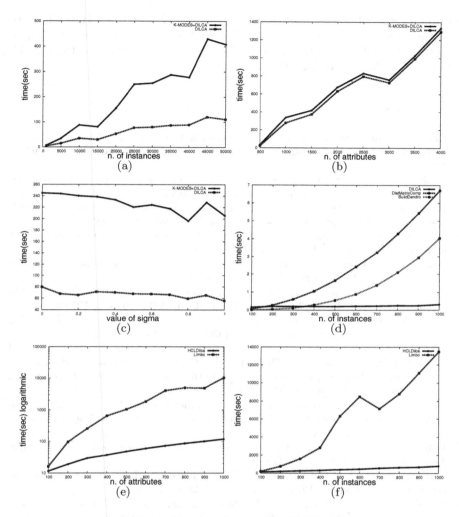

Fig. 2. Time performances of DILCA

5 Conclusion

We introduced a scalable approach to learn a context-based distance between values of categorical attributes. We showed the effective impact of this approach on two distance-based clustering approaches. We believe that the proposed method is general enough and it can be applied to any data mining task that involves categorical data and requires distance computations. As a future work we will investigate the application of our distance learning approach to different distance-based tasks such as: outlier detection and nearest neighbors classification. Moreover, using this distance it will be possible to compute distances between objects described by both numerical and categorical attributes.

Acknowledgments. The authors wish to thank Dr. Eui-Hong Han who provided the source code ROCK, Dr. Periklis Andritsos who provided the implementation of LIMBO, and Elena Roglia for stimulating discussions. Ruggero G. Pensa is co-funded by Regione Piemonte.

References

1. Han, J., Kamber, M.: Data Mining: Concepts and Techniques. The Morgan Kaufmann Series in Data Management Systems. Morgan Kaufmann, San Francisco (2000)
2. Kasif, S., Salzberg, S., Waltz, D., Rachlin, J., Aha, D.: Towards a framework for memory-based reasoning (manuscript, 1995) (in review)
3. Huang, Z.: Extensions to the k-means algorithm for clustering large data sets with categorical values. Data Min. Knowl. Discov. 2(3), 283–304 (1998)
4. Guha, S., Rastogi, R., Shim, K.: Rock: A robust clustering algorithm for categorical attributes. In: Proc. of IEEE ICDE 1999 (1999)
5. Andritsos, P., Tsaparas, P., Miller, R.J., Sevcik, K.C.: Scalable clustering of categorical data. In: Proc. of EDBT 2004, pp. 123–146 (2004)
6. Zaki, M.J., Peters, M.: Clicks: Mining subspace clusters in categorical data via k-partite maximal cliques. In: Proc. of IEEE ICDE 2005, pp. 355–356 (2005)
7. Ganti, V., Gehrke, J., Ramakrishnan, R.: Cactus-clustering categorical data using summaries. In: Proc. of ACM SIGKDD 1999, pp. 73–83 (1999)
8. Barbara, D., Couto, J., Li, Y.: Coolcat: an entropy-based algorithm for categorical clustering. In: Proc. of CIKM 2002, pp. 582–589. ACM Press, New York (2002)
9. Li, T., Ma, S., Ogihara, M.: Entropy-based criterion in categorical clustering. In: Proc. of ICML 2004, pp. 536–543 (2004)
10. Ahmad, A., Dey, L.: A method to compute distance between two categorical values of same attribute in unsupervised learning for categorical data set. Pattern Recogn. Lett. 28(1), 110–118 (2007)
11. Guyon, I., Elisseeff, A.: An introduction to variable and feature selection. J. Mach. Learn. Res. 3, 1157–1182 (2003)
12. Yu, L., Liu, H.: Feature selection for high-dimensional data: A fast correlation-based filter solution. In: Proc. of ICML 2003, Washington, DC (2003)
13. Quinlan, R.J.: C4.5: Programs for Machine Learning. Morgan Kaufmann Series in Machine Learning. Morgan Kaufmann, San Francisco (1993)
14. Strehl, A., Ghosh, J., Cardie, C.: Cluster ensembles - a knowledge reuse framework for combining multiple partitions. Journal of Machine Learning Research 3, 583–617 (2002)
15. Blake, C.L., Merz, C.J.: UCI repository of machine learning databases (1998), http://www.ics.uci.edu/~mlearn/MLRepository.html
16. Melli, G.: Dataset generator, perfect data for an imperfect world (2008), http://www.datasetgenerator.com
17. Witten, I.H., Frank, E.: Data Mining: Practical Machine Learning Tools and Techniques, 2nd edn. Data Management Systems. Morgan Kaufmann, San Francisco (2005)

Semi-supervised Text Classification Using RBF Networks

Eric P. Jiang

University of San Diego, Serra Hall 150, 5998 Alcala Park, San Diego, CA 92110, USA
jiang@sandiego.edu

Abstract. Semi-supervised text classification has numerous applications and is particularly applicable to the problems where large quantities of unlabeled data are readily available while only a small number of labeled training samples are accessible. The paper proposes a semi-supervised classifier that integrates a clustering based Expectation Maximization (EM) algorithm into radial basis function (RBF) neural networks and can learn for classification from a very small number of labeled training samples and a large pool of unlabeled data effectively. A generalized centroid clustering algorithm is also investigated in this work to balance predictive values between labeled and unlabeled training data and to improve classification accuracy. Experimental results with three popular text classification corpora show that the proper use of additional unlabeled data in this semi-supervised approach can reduce classification errors by up to 26%.

1 Introduction

Text classification is a process assigning textual documents into one or more predefined categories or classes, based on their contents. The process is typically carried out by applying machine learning algorithms to build models from some pre-labeled training samples and then by deploying the models to classify previously unseen documents. In general, given a sufficient set of labeled training data, this classification approach produces reasonably good results. However, it can perform poorly when there is only a limited quantity of labeled data on hand. In many applications, hand-labeling a large training dataset can be very time-consuming or even impractical. For instance, developing a software agent, which will automatically route the relevant news articles to online newsgroup readers based on their reading interests, likely requires at least a few hundred labeled articles to achieve acceptable accuracy [8]; a task can be very tedious and labor intensive. Web page classification is another example in this context. Given the rapid proliferation of online textual material, this classification task is very valuable. But, any attempt to classify Web pages (even it is limited to a small number of specific topics) would need a set of labeled training pages with the size that is simply way too large to be reasonably accomplishable.

This is one of the primary reasons why semi-supervised learning has been a very popular research topic in the last few years. Semi-supervised learning aims to develop techniques that improve model efficacy by augmenting labeled samples with

N. Adams et al. (Eds.): IDA 2009, LNCS 5772, pp. 95–106, 2009.
© Springer-Verlag Berlin Heidelberg 2009

additional unlabeled data in model training. It is particularly pertinent to the text classification domain because large quantities of online unlabeled textual data are readily available. Recently, a number of semi-supervised algorithms have been developed to learn for classification (for instance, see [2][7][8]). A comprehensive survey on learning algorithms with labeled and unlabeled data can be found in [9].

Successful applications of such semi-supervised techniques in text classification have been reported in recent years. With the careful use of unlabeled data, these algorithms are able to improve classification accuracy considerably over traditional supervised classifiers. However, some of the algorithms can still encounter difficulties when labeled training data are extremely small [11]. This could well be due to the fact that most of the algorithms start with an initial model based on a set of labeled data and then, iteratively, they build subsequent and presumably improved models by incorporating the labeled data with additional unlabeled documents. And when the labeled data are very scarce and carry a distribution that is not close to that of the entire data to be modeled, the process can lead to an inaccurate initial model and the inaccuracy from the initial model can be further propagated to the subsequent models through the iterations. This error propagation phenomenon is particularly relevant for text classification tasks because textual documents generally have very high feature dimensionality. Therefore, incorporating additional unlabeled data in model learning can actually increase or decrease classification accuracy, and the impact of the unlabeled data on learning needs to be carefully modulated to make them useful [8].

This paper proposes a semi-supervised text classifier that integrates a clustering based Expectation Maximization (EM) algorithm into radial basis function (RBF) networks and it can learn from a very small number of labeled samples and a large quantity of unlabeled data. The classifier starts with a feature selection procedure on training data to reduce their feature dimensionality, and then applies a clustering algorithm to both labeled and unlabeled data iteratively for computing RBF middle layer parameters. Finally, it uses a regression model to determine RBF output layer weights. Note that the approach utilizes class labels in the labeled training data to guide the clustering process and also applies a weighting scheme to incorporate the unlabeled data in estimating network parameters. This guided and weighted clustering process helps balance predictive values between the labeled and unlabeled data and improve classification accuracy.

2   RBF Networks for Text Classification

The Radial basis function (RBF) networks have many applications in science and engineering and they can also be applied to text classification. For a given document training set $T = \{d_1, d_2, ..., d_{|T|}\}$ with classes $C = \{c_1, c_2, ..., c_m\}$, where $|T|$ is the size of T, the networks can be used to build models that characterize principal correlations of terms (or features) and documents in T and then to classify previously unseen documents into the classes. Recently, a RBF based email filter has been proposed [6] that it is capable of filtering spam email effectively.

A typical RBF network has a feed-forward connected structure of three layers: an input layer, a hidden layer of nonlinear processing neurons and an output layer [1]. For text classification, the input layer of the network has n neurons and the size n

corresponds to the number of features in the input vectors d_i. The hidden layer contains k computational neurons; each of them can be mathematically described by a radial basis function ϕ_p that maps a distance in the Euclidean norm to a real value:

$$\phi_p(x) = \phi(\| x - a_p \|_2), \quad p = 1, 2, \ldots, k \tag{1}$$

where $a_p, p = 1, 2, \ldots, k$ are the RBF centers in input document space and in general, k is less than $|T|$. The output layer of the network has m neurons and it determines the document classes according to:

$$c_j = \sum_{p=1}^{k} w_{pj} \phi_p(x), \quad j = 1, 2, \ldots, m \tag{2}$$

where w_{pj} is the weight that connects the pth neuron in the hidden layer to the jth neuron in the output layer. A popular choice of RBF is the Gaussian function

$$\phi(x) = e^{-\frac{x^2}{2o^2}} \tag{3}$$

where σ is a width parameter that controls the smoothness properties of the function.

3 Semi-supervised Learning Process

In this section we describe major components of the semi-supervised RBF process, which include data preprocessing and representation, a two-stage network training procedure, a weighted centroid clustering method, and an integrated learning algorithm that uses both labeled and unlabeled data.

3.1 Feature Selection and Document Representation

In text classification, appropriate feature selection can be useful to aid in the classification task [10]. Features or terms occurring in documents are selected according to their contributions to profiling the documents and document classes. Feature selection can trim down feature dimensionality and is beneficial to model training for reduced computational cost. It can also help filter out irrelevant features. This is particularly valuable for RBF networks since their training algorithms generally treat every data feature equally in computing the neuron activations (Eq. (1)).

In this work, document features are first extracted by removing stop or common words and by applying a word stemming procedure. Then, the features with very low document frequencies or corpus frequencies are eliminated from model training and deploying. Our feature selection process also removes the features with very high corpus frequencies since these features can behave like common words in a specific textual domain, which can be undetected by a general common-word removal process based on a predefined word collection. Since the semi-supervised classifier proposed in this paper can potentially be applied to extremely small sets of labeled samples, we use the unsupervised document frequency (DF) as the criterion to carry out additional feature reduction. Document frequency has been reported as a simple and inexpensive alternative to other more sophisticated supervised feature selection measures [10].

Once the process of feature selection is completed, each training document is encoded as a numerical vector whose elements are the values of retained features. In this work, the traditional $log(tf)$-idf feature weighting is used.

3.2 Network Training

The training of a RBF network can be conducted by some optimization algorithms that determine network parameters using all connected neuron activations and weights between the layers. Alternatively, it can be done by a two-stage training procedure, which is more computationally efficient. More specifically, the first stage of training is to form a representation of density distribution of the input data in terms of the RBF parameters. It determines the centers a_p (Eq. (1)) and widths σ (Eq. (3)) by relatively fast clustering algorithms, clustering each document class independently to obtain k basis functions for the class. In this paper, we use a centroid clustering algorithm [6]. With the centers and widths for the hidden layer being estimated, the second stage of training is to select weights of the output layer by a logistic regression model. Once all network parameters are determined, the network can be deployed to new documents for classification and the classification outcomes from the network are computed numerically by a weighted sum of the hidden layer activations, as shown in Eq. (2).

3.3 Incorporating Unlabeled Data

For many text classification problems, especially those involving online textual databases, collecting unlabeled documents is quite easy while hand labeling a large number of document training samples can be labor intensive and also prone to human errors. Therefore, incorporating additional unlabeled data in RBF network learning is very valuable. It can be used to compensate for insufficient labeled training data in terms of achieving good classification performance of the system.

The two-stage training process on RBF networks is well structured for augmenting additional unlabeled data. In principle, the data used in the first training stage for determining the basis functions in the network hidden layer are not required to be labeled, since it is carried out by a clustering algorithm. The modeling in the second training stage, however, does depend on some labeled samples for finding the weights in the network output layer. Several experiments on clustering for the first training stage indicate that, while the training can be done solely by unlabeled data, the RBF classifier in general delivers substantially higher classification accuracy if the process is performed by using both labeled and unlabeled data. This is somewhat explainable because, in the context of text classification, the class information embedded in labeled samples is typically useful to produce accurate clusters for estimating network parameters.

In the first stage, the value of the labeled samples can be further utilized by incorporating the samples with unlabeled data in an iterative cluster refining process. This can be accomplished by the well-known Expectation Maximization (EM) algorithm. EM is typically used to iteratively estimate the maximum likelihood of hidden parameters for problems with incomplete data [5]. If we consider the labels of unlabeled data as unknown entries, EM can be applied to estimate these class labels. A similar approach that combines EM with the naïve Bayes classifier has been proposed in [8].

In the first stage of network training, we start with a number of initial clusters, which are constructed by the labeled samples. Then, the unlabeled data are classified by the initial clusters and these newly classified data, together with the labeled samples, are used to form the new clusters with the updated cluster centroids. The process is repeated until the clusters are stabilized.

3.4 Adjusting the Weights for Unlabeled Data

As we described earlier, the network training uses only a very small number of labeled samples and in comparison to the labeled, many order of magnitude more unlabeled data are used in the EM iterations to update the estimates of network parameters. This imbalance in count between the labeled and unlabeled data can make the process almost like performing unsupervised clustering [8] and the unlabeled data could potentially play a dominant role in parameter estimation. In general, when the natural clusters of the combined training data are in correspondence with the class labels, the unsupervised clustering with a large number of unlabeled data will produce the parameter estimates that are helpful for classification. However, when the natural clustering of the data generates parameter estimates that are not in correspondence with the class labels, then these estimates are likely destructive to classification accuracy.

In order to modulate the influence of the unlabeled data in parameter estimation, we introduce a weighted centroid algorithm for the EM iterations. Let β_1 be a parameter, $0 \leq \beta_1 \leq 1$, and L_p and U_p the labeled and unlabeled document set in the pth cluster, respectively, the cluster centroid a_p from the weighted centroid algorithm is computed as:

$$a_p = \frac{1}{\mid L_p \mid + \beta_1 \mid U_p \mid} (\sum_{d_i \in L_p} d_i + \beta_1 \sum_{d_i \in U_p} d_i) \qquad (4)$$

This equation is used in the iterations to update cluster centroids.

Once the iteration process is converged, we compute the mean and standard deviation values for each cluster and use another separate weighting parameter β_2, $0 \leq \beta_2 \leq 1$, to control the influence of the unlabeled data on the final estimation of network basis functions. The cluster mean value is computed by a similar formula as Eq. 4 and the cluster standard deviation is computed by

$$\sigma_p^2 = \frac{1}{\mid L_p \mid + \beta_2 \mid U_p \mid} (\sum_{d_i \in L_p} (d_i - a_p)^2 + \beta_2 \sum_{d_i \in U_p} (d_i - a_p)^2) \qquad (5)$$

Note that when both parameters β_1 and β_2 take small values that are close to zero, the unlabeled data will have little influence on parameter estimation of the basis functions. In particular, when both parameters are set to zero, the entire network training is performed only on the labeled data and effectively, it reduces to a (purely) supervised algorithm. On the other hand, when both parameters are set to one, each unlabeled document will carry the same weight as a labeled one, and it becomes the process that applies the traditional centroid clustering algorithm.

In this work, we have set the weighting parameters β_1 and β_2 to some fixed values and also the values that maximize classification accuracy through a limited number of trials. Experimental results presented in Section 4.3 indicate that, by setting both weighting parameters to some values between 0 and 1, additional unlabeled data can almost always be useful to aid in classification even their natural clustering, without weight adjusting, would produce less accurate classification.

As a summary of the major components that have been discussed in this section, we present a semi-supervised RBF training algorithm in Table 1.

Table 1. A semi-supervised RBF training algorithm

Inputs: A set of labeled document vectors L and a set of unlabeled document vectors U (both sets have gone through feature selection and feature weighting)

Output: A RBF text classifier

- Set values for weighting parameter β_1 and β_2, $0 \leq \beta_1, \beta_2 \leq 1$
- Assign all labeled documents of class c_j in L to set L_j, $j = 1, 2, ..., m$
- Compute the initial cluster centroids a_{jp} of L_j, $j = 1, 2, ..., m$, $p = 1, 2, ..., k$
- Loop
  - Set the unlabeled document set U_j of class c_j to empty, $j = 1, 2, ..., m$
  - For each unlabeled document d_i in U
    - Compute its normalized distances from all centroids a_{jp}, $j = 1, 2, ..., m$, $p = 1, 2, ..., k$
    - Include d_i in U_j if the distance between d_i and a_{jp} is minimal, $p = 1, 2, ..., k$
  - Update the cluster centroids a_{jp} of $L_j \cup U_j$ using Eq. (4), $j = 1, 2, ..., m$, $p = 1, 2, ..., k$
  - If there is no change on the cluster centroids, then forward to the next step; otherwise repeat the loop
- Estimate the parameters of network basis functions of $L_j \cup U_j$ using Eq. (4) (but substitute β_1 with β_2) and Eq. (5), $j = 1, 2, ..., m$
- Determine the weights of the network output layer by logistic regression on set L

---

4 Experiments

In this section, we provide empirical results and analysis of the proposed semi-supervised RBF classifier with three different text corpora: UseNet news articles (20 Newsgroups), Web pages (WebKB) and newswire articles (Reuters 21578). We show that incorporating unlabeled data into RBF training can improve classification accuracy when the number of labeled training samples is very small. With additional adjusted weighting on the unlabeled data, the semi-supervised RBF classifier can outperform the traditional supervised RBF model trained only by the labeled samples. We also discuss the impact of document feature size on classification accuracy.

4.1   Datasets and Experiment Settings

The 20 Newsgroups dataset is a collection of 20,017 news postings divided almost evenly among 20 different UseNet newsgroups and several of these newsgroups share closely related topics. The standard *byDate* split is used to order articles by date and then to take the first two-thirds for training and the remaining one-third for testing. In our experiments, we have further selected at random 5,842 documents for training, and 3,560 documents for testing.

The second dataset, WebKB is a collection of 8,145 web pages gathered from university computer science departments. It has 7 different classes: student, faculty, project, course, department, staff and other. The data from the first four classes are used in our experiments and they consist of 4,199 pages. With a random partition, it has 2,803 pages for training (two-thirds) and 1,296 pages for testing (one-third) [3].

The third data corpus, Reuters 21578, contains 21,578 articles and 135 topic categories from the Reuters newswire in 1978, and the standard *ModApte* training-testing split is used [4]. Note that in this corpus the distribution of the documents across the categories is highly skewed and some of the documents are assigned to multiple categories or no category at all. For our experiments, we use the subset R8 [3] from the corpus that contains the articles from the top eight most frequent categories and with a single label (i.e., belong to one category). R8 has been further partitioned with 5,485 documents for training and 2,189 documents for testing, and they belong to the categories: acq, crude, earn, grain, interest, money-fx, ship and trade.

Text classification performance can be evaluated by precision and recall or other related measures such as F_1. Since some datasets included in our experiments have very unevenly distributed classes, we use the micro-averaged F_1 measure, or a weighed average over all classes [11]. Note that for single-label classification, the total number false positive decisions is the same as the total number of false negative decisions and hence, the micro-averaged F_1 is identical to several other commonly used measures: accuracy, micro-averaged precision and micro-averaged recall.

In this paper we are primarily concerned with semi-supervised learning using very small numbers of labeled samples. In our experiments, we varied the size of labeled training samples per class from 1 to 5 for all three datasets and then from 10 to 70 for WebKB and 20 Newsgroups, and from 10 to 40 for R8, with an increment of 10. The reason for having a smaller range with R8 is that the category *grain* in the dataset contains only 41 labeled samples.

For a given training data set, we first randomly selected the labeled samples with a specified size from the set. Then, from the remaining training data, we selected a number of unlabeled documents at random. The labels of these designated unlabeled data are ignored in model learning. In the experiments, we used a fixed size for the unlabeled data set, namely, 5,000 unlabeled documents with 20 Newsgroups and R8, and 2,500 with WebKB. It should be noted that all results presented in this section are the average classification accuracy values over five repeated experiments; each experiment independently selects its own labeled and unlabeled data sets.

4.2   Comparing Performance with and without Using Labeled Data

In this subsection we evaluate the use of the semi-supervised RBF classifier to incorporate additional unlabeled data and assume each unlabeled document carries an

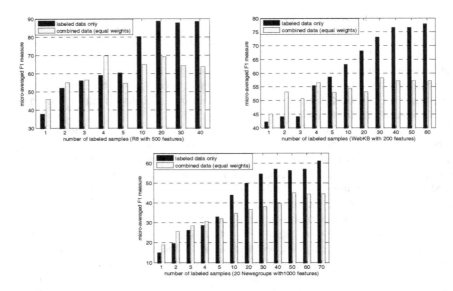

Fig. 1. Classification accuracy results of RBF on R8, WebKB and 20 Newsgroups with and without using unlabeled data. When using the combined data set, an equal weight scheme is applied to the unlabeled documents.

equal weight (i.e., $\beta_1 = \beta_2 = 1$) as a labeled sample in estimating network parameters. Fig.1 shows that the classification accuracy (or micro-averaged F_1) results of the classifier with (as *combined data* (*equal weights*)) and without (as *labeled data only*) using unlabeled training documents on R8, WebKB and 20 Newsgroups. The horizontal axis in the graphs is the number of labeled samples per class used in the training, and the vertical axis is the micro-averaged F_1 measure on the test set.

Fig. 1 indicates that additional unlabeled documents are useful to boost classification accuracy at small labeled data sizes that are less than five. However, as the data size becomes larger, the usefulness of the unlabeled data seems to be vanishing and comparing to the training with the combined data, the traditional supervised RBF networks (using only the labeled data) actually achieve significantly better accuracy results. This reflects some similar observations made in previous studies [8] that unlabeled data may not always help improve classification performance. It can be hypothesized that when the labeled sample set for a class is extremely small, the set may not accurately represent the data distribution of the class and it can carry a large variance towards parameter estimates of the network model. Using a large pool of unlabeled data to augment the very limited labeled could improve the estimates. On the other hand, when the labeled sample set becomes larger, a stronger data representation of the document classes generated from the labeled samples is anticipated and in this case, the clustering of the unlabeled data might not be in good correspondence with the labeled samples. Therefore this inconsistency between the labeled and unlabeled data can cause problems in parameter estimation and consequently decrease classification accuracy.

4.3 Varying the Weights for Unlabeled Data

As we discussed in the previous subsection, unlabeled data can have some significant influence on performance of the semi-supervised RBF classifier. In comparison to unlabeled data, the numbers of labeled training samples are very small, and in the first stage of network training, they are primarily used in computing the initial parameter estimates. In order to coordinate both types of data (labeled and unlabeled) and to modulate the impact of the unlabeled data on the training, we introduced two weighting parameters β_1, β_2 that are used in updating cluster centroids in the EM iterations and in computing the final cluster means and standard deviations, respectively.

In the experiments we set the values for β_1 and β_2 by two weighting schemes: *fixed weights* and *selected weights*. The *fixed weights* scheme simply sets the value of one to both parameters when the number of labeled samples per class is less than five and then, as the size of labeled data increases, it assigns a systematically deceased value to β_1 while keeping a fixed value for β_2 (0.001). The decreased parameter value for β_1 intends to reduce the influence of unlabeled data on network training. The detailed parameter setting of this weighting scheme is described in the following table:

Number of labeled	1-4	5	10	20	30	40	50-70 (if applicable)
B_1 / β_2 values	1/1	.1/.001	.05/.001	.01/.001	.005/.001	.001/.001	.0005/.001

The second *selected weights* scheme selects the values for β_1 and β_2 over a limited number of trials and uses the values that correspond to the best accuracy results obtained from the trials. More specifically, each of the weighting parameters can take one of the seven possible values: 0.0005, 0.001, 0.005, 0.01, 0.05, 0.1 and 1, and it creates a total of 14 different combinations or trials. Note that this scheme is only used to demonstrate if there exist some values for the weighting parameters that can further increase performance of the semi-supervised RBF classifier. In practice, the values of the weighting parameters can be set by the *fixed weights* scheme, or as an alternative, by cross-validation or other improved weighting methods.

Fig. 2 compares classification accuracy of the semi-supervised RBF classifier, which is equipped with the adjusted weighting schemes (*fixed* and *selected*) on unlabeled data, with that of the supervised counterpart on all three datasets. We observe that the semi-supervised classifier can achieves better classification performance, in most cases if the *fixed* scheme is used and in all cases if the *selected* scheme is used. For instance, using just four labeled samples per class with R8, the supervised RBF reaches 59% accuracy while the semi-supervised RBF achieves 69.7%. This represents a 26% reduction in classification error.

In comparison to the results from Fig. 1, the adjusted weighting on unlabeled data can help the semi-supervised RBF avoid its degradation in accuracy at large labeled data sizes, while still preserving the benefits of the algorithm with small labeled sets. In addition, the consistently superior accuracy values with the *selected* scheme suggest that improved methods of setting weighting parameters have the potential to further increase the practical performance of the classifier.

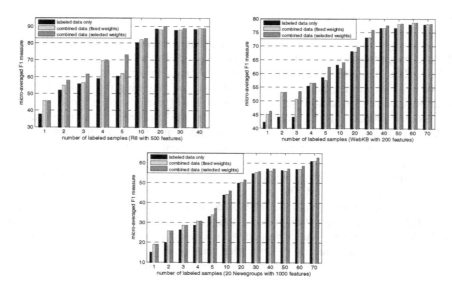

Fig. 2. Classification accuracy results of RBF on R8, WebKB and 20 Newsgroups with and without using the unlabeled data. When using the combined data set, the fixed and selected weighting schemes are applied to the unlabeled documents, respectively.

4.4 Impact of Feature Size on Performance

In the empirical results reported in previous sections (Fig. 1 and 2), we used a fixed document feature size, one for each dataset, which is somewhat proportional to the retained feature dimensionality of the dataset after its data preprocessing procedures. As we discussed earlier, in the cases of very small labeled sample sets, the use of unlabeled data in the RBF network training can noticeably improve the network parameter estimates. Since the variance of a training set depends on the number of samples in the set and it can also be influenced by the dimensionality of document feature space, we presume that in such cases some additional feature reduction may further help improve network parameter estimation and classification accuracy. Of course, this additional reduction needs to be performed reasonably and any excessive feature reduction could produce a feature space that is simply incapable of profiling documents and document classes.

Experiments in this regard were conducted with the WebKB dataset. Instead of using 200 features as seen in Fig. 2, we used only the top 100 features and applied the semi-supervised RBF with the weighting parameters being set by the *fixed* scheme. The comparison results are shown in Fig. 3 and they seem to support our hypothesis that appropriate document feature size can have a significant impact on classification accuracy and, with a very small number of labeled samples, some further yet reasonable feature reduction could assist the semi-supervised RBF classifier in its classification task. For instance, when only one labeled training sample for each class is used, Fig. 3 indicates that the semi-supervised RBF network attains the classification accuracy of 57.4% by using 100 selected features and 45.1% by using 200 selected

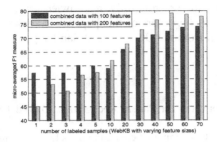

Fig. 3. Classification accuracy results of the semi-supervised RBF classifier with two different feature sizes on WebKB. The combined data are used and the fixed weighting scheme is applied to the unlabeled data.

features. This suggests that in this case, a 27.3% accuracy improvement can be achieved if a relatively smaller feature size is used.

The results from Fig. 3 also indicate that a relatively larger feature size will be beneficial to the classifier in accuracy at large labeled set sizes. In practice, we can adaptively adjust the network's document feature dimensionality with the size of available labeled samples to be trained. An optimal algorithm that coordinates these two quantities is an area of future work.

5   Conclusions and Future Work

This paper has presented a semi-supervised text classifier that integrates a clustering based EM algorithm into radial basis function (RBF) networks and can learn for classification from a very small set of labeled samples and a large volume of unlabeled data effectively. Since the number of the unlabeled data is so large in comparison to the labeled, their influence on the network parameter estimation needs to be appropriately modulated in order to make them useful. Towards this direction, a generalized centroid clustering algorithm has been investigated in this work to balance predictive values between the labeled and unlabeled training data and to improve classification accuracy. In this paper, we have also studied the effect of document feature size on classification performance.

Experiments of the proposed semi-supervised RBF classifier with three popular text classification corpora have shown that, with proper use of additional unlabeled data, the classifier can achieve classification error reduction by up to 26%. In addition, appropriate settings on the selected document feature size can further improve classification accuracy.

As future work we plan to improve the semi-supervised RBF classifier in several directions. Currently in our system, we are using simple and fixed weighting parameters to modulate the impact of unlabeled data in network learning. We plan to develop methods for dynamically adjusting the parameter values in EM iterations and improving the estimates of network basis functions. We will also investigate the coordination between the dimensionality of document feature space and the size of labeled training data set.

Acknowledgments. The author gratefully acknowledges the valuable comments provided by the anonymous reviewers of this paper. This work was in part supported by a faculty research grant from the University of San Diego.

References

1. Bishop, C.: Neural Networks for Pattern Recognition. Oxford University Press, Oxford (1995)
2. Blum, A., Mitchell, T.: Combining labeled and unlabeled data with Co-Training. In: 11th COLT conference, pp. 92–100 (1998)
3. Cardoso-Cachopo, A., Oliveira, A.: Semi-supervised Single-label Text Categorization Using Centroid-based Classifiers. In: ACM Symposium on Applied Computing, pp. 844–851 (2007)
4. Cohen, F., Sebastiani, F.: An analysis of the relative hardness of reuters-21578 subsets. J. American Society for information Science and Technology 56(6), 584–596 (2004)
5. Dempster, A., Laird, N., Rubin, D.: Maximum likelihood from incomplete data via the EM algorithm. J. Royal Statistical Society, Series B 39, 1–38 (1977)
6. Jiang, E.: Detecting spam email by radial basis function networks. International J. Knowledge based and Intelligent Engineering Systems 11, 409–418 (2007)
7. Joachims, T.: Transductive inference for text classification using support vector machines. In: 16th ICML conference, pp. 200–209 (1999)
8. Nigam, K., McCallum, A., Thurn, S., Mitchell, T.: Text classification from labeled and unlabeled documents using EM. Machine Learning 39(2/3), 103–134 (2000)
9. Seeger, M.: Learning with labeled and unlabeled data. Technical report, Edinburgh University (2001)
10. Yang, Y., Pederson, J.O.: A Comparative Study on Feature Selection in Text Classification. In: 14th International Conference on Machine Learning, pp. 412–420 (1997)
11. Zeng, H., Wang, X., Chen, Z., Lu, H., Ma, W.: CBC-clustering based text classification requiring minimal labeled data. In: 3rd International Conference on Data Mining, pp. 443–450 (2003)

Improving k-NN for Human Cancer Classification Using the Gene Expression Profiles

Manuel Martín-Merino[1] and Javier De Las Rivas[2]

[1] Universidad Pontificia de Salamanca
C/Compañía 5, 37002, Salamanca, Spain
mmartinmac@upsa.es
[2] Cancer Research Center (CIC-IBMCC, CSIC/USAL)
Salamanca, Spain
jrivas@usal.es

Abstract. The k Nearest Neighbor classifier has been applied to the identification of cancer samples using the gene expression profiles with encouraging results. However, k-NN relies usually on the use of Euclidean distances that fail often to reflect accurately the sample proximities. Non Euclidean dissimilarities focus on different features of the data and should be integrated in order to reduce the misclassification errors.

In this paper, we learn a linear combination of dissimilarities using a regularized kernel alignment algorithm. The weights of the combination are learnt in a HRKHS (Hyper Reproducing Kernel Hilbert Space) using a Semidefinite Programming algorithm. This approach allow us to incorporate a smoothing term that penalizes the complexity of the family of distances and avoids overfitting.

The experimental results suggest that the method proposed outperforms other metric learning strategies and improves the classical k-NN algorithm based on a single dissimilarity.

1   Introduction

DNA microarrays allow us to monitor the expression levels of thousands of genes simultaneously across a collection of related samples. This technology has been applied to the identification of cancer samples with encouraging results [3].

The k Nearest Neighbor (k-NN) classifier has been applied to the identification of cancer samples using the gene expression profiles. However, k-NN is based usually on Euclidean distances that fail often to model accurately the sample proximities [1]. Non Euclidean dissimilarities reflect complementary features of the data and misclassify frequently different subsets of patterns. Therefore, they should be integrated in order to reduce the misclassification errors.

Several authors have proposed techniques to learn the metric from the data [22,23]. Some of them, are based on a linear transformation of the Euclidean metric [20,22] that fails often to reflect the proximities among the sample profiles [1]. Other approaches such as [23] are more general, but are prone to overfitting when the sample size is small because they learn the metric without taking into

N. Adams et al. (Eds.): IDA 2009, LNCS 5772, pp. 107–118, 2009.
© Springer-Verlag Berlin Heidelberg 2009

account the generalization ability of the classifier. Besides, they rely on complex non-linear optimization algorithms.

Our approach considers that the integration of dissimilarities that reflect different features of the data should help to reduce the classification errors. To this aim, a linear combination of dissimilarities is learnt considering the relation between kernels and distances. Each dissimilarity is embedded in a feature space using the Empirical Kernel Map [18]. Next, learning the dissimilarity is equivalent to optimize the weights of the linear combination of kernels. The combination of kernels is learnt in the literature [2,8] maximizing the alignment between the input kernel and an idealized kernel. However, this error function does not take into account the generalization ability of the classifier and is prone to overfitting.

In this paper, we consider a regularized version of the kernel alignment proposed by [2]. The linear combination of kernels is learnt in a HRKHS (Hyper Reproducing Kernel Hilbert Space) following the approach of hyperkernels proposed in [15]. This formalism exhibits a strong theoretical foundation, is less sensitive to overfitting and allow us to work with infinite families of distances.

The algorithm has been applied to the identification of human cancer samples using the gene expression profiles with remarkable results.

This paper is organized as follows: Section 2 introduces briefly the idea of Kernel Alignment, section 3 presents the algorithms considered to learn a linear combination of dissimilarities. Section 4 illustrates the performance of the algorithm in the challenging problem of gene expression data analysis. Finally, Section 5 gets conclusions and outlines future research trends.

2   Kernel Target Alignment

Given two kernels k_1 and k_2 and a sample $\mathcal{S}$, the empirical alignment evaluates the similarity between the corresponding kernel matrices. Mathematically it is defined as:

$$A(\mathcal{S}, k_1, k_2) = \frac{\langle K_1, K_2 \rangle_F}{\sqrt{\langle K_1, K_1 \rangle_F \langle K_2, K_2 \rangle_F}}, \tag{1}$$

where K_1 denotes the kernel matrix for the kernel k_1, and $\langle K_1, K_2 \rangle_F = \sum_{ij} K_{ij}^1 K_{ij}^2 = Tr(K_1 K_2)$ is the Frobenius product between matrices. If the kernel matrices K_1 and K_2 are considered as bidimensional vectors, the alignment evaluates the cosine of the angle and is a similarity measure.

For classification purposes we can define an ideal target matrix kernel as $K_2 = yy^T$, where y is the vector of labels for the sample $\mathcal{S}$. $K_2(x_i, x_j) = 1$ if $y(x_i) = y(x_j)$ and -1 otherwise. Substituting K_2 in equation (1) the empirical alignment between the matrix kernel K_1 and the target labels for the sample $\mathcal{S}$ can be written as:

$$A(\mathcal{S}, k_1, k_2) = \frac{y^T K_1 y}{m \|K_1\|_F}, \tag{2}$$

where m is the size of the training set $\mathcal{S}$.

It has been shown in [2] that the empirical alignment is stable with respect of different splits of the data and that larger values for the alignment increase the separability among the classes.

2.1   Empirical Kernel Map

Now we introduce shortly the Empirical Kernel Map that allow us to incorporate non-Euclidean dissimilarities into any kernel classifier [18,11].

Let d: $\mathcal{X} \times \mathcal{X} \to \mathbb{R}$ be a dissimilarity and $R = \{p_1, \ldots, p_n\}$ a subset of representatives drawn from the training set. Define the mapping $\phi : \mathcal{F} \to \mathbb{R}^n$ as:

$$\phi(z) = D(z, R) = [d(z, p_1), d(z, p_2), \ldots, d(z, p_n)] \tag{3}$$

This mapping defines a dissimilarity space where feature i is given by $d(., p_i)$.

The set of representatives R determines the dimensionality of the feature space. The choice of R is equivalent to select a subset of features in the dissimilarity space. Due to the small number of samples in our application, we have considered the whole training set as representatives. Notice that it has been suggested in the literature [11] that for small samples reducing the set of representatives does not help to improve the classifier performance.

3   Learning the Metric in a HRKHS Using Kernel Alignment

In order to incorporate a linear combination of dissimilarities into k-NN, we follow the approach of Hyperkernels developed by [15]. To this aim, each distance is embedded in a RKHS via the Empirical Kernel Map introduced in section 2.1. Next, a regularized version of the alignment is presented that incorporates a L_2-penalty over the complexity of the family of distances considered. The solution to this regularized quality functional is searched in a Hyper Reproducing Kernel Hilbert Space. This allows to minimize the quality functional using a semidefinite programming approach (SDP).

Let $X_{train} = \{x_1, x_2, \ldots, x_m\}$ and $Y_{train} = \{y_1, y_2, \ldots, y_m\}$ be a finite sample of training patterns where $y_i \in \{-1, +1\}$. Let $\mathcal{K}$ be a family of semidefinite positive kernels. Our goal is to learn a kernel of dissimilarities [11] $k \in \mathcal{K}$ that represents the combination of dissimilarities and that minimizes the empirical quality functional defined by:

$$Q_{emp}^{align} = 1 - A(\mathcal{S}, k_1, k_2) = 1 - \frac{y^T K_1 y}{m \|K_1\|_F} \tag{4}$$

However, if the family of kernels $\mathcal{K}$ is complex enough it is possible to find a kernel $(k^* = y^T y)$ that achieves training error equal to zero overfitting the data. To avoid this problem, we introduce a term that penalizes the kernel complexity in a Hyper Reproducing Kernel Hilbert Space (HRKHS):

$$Q_{reg}(k, X, Y) = Q_{emp}^{align}(k, X, Y) + \frac{\lambda_Q}{2} \|k\|_{\underline{H}}^2 \tag{5}$$

where $\| \; \|_{\mathcal{H}}$ is the L_2 norm defined in the Hyper Reproducing Kernel Hilbert space generated by the hyperkernel $\underline{k}$. λ_Q is a regularization parameter that controls the complexity of the resulting kernel. The definition of Hyper Reproducing Kernel Hilbert Spaces (HRKHS) is provided in appendix A.

The following theorem allows us to write the solution to the minimization of this regularized quality functional as a linear combination of hyperkernels in a HRKHS.

Theorem 1 (Representer Theorem for Hyper-RKHS [15]). *Let X, Y be the combined training and test set, then each minimizer $k \in \underline{\mathcal{H}}$ of the regularized quality functional $Q_{reg}(k, X, Y)$ admits a representation of the form:*

$$k(x, x') = \sum_{i,j=1}^{m} \beta_{ij} \underline{k}((x_i, x_j), (x, x')) \tag{6}$$

for all x, $x' \in X$, where $\beta_{ij} \in \mathbb{R}$, for each $1 \leq i, j \leq m$.

However, we are only interested in solutions that give rise to positive semidefinite kernels. The following condition over the hyperkernels [15] allow us to guarantee that the solution is a positive semidefinite kernel.

Property 1. Given a hyperkernel $\underline{k}$ with elements such that for any fixed $\underline{x} \in X$, the function $k(x_p, x_q) = \underline{k}(\underline{x}, (x_p, x_q))$, with $x_p, x_q \in \mathcal{X}$, is a positive semidefinite kernel, and $\beta_{ij} \geq 0$ for all $i, j = 1, \ldots, m$, then the kernel

$$k(x_p, x_q) = \sum_{i,j=1}^{m} \beta_{ij} \underline{k}(x_i, x_j, x_p, x_q) \tag{7}$$

is positive semidefinite.

Now, we address the problem of combining a finite set of dissimilarities. As we mentioned earlier, each dissimilarity can be represented by a kernel using the Empirical Kernel Map. Next, the hyperkernel is defined as:

$$\underline{k}(\underline{x}, \underline{x}') = \sum_{i=1}^{n} c_i k_i(\underline{x}) k_i(\underline{x}') \tag{8}$$

where each k_i is a positive semidefinite kernel of dissimilarities and c_i is a constant ≥ 0.

Now, we show that $\underline{k}$ is a valid hyperkernel: First, $\underline{k}$ is a kernel because it can be written as a dot product $\langle \underline{\Phi}(\underline{x}), \underline{\Phi}(\underline{x}') \rangle$ where

$$\underline{\Phi}(\underline{x}) = (\sqrt{c_1}\, k_1(\underline{x}), \sqrt{c_2}\, k_2(\underline{x}), \ldots, \sqrt{c_n}\, k_n(\underline{x})) \tag{9}$$

Next, the resulting kernel (7) is positive semidefinite because for all $\underline{x}, \underline{k}(\underline{x}, (x_p, x_q))$ is a positive semidefinite kernel and β_{ij} can be constrained to be ≥ 0. Besides, the linear combination of kernels is a kernel and therefore is positive semidefinite. Notice that $\underline{k}(\underline{x}, (x_p, x_q))$ is positive semidefinite if $c_i \geq 0$ and

k_i are pointwise positive for training data. Both Laplacian and multiquadratic kernels verify this condition.

Finally, we show that the resulting kernel is a linear combination of the original k_i. Substituting the expression of the hyperkernel (8) in equation (7), the kernel is written as:

$$k(x_p, x_q) = \sum_{i,j=1}^{m} \beta_{ij} \sum_{l=1}^{n} c_l k_l(x_i, x_j) k_l(x_p, x_q) \tag{10}$$

Now the kernel can be expressed as a linear combination of base kernels.

$$k(x_p, x_q) = \sum_{l=1}^{n} \left[c_l \sum_{i,j=1}^{m} \beta_{ij} k_l(x_i, x_j) \right] k_l(x_p, x_q) \tag{11}$$

Therefore, the above kernel introduces into the k-NN a linear combination of base dissimilarities represented by k_l with coefficients $\gamma_l = c_l \sum_{i,j=1}^{m} \beta_{ij} k_l(x_i, x_j)$.

The previous approach can be extended to an infinite family of distances. In this case, the space that generates the kernel is infinite dimensional. Therefore, in order to work in this space, it is necessary to define a hyperkernel and to optimize it using a HRKHS. Let k be a kernel of dissimilarities. The hyperkernel is defined as follows [15]:

$$\underline{k}(\underline{x}, \underline{x}') = \sum_{i=0}^{\infty} c_i (k(\underline{x}) k(\underline{x}'))^i \tag{12}$$

where $c_i \geq 0$ and $i = 0, \ldots, \infty$. In this case, the non-linear transformation to feature space is infinite dimensional. Particularly, we are considering all powers of the original kernels which is equivalent to transform non-linearly the original dissimilarities.

$$\Phi(\underline{x}) = (\sqrt{c_1}\, k(\underline{x}), \sqrt{c_2}\, k^2(\underline{x}), \ldots, \sqrt{c_n}\, k^n(\underline{x})) \tag{13}$$

where n is the dimensionality of the space which is infinite in this case.

As for the finite family, it can be easily shown that $\underline{k}$ is a valid hyperkernel provided that the kernels considered are pointwise positive. The inverse multiquadratic and Laplacian kernels satisfy this condition. The following proposition allow us to derive the hyperkernel expression for any base kernel.

Proposition 1 (Harmonic Hyperkernel). *Suppose k is a kernel with range $[0, 1]$ and $c_i = (1 - \lambda_h)\lambda_h^i$, $i \in \mathbb{N}$, $0 < \lambda_h < 1$. Then, computing the infinite sum in equation (12), we have the following expression for the harmonic hyperkernel:*

$$\underline{k}(\underline{x}, \underline{x}') = (1 - \lambda_h) \sum_{i=0}^{\infty} (\lambda_h k(\underline{x}) k(\underline{x}'))^i = \frac{1 - \lambda_h}{1 - \lambda_h k(\underline{x}) k(\underline{x}')}, \tag{14}$$

λ_h is a regularization term that controls the complexity of the resulting kernel. Particularly, larger values for λ_h give more weight to strongly non-linear kernels while smaller values give coverage for wider kernels.

3.1 Kernel Alignment k-NN in a HRKHS

We start with some notation that is used in the kernel alignment algorithm. For p,q,r $\in \mathbb{R}^n$, n $\in \mathbb{N}$ let $r = p \circ q$ be defined as element by element multiplication, $r_i = p_i \times q_i$. The pseudo-inverse of a matrix K is denoted by $K^\dagger$. Define the hyperkernel Gram matrix $\underline{K}$ by $\underline{K}_{ijpq} = \underline{k}((x_i, x_j), (x_p, x_q))$, the kernel matrix $K = reshape(\underline{K}\beta)$ (reshaping an m^2 by 1 vector, $\underline{K}\beta$, to an $m \times m$ matrix), Y = diag(y) (a matrix with y on the diagonal and zero otherwise), $G(\beta) = YKY$ (the dependence on β is made explicit) and $\mathbf{1}$ a vector of ones.

The optimization of the regularized quality functional (4) for the kernel alignment in a HRKHS can be written as:

$$\max_{k \in \underline{H}} \quad tr(Kyy^T) + \frac{\lambda_Q}{2}\|k\|^2_{\underline{H}} \tag{15}$$

$$\text{subject to} \quad \|K\|^2_F = C \tag{16}$$

where λ_Q is a parameter that penalizes the complexity of the family of kernels considered and $\|K\|^2_F = tr(KK^T) = \sum_{ij}(K_{ij})^2$ is the Frobenius norm of the kernel.

The minimization of the previous equation leads to the following SDP optimization problem [10].

$$\min_{\beta} \quad \frac{1}{2}t_1 + \frac{\lambda_Q}{2}t_2 \tag{17}$$

$$\text{subject to} \quad \beta \geq 0 \tag{18}$$

$$\|\underline{K}^{\frac{1}{2}}\beta\| \leq t_2, \ \mathbf{1}^T\beta = 1 \tag{19}$$

$$\begin{bmatrix} K & y \\ y^T & t_1 \end{bmatrix} \succeq 0 \tag{20}$$

Once the kernel is learnt, the first k nearest neighbors are identified considering that the Euclidean distance in feature space can be written exclusively in terms of kernel evaluations:

$$d_e^2(x_i, x_j) = k(x_i, x_i) + k(x_j, x_j) - 2k(x_i, x_j) \tag{21}$$

where k is the kernel of dissimilarities learnt by the regularized kernel alignment algorithm introduced previously.

Now we comment shortly some issues about the implementation. The optimization problem (17) were solved using SeDuMi 1.1R3 [17] and YALMIP [9] SDP optimization packages running under MATLAB.

As in the SDP problem there are m^2 coefficients β_{ij}, the computational complexity is high. However, it can be significantly reduced if the Hyperkernel $\{\underline{k}((x_i, x_j), .)|1 \leq i, j \leq m^2\}$ is approximated by a small fraction of terms, $p \ll m^2$ for a given error. In particular, we have chosen an $m \times p$ truncated lower triangular matrix G which approximate the hyperkernel matrix to an error $\delta = 10^{-6}$ using the incomplete Cholesky factorization method [4].

4   Experimental Results

The algorithms proposed have been applied to the identification of several cancer human samples using microarray gene expression data.

The gene expression datasets considered in this paper exhibit different features as shown in table 1. We have chosen problems with a broad range of signal to noise ratio (Var/Samp.), different number of samples and varying priors for the larger category. All the datasets are available from the Broad Institute of MIT and Harvard www.broad.mit.edu/cgi-bin/cancer/datasets.cgi. Next we detail the features and preprocessing applied to each dataset.

Table 1. Features of the different cancer datasets

	Samples	Genes	Var/Samp.	Priors %
Lymphoma MLBCL/DLBCL	210	44928	213	84
Breast Cancer LN	49	7129	145	51
Medulloblastoma	60	7129	119	65

The first dataset consists of frozen tumors specimens from newly diagnosed, previously untreated MLBCL patients (34 samples) and DLBCL patients (176 samples). They were hybridized to Affymetrix $hgu133b$ gene chip containing probes for 44000 genes [13]. The raw intensities have been normalized using the rma algorithm [5]. The second problem we address concerns the clinically important issue of metastatic spread of the tumor. The determination of the extent of lymph node involvement in primary breast cancer is the single most important risk factor in disease outcome and here the analysis compares primary cancers that have not spread beyond the breast to ones that have metastasized to axillary lymph nodes at the time of diagnosis. We identified tumors as 'reported negative' (24) when no positive lymph nodes were discovered and 'reported positive' (25) for tumors with at least three identifiably positive nodes [21]. All assays used the human HuGeneFL Genechip microarray containing probes for 7129 genes. The third dataset [12] addresses the clinical challenge concerning medulloblastoma due to the variable response of patients to therapy. Whereas some patients are cured by chemotherapy and radiation, others have progressive disease. The dataset consists of 60 samples containing 39 medulloblastoma survivors and 21 treatment failures. Samples were hybridized to Affymetrix HuGeneFL arrays containing 5920 known genes and 897 expressed sequence tags.

All the datasets have been standardised subtracting the median and dividing by the Inter-quantile range. The rescaling were performed based only on the training set to avoid bias.

In order to assure a honest evaluation of all the classifiers we have performed a double loop of crossvalidation [16]. The outer loop is based on stratified ten fold cross-validation that iteratively splits the data in ten sets, one for testing and the others for training. The inner loop performs stratified nine fold cross-validation over the training set and is used to determine the optimal parameters avoiding bias in the error estimation. The stratified variant of cross-validation keeps the same proportion of patterns for each class in training and test sets. This is necessary in our problem because the class proportions are not equal. Finally, the error measure considered to evaluate the classifiers has been accuracy. This metric computes the proportion of samples misclassified. The accuracy is easy to interpret and allow us to compare with the results obtained by previously published studies.

Regarding the value of the parameters, $c_i = 1/M$ for the finite family of distances where M is the number of dissimilarities considered, and the regularization parameter $\lambda_Q = 1$ which gives good experimental results for all the problems considered in this paper. Finally, for the infinite family of dissimilarities, the regularization parameter λ_h in the Harmonic hyperkernel (14) has been set up to 0.6 which gives an adequate coverage of various kernel widths. Smaller values emphasize only wide kernels. All the base kernel of dissimilarities have been normalized so that all ones have the same scale. Three different kernels have been considered, linear, inverse multiquadratic and Laplacian.

The optimal values for the kernel parameters, the number of genes and the nearest neighbors considered have been set up by crossvalidation and using a grid search strategy.

Gene selection can improve significantly the classifier performance [6]. Therefore, we have evaluated the classifiers for subsets of $280, 146, 101, 56$ and 34 top ranked genes. k-NN is quite sensitive to the 'curse of dimensionality'. Thus, considering a larger number of genes or even the whole set of genes does not help to reduce the misclassification errors.

The genes are ranked according to the ratio of between-group to within-group sums of squares defined in [3]

$$BW(j) = \frac{\sum_i \sum_k I(y_i = k)(\bar{x}_{.j}^{(k)} - \bar{x}_{.j})^2}{\sum_i \sum_k I(y_i = k)(x_{ij} - \bar{x}_{.j}^{(k)})^2} \tag{22}$$

where $\bar{x}_{.j}^{(k)}$ and $\bar{x}_{.j}$ denote respectively the average expression level of gene j for class k and the overall average expression level of gene j across all samples, y_i denotes the class of sample i, and $I(\cdot)$ is the indicator function. Next, the top ranked genes are chosen. This feature selection method is simple but compares well with more sophisticated methods. For a discussion of other approaches considered in gene expression data analysis the reader is referred to [6]. Finally, the ranking of genes has been carried out considering only the training set to avoid bias. Therefore, feature selection is repeated in each iteration of cross-validation.

We have compared our method with the Lanckriet formalism [10] that allow us to incorporate a linear combination of dissimilarities into the SVM considering the connection between kernels and dissimilarities, the Large Margin Nearest Neighbor algorithm [20] that learns a Mahalanobis metric maximizing the k-NN margin in input space and the classical k-NN with the best dissimilarity for a subset of six measures widely used in the Microarray literature.

Table 2. Empirical results for the k-NN classifier considering different dissimilarities. The ν-SVM based on coordinates and the best dissimilarity have also been considered.

Technique	DLBCL-MLBCL	Breast LN	Medulloblastoma
ν-SVM (Coordinates)	16%	8.16%	16.6%
ν-SVM (Best Distance)	11%	8.16%	13.3%
k-NN Euclidean	10%	10%	10%
k-NN Cosine	15.1%	6%	10%
k-NN Manhattan	10%	12%	16.6%
k-NN Correlation	23%	18%	15%
k-NN χ^2	16%	6%	10%
k-NN Spearman	31%	28%	23.3%

Table 3. Empirical results for the kernel alignment k-NN based on a combination of dissimilarities. For comparison we have included two learning metric strategies proposed in the literature.

Technique	DLBCL-MLBCL	Breast LN	Medulloblastoma
Kernel align. k-NN (Finite family, linear kernel)	10%	6%	11.66%
Kernel align. k-NN (Infinite family, linear kernel)	10%	4%	10%
Kernel align. k-NN (Finite family, inverse kernel)	10%	8%	10%
Kernel align. k-NN (Infinite family, inverse kernel)	9%	4%	10%
Kernel align. k-NN (Finite family, laplacian kernel)	9%	6%	8.33%
Kernel align. k-NN (Infinite family, laplacian kernel)	9%	4%	10%
Lanckriet SVM	11%	8.16%	11.66%
Large Margin NN	17%	8.50%	13.3%

From the analysis of tables 2 and 3, the following conclusions can be drawn:

- Kernel alignment k-NN outperforms two widely used strategies to learn the metric such as Large Margin NN and Lanckriet SVM. The first one is prone to overfitting and does not help to reduce the error of k-NN based on the best dissimilarity. Similarly, our method improves the Lanckriet formalism particularly for Breast LN problem in which the sample size is smaller.

    Kernel alignment k-NN is quite insensitive to the kind of non-linear kernel employed.
- Kernel alignment k-NN considering an infinite family of distances outperforms k-NN with the best distance and the ν-SVM, particularly for breast cancer and Leukemia DLBCL-MLBCL. The infinite family of dissimilarities helps to reduce the errors of the finite counterpart particularly for breast cancer. This suggests that for certain complex non-linear problems, the non-linear transformation of the original dissimilarities helps to improve the

classifier accuracy. We report, that only for the Medulloblastoma and with Laplacian base kernel the error is slightly larger for the infinite family. This suggests that the regularization term controls appropriately the complexity of the resulting dissimilarity.

- Table 2 shows that the best distance depends on the dataset considered and that the performance of k-NN depends strongly on the particular measure employed to evaluate the sample proximities. Finally, an interesting result is that k-NN outperforms the ν-SVM algorithm for all the datasets.

5 Conclusions

In this paper, we propose two methods to incorporate in the k-NN algorithm a linear combination of non-Euclidean dissimilarities. The family of distances is learnt in a HRKHS (Hyper Reproducing Kernel Hilbert Space) using a Semi-definite Programming approach. A penalty term has been added to avoid the overfitting of the data. The algorithm has been applied to the classification of complex cancer human samples.

The experimental results suggest that the combination of dissimilarities in a Hyper Reproducing Kernel Hilbert Space improves the accuracy of classifiers based on a single distance particularly for non-linear problems. Besides, this approach outperforms other learning metric strategies widely used in the literature and is robust to overfitting.

Future research trends will apply this formalism to integrate heterogeneous data sources.

References

1. Blanco, A., Martín-Merino, M., De Las Rivas, J.: Combining dissimilarity based classifiers for cancer prediction using gene expression profiles. BMC Bioinformatics, 1–2 (2007); ISMB/ECCB 2007
2. Cristianini, N., Kandola, J., Elisseeff, J., Shawe-Taylor, A.: On the kernel target alignment. Journal of Machine Learning Research 1, 1–31 (2002)
3. Dudoit, S., Fridlyand, J., Speed, T.P.: Comparison of Discrimination Methods for the Classification of Tumors Using Gene Expression Data. Journal of the American Statistical Association 97(457), 77–87 (2002)
4. Fine, S., Scheinberg, K.: Efficient svm training using low-rank kernel representations. Journal of Machine Learning Research 2, 243–264 (2001)
5. Gentleman, R., Carey, V., Huber, W., Irizarry, R., Dudoit, S.: Bioinformatics and Computational Biology Solutions Using R and Bioconductor. Springer, Berlin (2006)
6. Jeffery, I.B., Higgins, D.G., Culhane, A.C.: Comparison and Evaluation Methods for Generating Differentially Expressed Gene List from Microarray Data. BMC Bioinformatics 7(359), 1–16 (2006)
7. Jiang, D., Tang, C., Zhang, A.: Cluster Analysis for Gene Expression Data: A Survey. IEEE Transactions on Knowledge and Data Engineering 16(11), 1370–1386 (2004)
8. Kandola, J., Shawe-Taylor, J., Cristianini, N.: Optimizing kernel alignment over combinations of kernels. NeuroCOLT, Tech. Rep. (2002)

9. Löfberg, J.: YALMIP, yet another LMI parser (2002),
   www.control.isy.liu.se/~johanl/yalmip.html
10. Lanckriet, G., Cristianini, N., Barlett, P., El Ghaoui, L., Jordan, M.: Learning
    the kernel matrix with semidefinite programming. Journal of Machine Learning
    Research 3, 27–72 (2004)
11. Pekalska, E., Paclick, P., Duin, R.: A generalized kernel approach to dissimilarity-
    based classification. Journal of Machine Learning Research 2, 175–211 (2001)
12. Pomeroy, S.E.A.: Prediction of central nervous system embryonal tumour outcome
    based on gene expression. Nature 415 (2002)
13. Savage, K., et al.: The molecular signature of mediastinal large B-cell lymphoma
    differs from that of other diffuse large B-cell lymphomas and shares features with
    classical hodgkin lymphoma. Blood 102(12) (December 2003)
14. Scholkopf, B., Tsuda, K., Vert, J.: Kernel Methods in Computational Biology. MIT
    Press, Cambridge (2004)
15. Soon Ong, C., Smola, A., Williamson, R.: Learning the kernel with hyperkernels.
    Journal of Machine Learning Research 6, 1043–1071 (2005)
16. Statnikov, A.: A comprehensive evaluation of multicategory classification methods
    for microarray gene expression cancer diagnosis. Bioinformatics 21(5), 631–643
    (2004)
17. Sturm, J.F.: Using SeDuMi 1.02, a MATLAB toolbox for optimization over sym-
    metric cones. Optimization Methods and Software 11/12(1-4), 625–653 (1999)
18. Tsuda, K.: Support Vector Classifier with Assymetric Kernel Function. In: Pro-
    ceedings of ESANN, Bruges, pp. 183–188 (1999)
19. Vapnik, V.: Statistical Learning Theory. John Wiley & Sons, New York (1998)
20. Weinberger, K.Q., Saul, L.K.: Distance Metric Learning for Large Margin Nearest
    Neighbor Classification. J. Machine Learning Research 10, 207–244 (2009)
21. West, M., et al.: Predicting the clinical status of human breast cancer by using
    gene expression profiles. PNAS 98(20) (2001)
22. Wu, G., Chang, E.Y., Panda, N.: Formulating distance functions via the kernel
    trick. In: ACM SIGKDD, Chicago, pp. 703–709 (2005)
23. Xiong, H., Chen, X.-W.: Kernel-Based Distance Metric Learning for Microarray
    Data Classification. BMC Bioinformatics 7(299), 1–11 (2006)

Appendix A

In this section we define shortly the Hyper-Reproducing Kernel Hilbert Spaces.
First, we define a Reproducing Kernel Hilbert Space.

Definition 1 (Reproducing Kernel Hilbert Space). *Let $\mathcal{X}$ be a nonempty
set and $\mathcal{H}$ be a Hilbert space of functions $f : \mathcal{X} \to \mathbb{R}$. Let $\langle \cdot, \cdot \rangle$ be a dot product
in $\mathcal{H}$ which induces a norm as $\|f\| = \sqrt{\langle f, f \rangle}$. $\mathcal{H}$ is called a RKHS if there is a
function $k : \mathcal{X} \times \mathcal{X}$ with the following properties:*

- *k has the reproducing property $\langle f, k(x, \cdot) \rangle = f(x)$ for all $f \in \mathcal{H}$, $x \in \mathcal{X}$*
- *k spans $\mathcal{H}$, i.e. $\mathcal{H} = \overline{span\{k(x, \cdot)|x \in \mathcal{X}\}}$, where $\overline{X}$ is the completion of the
  set X.*
- *k is symmetric, i.e, $k(x, y) = k(y, x)$*

Next, we introduce the Hyper Reproducing Kernel Hilbert Space.

Definition 2 (Hyper-Reproducing Kernel Hilbert Space). *Let $\mathcal{X}$ be a nonempty set and $\underline{\mathcal{X}} = \mathcal{X} \times \mathcal{X}$ be the Cartesian product. Let $\underline{\mathcal{H}}$ be the Hilbert space of functions $k : \underline{\mathcal{X}} \to \mathbb{R}$ with a dot product $\langle \cdot, \cdot \rangle$ and a norm $\|k\| = \sqrt{(\langle k, k \rangle)}$. $\underline{\mathcal{H}}$ is a Hyper Reproducing Kernel Hilbert Space if there is a hyperkernel $\underline{k} : \underline{X} \times \underline{X} \to \mathbb{R}$ with the following properties:*

- *$\underline{k}$ has the reproducing property $\langle k, \underline{k}(\underline{x}, \cdot) \rangle = k(\underline{x})$ for all $k \in \underline{\mathcal{H}}$*
- *$\underline{k}$ spans $\underline{H} = \overline{span\{\underline{k}(\underline{x}, \cdot) | \underline{x} \in \underline{X}\}}$*
- *$\underline{k}(x, y, s, t) = \underline{k}(y, x, s, t)$ for all $x, y, s, t \in \mathcal{X}$.*

Subgroup Discovery for Test Selection: A Novel Approach and Its Application to Breast Cancer Diagnosis

Marianne Mueller[1], Rómer Rosales[2], Harald Steck[2],
Sriram Krishnan[2], Bharat Rao[2], and Stefan Kramer[1]

[1] Technische Universität München, Institut für Informatik, 85748 Garching, Germany
[2] IKM CAD and Knowledge Solutions, Siemens Healthcare, Malvern PA 19335, USA

Abstract. We propose a new approach to test selection based on the discovery of subgroups of patients sharing the same optimal test, and present its application to breast cancer diagnosis. Subgroups are defined in terms of background information about the patient. We automatically determine the best t subgroups a patient belongs to, and decide for the test proposed by their majority. We introduce the concept of prediction quality to measure how accurate the test outcome is regarding the disease status. The quality of a subgroup is then the best mean prediction quality of its members (choosing the same test for all). Incorporating the quality computation in the search heuristic enables a significant reduction of the search space. In experiments on breast cancer diagnosis data we showed that it is faster than the baseline algorithm APRIORI-SD while preserving its accuracy.

1 Introduction

Diagnosis is the art or act of identifying a disease from its signs and symptoms. This implies that the more information is available about a patient, the easier it is to pose an accurate diagnosis. Information can be obtained by a variety of tests including questioning the patient, physical examinations, imaging modalities, or laboratory tests. However, due to costs, time, and risks for the patient, in clinical routine it is often preferable for patients to undergo as few tests as needed. Consequently, there is a trade-off between the costs (and number) of tests and the accuracy of the diagnosis. Therefore, optimal test selection plays a key role for diagnosis. The goal of this paper is to find the optimal set of tests to choose for a patient in a given situation, where the definition of optimality is also provided in this paper. Existing work on test selection [1,2] mostly addresses the problem of finding global solutions for all patients. However, it is not likely that for each patient the same test is the most informative one. Therefore, we believe that it is a better approach to concentrate on the task of identifying subgroups of patients for which the optimal test is the same. In this paper, we present a novel solution to this problem based on subgroup discovery (SD) [3,4], a family of data mining algorithms. Subgroup discovery methods compute all

N. Adams et al. (Eds.): IDA 2009, LNCS 5772, pp. 119–130, 2009.
© Springer-Verlag Berlin Heidelberg 2009

subgroups of a population that are statistically most interesting with respect to a specified property of interest. Consider, for instance, a population described by general demographic attributes and a target variable (property/attribute of interest) representing a disease status (disease, non-disease). Let us assume a distribution of 43% disease and 57% non-disease in the entire population. Then, an SD algorithm might come up with a subgroup identified by two conditions, $age > 75$ and $gender = female$ in which the distribution is 85% disease and 15% non-disease. Here, the subgroup description consists of two attribute-value tests, and it selects a set of persons with a particularly high prevalence of the disease (85% instead of 43% in the entire population). Standard SD approaches are designed for a single target variable. However, in the setting of test selection, a single variable seems not sufficient. In fact, we want to target the relation between two variables: the outcome of a test and the actual state of disease. Therefore, the quality of a subgroup should correspond to the value of the result of a selected test with respect to the actual state of disease, e.g., a biopsy result. To quantify the value of a test result, we define a so-called *prediction quality* function in Section 2.1. The function gives high scores to a pair of a subgroup and a test if the result is close to the actual state of disease, and therefore leads to an accurate diagnosis. Since standard SD does not take into account complex scenarios like this, including benefits or costs of subgroups, we developed a new, cost-sensitive variant. Throughout the paper, we will use the term *prediction quality*, which corresponds to the *benefits* of a prediction rather than to its *costs*. However, as it is easy to transform one into the other, we can also speak of *cost-sensitive subgroup discovery*. The algorithm outputs subgroup descriptions consisting of background information about the patients. The overall goal is to compute an optimal test selection for a new patient. More precisely, our proposed solution is to identify subgroups of the data for which the same test is the optimal selection, to arrive at a correct diagnosis. In a second step, analyzing the subgroups will help to find out which features determine the performance of the tests. Hence, it will be possible to decide for a new patient, given its features, which test is the best to choose. We apply and validate this approach on a data set from breast cancer diagnosis, where for each patient four different tests are possible.

2   Background and Data

Our study was conducted in the area of breast cancer diagnosis. In breast cancer diagnosis, different imaging modalities are used routinely, in particular, Film Mammography (FMAM), Digital Mammography (DMAM), Ultrasound (USND), and Magnetic Resonance Imaging (MRI). Each modality has its own specific characteristics. When a patient is under scrutiny for breast cancer, it is often not clear which of these modalities is best suited to answer the basic question to whether the patient has or does not have cancer. The choice of a modality usually requires considerable experience of the health care workers. In this paper we show how to support the optimal test selection for a new patient

Table 1. Prediction score for agreement between the overall assessment (OA_m = BIRADS) of a modality m and the biopsy finding BIO

pscr		OA_m					
		0	1	2	3	4	5
	Malignant	75	0	0	25	100	100
BIO	Atypia	75	75	90	90	90	75
	Benign	75	100	100	100	75	50

by retrospectively analyzing the performance of the tests on subgroups of previously examined patients with similar features. The basis of our work is a dataset collected in a breast cancer study of a large University Hospital, which comprises patients that had a suspicious finding in a screening. The study gathers patient specific information like medical history, demographic information, and a breast cancer risk summary. Each patient in the study underwent all four above mentioned modality tests. Each of these tests was independently analyzed by the appropriate specialist to judge for the occurrence of breast cancer. For each lesion detected, the specialist determines in which category it falls. The categories are called BIRADS score and range from 0 to 5: The higher the BIRADS, the higher the probability (assessed by the medical expert) for the lesion to be malignant. (0 = incomplete, i.e., needs additional imaging evaluation, 1 = no finding, 2 = benign finding, 3 = probably benign, 4 = suspicious abnormality, 5 = highly suggestive of malignancy) [5]. To obtain evidence of the initial assessments, a biopsy has to be performed. A pathologic examination of a biopsy determines whether the lesion is benign, atypia benign, or malignant. In this study at least one lesion for each patient is subject to biopsy.

2.1 Definition of Prediction Quality

To quantify the accuracy of a diagnosis, we propose a measure of prediction quality. Each test m results for each lesion l in an overall assessment $OA_m(l)$ of the physician. This is defined by the BIRADS score (see above). Each lesion has a biopsy $BIO(l)$ proving the status of the lesion (malignant, benign or atypia benign). The prediction quality expresses how close the assessment comes to the biopsy finding. Therefore, we define a prediction score $pscr$ that evaluates the performance of a test for a single lesion. Table 1 gives the $pscr$ for each pair (Overall Assessment, Biopsy). The values in the table were proposed by a domain expert in the field of breast cancer diagnosis.[1] The higher the prediction score, the more accurate is the prediction.

Having defined $pscr$ for a single lesion l, we can easily obtain the prediction quality $pq(S, m)$ of a modality m for an example set S by averaging over the prediction scores of m and all lesions in S:

---

[1] However, they are certainly not the only possible values for the prediction score. For instance, as not all types of malignant findings are equally harmful, it might be more accurate to distinguish between invasive and non-invasive types of cancer.

$$pq(S, m) = \frac{1}{|S|} \sum_{l \in S} \cdot pscr(OA_m(l), BIO(l))$$

In our data set we have 138 lesions (of 72 patients) with biopsy and four modalities (Digital Mammography (DMAM), Film Mammography (FMAM), Magnet Resonance Imaging (MRI), and Ultrasound (USND)) to choose from. The prediction quality for the entire dataset separated for each modality is 77.9 for DMAM, 78.0 for FMAM, 78.4 for MRI, and 80.2 for USND. It shows that the prediction qualities of the different modalities over all lesions are quite similar (Entropy = 1.999 bits of a maximum 2 bits), with USND performing slightly better. By considering subgroups of patients we expect to increase the prediction quality for at least one modality per subgroup. Then, we apply this modality to all lesions in the subgroup to obtain the most accurate diagnosis.

3   Method

The general idea is to determine subgroups of lesions with an unusual modality performance. Let X be a training set of observed examples and n the number of tests $\{m_1, \ldots, m_n\}$ that can be performed. For each group[2] of lesions $S \subseteq X$ we consider the prediction qualities $pq(S, m_i)$ of the possible modalities and decide for the modality $m^*(S)$ with the highest pq-value[3]: $m^*(S) = \text{argmax}_m\, pq(S, m)$. The optimal prediction quality of S is then defined as $pq^*(S) = \max_m pq(S, m)$.

We introduce an algorithm called $SD4TS$ (Subgroup Discovery for Test Selection). The task of the algorithm is defined in the following way:

Given: X, n, $minsupport$, the minimal number of examples that have to be covered by a subgroup description, t, the number of best subgroups we want to obtain from the algorithm, and a set of pq-values $\{pscr(s, m_i) | s \in X, m_i \in tests\}$ (in a more general setting a set of cost/benefit values).

Find: The t subgroups with the highest pq^* values (best costs/benefit) and at least $minsupport$ examples.

We base our algorithm on APRIORI-SD [3], an adaptation of the association rule learning algorithm APRIORI [6] to subgroup discovery. APRIORI-SD starts with generating subgroups described by a single attribute-value-pair. Subsequently, it generates subgroups with longer (and thus more specific) descriptions. Subgroups are only kept if they contain more examples than $minsupport$. All smaller subgroups are pruned, and no subgroups more specific than these are generated. For our task, we are interested in the t subgroups that are cost-efficient for at least one modality. Therefore, we can prune the search space even further, namely in a way that only the promising subgroups are kept. That means, during

---

[2] As in other SD algorithms we consider only groups that can be described by a conjunction of attribute-value pairs.

[3] In a more general setting, instead of pq we can assume any types of costs (where max should be replaced by min) or benefits that rate the performance of the tests.

the generation of subgroups, candidates are immediately evaluated and checked whether they have the potential to lead to improved costs.

3.1   Quality Pruning

Pruning is possible when a subgroup and all specializations of the subgroup will not outperform the quality of the already discovered subgroups. Specialization of a subgroup means adding an attribute-value pair to the subgroup description of a subgroup sg. This can cause changes of both the frequency and the quality. The frequency can only decrease. The defined quality, however, can change in both directions.

The critical point is hence to recognize when the quality of subgroup sg can not outperform at least one of the best t subgroups found so far. Thus, it is not enough to consider the actual $pq(sg)$ to determine if sg can be pruned. Furthermore, it is necessary to consider what we call the *coreGroup* of sg. The *coreGroup* is a group consisting of the *minsupport* examples covered by sg with the highest quality. The cost of the *coreGroup* upperbounds the costs of all possible specializations of sg, because the overall score is defined as an average of the elements of the group.

The example in Figure 1 demonstrates the discussed characteristics. The seven dots represent the generated subgroup sg, with $pq(sg) = 0.5$. Assume we have generated already a subgroup sg_{best} with $pq(sg_{best}) = 0.6$. In this case, sg has a worse pq value and seems to be not promising. However, pruning sg will inhibit finding an optimal subgroup sg_{new} (the four black dots) contained in sg with $pq(sg_{new}) = 0.625$.

Considering the pq-value of the *coreGroup* of sg will circumvent this mistake by providing the upper bound of the pq-values of any specialization of sg: For the given example, we assume a *minsupport* of 3. Then $pq(coreGroup(sg)) = 1$. Since $pq(coreGroup(sg)) > pq(sg_{best})$, sg is not pruned and keeps the option of generating the improved subgroup sg_{new} in a later iteration of the algorithm.

3.2   The *SD4TS* Algorithm

The pseudo-code of the algorithm is shown in Algorithm 1. *SD4TS* starts with generating 1-itemset candidates (described by one attribute-value-pair).

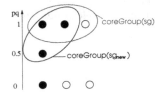

Fig. 1. Simple pruning fails. All dots are examples in sg. The black dots are also covered by sg_{new}. The y-direction corresponds to the pq-value of each example.

Candidates are pruned if they are not frequent or if they can not outperform (not even through specialization) one of the the best t subgroups generated so far. All remaining candidates are stored in *optimizableCandidates*. The best t subgroups are stored in *topCandidates*. For efficiency, we store all created subgroups (including the list of transactions that are covered, costs, *bestPossibleCosts*, support and a list *still2test* of 1-itemsets that have not been tested yet to specialize the subgroup) in an array *allSubgroups* in the order they were created. The sorted lists *topCandidates* and *optimizableCandidates* contain only pointers (the indices of the array) to the subgroups stored in *allSubgroups*. The list *topCandidates* is sorted according to the actual costs of the subgroups. This facilitates removing the worst subgroup, whenever a newly generated subgroup has better costs. The list *optimizableCandidates* is sorted according to the *bestPossibleCosts* a *coreGroup* can achieve. In that way, we explore always the subgroup with the highest potential first. That means specializing this subgroup

---

Algorithm 1. $SD4TS$ (subgroup discovery for test selection)

Input: Training set, set of costs, *minsupport*, number t of best subgroups to be produced

Output: list of *topCandidates*, including the proposed test(s) for each candidate

---

1: *optimizableCandidates* $= \{c | c$ frequent subgroup defined by 1 attribute-value-pair$\}$;
2: *topCandidates* $= \{c | c$ belongs to the t best candidates in *optimizableCandidates*$\}$;
3: *minpq* = worst quality of *topCandidates*;
4: remove all c from *optimizableCandidates* with $c.bestPossibleCosts < minpq$
5: **while** *optimizableCandidates* not empty **do**
6:    c_1 = *optimizableCandidates*.removeFirst();//candidate with bestPossibleCosts
7:    **for all** $c_2 \in c_1.still2test$ **do**
8:       remove c_2 from $c_1.still2test$
9:       c_{new} = generate_new_candidates(c_1, c_2);
10:      **if** c_{new} frequent and $c_{new}.bestPossibleCosts > minpq$ **then**
11:         add c_{new} to *optimizableCandidates*
12:         **if** c_{new} better than worst c_t of *topCandidates* **then**
13:            add c_{new} to *topCandidates*
14:            **if** $size(topCandidates) > t$ **then**
15:               remove worst c of *topCandidates*
16:            **end if**
17:            *minpq* = worst quality of *topCandidates*;
18:            remove all c from *optimizableCandidates* with $c.bestPossibleCosts < minpq$
19:         **end if**
20:      **end if**
21:   **end for**
22: **end while**
23: **return** *topCandidates*

is likely to lead to a subgroup that falls into the top t candidates and therefore raises $minpq$ which reduces the search space.

Safe pruning: A new subgroup candidate sg_{new} is only accepted if sg_{new} is frequent and at least one of the following holds:

1. there are less than t subgroups stored in $topCandidates$, or
2. sg_{new} has a better performance than the worst subgroup of $topCandidates$, or
3. at least one frequent subset of examples in sg_{new} (i.e., $coreGroup(sg_{new})$) leads to a better performance than the worst subgroup of $topCandidates$. This can be tested in $O(n * |sg_{new}| \log |sg_{new}|)$: For each test m determine the set CG_m of $minsupport$ examples covered by sg_{new} that have the best costs. $sg_{new}.BestPossibleCosts = \max_m pq(CG_m, m)$

In all cases we add the candidate to $optimizableCandidates$. In case 1 and 2 we also add the candidate to $topCandidates$. In the second case we additionally remove the worst stored subgroup from $topCandidates$.

3.3 Analysis of Runtime and Search Space

Figure 2 shows how the search space (i.e., the number of search nodes) depends on the parameters $minsupport$ and t. The higher $minsupport$, the smaller the search space. This is caused by the frequency pruning. We also see that a low t-value results in a small search space, which is the expected effect of quality pruning. For small values of t fewer subgroups are kept in $topCandidates$, which increases the threshold of costs below which subgroups are pruned. The right diagram in Figure 2 displays the runtime of the two algorithms. For $minsupport$ values below 25, $SD4TS$ is faster than APRIORI-SD, as frequency pruning is only effective for larger $minsupport$ values.

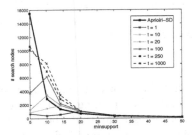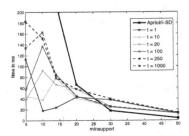

Fig. 2. Complexity of $SD4TS$. The left (right) diagram shows how the search space (runtime) depends on $minsupport$ (x-axis) for different t-values. In comparison, the black solid line shows the search space of APRIORI-SD.

4   Validation and Results

To evaluate the approach, we tested it in a predictive setting[4], more specifically, in a leave-one-out cross-validation. For each test lesion l, we generate only subgroups with attribute-value pairs contained in l. Table 2 shows the best $t = 5$ subgroups for 3 example lesions. From the resulting best t subgroups, we decide for the test proposed by the majority of the identified subgroups (for test lesion 9 it is USND). A test is proposed if it has the best costs averaged over all examples in the subgroup (for subgroup S1 it is USND). If more than one test has optimal costs, all of them are proposed (for subgroup S9 it is DMAM and USND). If more than one test is proposed most often by the subgroups, the cost for the test lesion l is determined by the mean of their costs.

4.1   Analysis of Performance Compared to Random Selection

For each lesion l, a vector of prediction qualities is given by

$$\vec{pq}(l) = (pq(l, FMAM), pq(l, DMAM), pq(l, MRI), pq(l, USND)).$$

This can be interpreted as a ranking of the modalities for each lesion. For instance, $\vec{pq}(l) = (0.8, 0.7, 0.9, 0.8)$ leads to the ranking $MRI > USND = DMAM > FMAM$. We encode this ranking as 1224. There is one modality ranked first, followed by two modalities ranked second, and one modality ranked fourth. In total, there are seven possible encodings: from 1111 (all modalities have the same prediction quality) to 1234 (all modalities have different prediction qualities).

Table 3 shows the distribution of codes of our dataset. It is remarkable that in 29% of the cases all modalities perform equally well. This implies that for those cases a random choice is as effective as a more informed choice. To have fairer conditions, we additionally validated the algorithm on two restricted subsets of test lesions (results are shown in Table 6). Set 1 consists of all 138 lesions. Set 2 is a subset of Set 1 containing only the 98 lesions whose costs are not the same over all modalities (all codes except 1111). Set 3 comprises 32 lesions, where one modality outperforms the other modalities (code 1222, 1224, and 1234). Note that the differences between the best and the worst, and between the best and the random choice improve significantly from Set 1 to Set 3.

4.2   Results

The results in Table 4 (column Set 1) show that the algorithm achieves in general better costs than picking a modality at random or picking always the same modality (compare with Table 3). It also becomes clear that the best results are achieved with low *minsupport* and high t values (considering even small subgroups), or, vice versa, high *minsupport* (50) and low t values (few large subgroups).

---

[4] Note that prediction is not our main goal. Additionally, we are interested in the discovery of new medical knowledge. Therefore, we prefer subgroup discovery over standard classifiers.

Table 2. Example of best 5 subgroups for 3 selected input test lesions. The shaded rows indicate the actual prediction scores of the modalities for the current input test lesion. The bold prediction qualities indicate the image modality proposed by $SD4TS$. For example, test lesion 9 will be assessed best by USND (pscr =100), the other three modalities fail to assess the lesion correctly (pscr = 0).

top 5 subgroups for selected test lesions	subgr. size	DMAM	FMAM	MRI	USND
test lesion 9		0	0	0	**100**
S1	Highschool or less + has past Mammo	17	76.5	76.5	57.4	**98.5**
S2	Highschool or less + has past Mammo + no relatives with cancer	16	75.0	75.0	56.3	**98.4**
S3	Highschool or less + has past Mammo + has past breast USND	14	85.7	78.6	62.5	**98.2**
S4	Highschool or less + has past Mammo + Race = white	14	85.7	78.6	66.1	**98.2**
S5	age 40-59 + no relatives with cancer + pre menopausal + Race=white	28	76.8	72.3	76.8	**93.8**
test lesion 19		75	100	100	**100**
S6	Graduate School after college + age 40-59 + no relatives with cancer + Race=white	14	76.8	75.0	**98.2**	82.1
S7	Graduate School after college + has no breast USND + no relatives with cancer	21	94.1	82.1	92.9	**96.4**
S8	Graduate School after college + has no breast USND + no relatives with cancer + Race = white	19	93.4	80.3	92.1	**96.1**
S9	Graduate School after college + age 40-59 + no relatives with cancer + has no breast USND	18	**95.8**	81.9	91.7	**95.8**
S10	Graduate School after college + has no breast USND+ no relatives with cancer + Race = white + age 40-59	16	**95.3**	79.7	90.6	**95.3**
test lesion 23		100	100	75	100
S11	no relatives with cancer + age ≥60	14	**94.6**	87.5	87.5	76.8
S12	Graduated from College + post menopausal status	16	78.1	82.8	**93.8**	70.3
S13	post menopausal status + age ≥ 60	15	**93.3**	86.7	86.7	76.7
S14	age ≥60	15	**93.3**	86.7	86.7	76.7
S15	Graduated form College + no relatives with cancer + post menopausal status	15	76.7	81.7	**93.3**	70.0

Table 3. Left: Distribution of lesions according to the ranking of modality performances. Right side shows the costs that arise for always picking the same modality (separated in always picking DMAM, always picking FMAM, etc), or for always picking the modality with the best costs, the worst costs, or picking one modality at random. The choice of parameters is good, if the costs are greater than the costs for random selection.

# best modalities	# cases	code	# cases		
4 or 0	40 (29%)	1111	40		
3	45 (33%)	1114	45		Set 1
2	21 (18%)	1133	16		
		1134	5	Set 2	
1	32 (23%)	1222	27		
		1224	3	Set 3	
		1234	2		

costs	Set 1	Set 2	Set 3
Best	99.45	99.49	99.22
Random	**78.6**	**70.22**	**40.63**
Worst	58.03	41.33	16.41
DMAM	77.92	69.13	30.47
FMAM	78.1	69.39	41.41
MRI	78.28	69.9	44.53
USND	80.11	72.45	46.09

Table 4. Results of leave-one-out cross-validation of $SD4TS$ with varying minsupport and t parameters over three different test sets (best parameter settings in **bold**). The displayed costs are averaged over all test lesions. A cost for a single lesion is derived by taking the costs of the test proposed by the majority of the returned subgroups. If two or more tests are proposed equally often, we take the average of their costs.

costs		Set 1					Set 2					Set 3			
min-s.	5	10	20	30	50	5	10	20	30	50	5	10	20	30	50
1	80.7	**82.6**	77.8	79.2	**81.5**	73.1	**75.8**	69.1	70.9	**74.2**	45.6	48.4	37.9	42.6	47.3
5	79.5	81.3	76.6	78.2	**82.7**	71.4	73.9	67.4	70.2	**75.9**	42.8	42.1	35.6	43.0	**51.6**
10	**82.7**	80.7	76.1	79.0	80.1	**75.9**	73.1	66.6	70.7	72.2	47.3	41.8	37.1	46.9	42.5
t 20	**82.6**	79.6	77.8	80.4	80.5	**75.8**	71.5	69.0	72.7	72.8	44.5	42.6	42.6	49.6	47.1
30	**82.5**	79.3	80.1	79.2	80.5	**75.6**	71.1	72.2	70.9	72.8	49.6	46.1	49.2	41.4	47.1
100	**82.8**	80.3	79.7	79.1	80.5	**76.0**	72.6	71.7	70.8	72.8	**51.0**	47.7	46.1	41.4	47.1
250	**82.2**	80.7	79.7	79.1	80.5	**75.2**	73.1	71.7	70.8	72.8	**51.0**	47.7	46.1	41.4	47.1

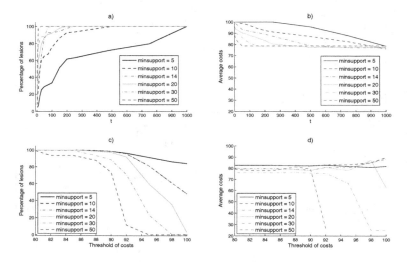

Fig. 3. a) Percentage of lesions that are covered by at least one subgroup for different values of t (x-axis) and *minsupport*. b) Average costs (prediction quality) of lesions when choosing the modality proposed by the best subgroup ignoring lesions that are not covered by any subgroup. c) Percentage of lesions covered by at least one subgroup with pq above a certain threshold (x-axis). d) Average quality of lesions when choosing modality proposed by the majority of the (maximal 10) best subgroups above the threshold. Lesions covered by no subgroup above the threshold are ignored.

We further validate the overall coverage of the lesions by the generated t subgroups. Figure 3a shows the percentage of lesions that are covered by at least one subgroup. With an increasing number of generated subgroups t, the coverage increases and the average quality (Figure 3b) decreases. It also shows that a higher minsupport induces a higher coverage, even for low values of t. Also for those cases the quality decreases. Figure 3c shows the behavior of the proportion of lesions covered by a subgroup when introducing a threshold, which has to be overcome by the prediction quality of the subgroups. Subgroups with lower prediction qualities are ignored in this setting. Of course with raising the threshold the number of uncovered lesions increases. Small *minsupport* allows more lesions to be covered. The average quality increases with a higher threshold for low *minsupport* and decreases for high *minsupport* and a higher threshold. The larger subgroups seem to be not specific enough.

5 Related Work

Test selection has been investigated extensively in the medical and the statistical literature. Andreassen [1] presents a valid framework for test selection based on conditional probabilistic networks. However, it does not allow identifying all subgroups with optimal costs. Doubilet [2] offers a mathematical approach to test selection, however, it assumes that prior probabilities can be computed or estimated, which is problematic for small training sets. Furthermore it is not clear how background knowledge can be incorporated. It proposes only models for the entire population instead of individual models for smaller subgroups. Other subgroup discovery algorithms [3,4,7,8,9] mostly focus on finding subgroups that are interesting or unusual with respect to a single target variable (mostly class membership; for numerical variables see [7]). In our problem setting we need a more complex target variable that expresses the relation between the test outcome and the biopsy. Exceptional Model Mining [10] provides an approach that is able to discover subgroups with a more complex target concept: a model and its fitting to a subgroup. It performs a level-wise beam search and explores the best t subgroups of each level. In contrast, $SD4TS$ does not require the definition of a model and is guaranteed to find the globally optimal subgroup. While subgroup discovery usually aims for a descriptive exploration of the entire population, we discover for each patient only subgroups that are supported by the patients features. Therefore, we do not need a covering heuristic. With the introduction of prediction quality we have a measure that enables quality pruning of the search space (comparable to optimistic estimate pruning [9]), whereas existing algorithms quite often only offer pruning according to frequency [3]. While test selection and subgroup discovery are well-investigated areas of research, their combination has not yet been considered in the literature.

6 Conclusion

Many questions in medical research are related to the discovery of statistically interesting subgroups of patients. However, subgroup discovery with a single target variable is rarely sufficient in practice. Rather, more complex variants, e.g., handling costs, are required. In this study, we considered such variants of subgroup discovery in the context of the clinical task of test selection and diagnosis: For our breast cancer diagnosis scenario, the task is to detect subgroups for which single modalities should be given priority over others, as indicated by a cost function. We designed an algorithm that handles costs and finds the most cost-efficient subgroups of a population. By limiting the output size to the best t subgroups, it is possible to prune the search space considerably, especially for lower values of the minimum frequency (i.e., support) parameter. Consequently, the proposed algorithm clearly outperforms the baseline algorithm used for comparison (APRIORI-SD) in our experiments. The main problems and limitations in the context of our study were caused by the small sample size (138 examples, i.e., lesions) and the non-unique solution for optimal test selection. In other words, for many cases, two or more tests perform equally well in practice. In future work we will investigate how to solve our task by applying slightly modified methods from Section 5. However, we expect this to result in longer runtimes. Moreover, we are planning to compare the method with a recently poposed information-theoretic approach of test selection [11]. Overall, we showed that subgroup discovery can be adapted for test selection. We believe that similar techniques should be applicable successfully in other areas as well.

References

1. Andreassen, S.: Planning of therapy and tests in causal probabilistic networks. Artifical Intelligence in Medicine 4, 227–241 (1992)
2. Doubilet, P.: A mathematical approach to interpretation and selection of diagnostic tests. Medical Decision Making 3, 177–195 (1983)
3. Kavšek, B., Lavrač, N.: APRIORI-SD: Adapting association rule learning to subgroup discovery. Applied Artificial Intelligence 20(7), 543–583 (2006)
4. Atzmüller, M., Puppe, F.: SD-map – A fast algorithm for exhaustive subgroup discovery. In: Fürnkranz, J., Scheffer, T., Spiliopoulou, M. (eds.) PKDD 2006. LNCS (LNAI), vol. 4213, pp. 6–17. Springer, Heidelberg (2006)
5. BI-RADS Breast Imaging Reporting and Data System, Breast Imaging Atlas. 4th edn. American College of Radiology (2003)
6. Agrawal, R., Srikant, R.: Fast algorithms for mining association rules. In: Proceedings of the 20th VLDB Conference, pp. 487–499 (1994)
7. Klösgen, W.: Explora: a multipattern and multistrategy discovery assistant, 249–271 (1996)
8. Lavrač, N., Kavšek, B., Flach, P., Todorovski, L.: Subgroup discovery with CN2-SD. Journal of Machine Learning Research (2004)
9. Wrobel, S.: An algorithm for multi-relational discovery of subgroups. In: Komorowski, J., Żytkow, J.M. (eds.) PKDD 1997. LNCS, vol. 1263, pp. 78–87. Springer, Heidelberg (1997)
10. Leman, D., Feelders, A., Knobbe, A.J.: Exceptional model mining. In: Daelemans, W., Goethals, B., Morik, K. (eds.) ECML PKDD 2008, Part II. LNCS (LNAI), vol. 5212, pp. 1–16. Springer, Heidelberg (2008)
11. Mueller, M., Rosales, R., Steck, H., Krishnan, S., Rao, B., Kramer, S.: Data-efficient information-theoretic test selection. In: Proceedings of the 12th Conference on Artificial Intelligence in Medicine (AIME 2009), pp. 410–415 (2009)

Trajectory Voting and Classification Based on Spatiotemporal Similarity in Moving Object Databases

Costas Panagiotakis[1], Nikos Pelekis[2], and Ioannis Kopanakis[3]

[1] Dept. of Computer Science, University of Crete, P.O. Box 2208, Greece
cpanag@csd.uoc.gr
[2] Dept. of Informatics, University of Piraeus, Greece
npelekis@unipi.gr
[3] E-Business Intelligence Lab, Dept. of Marketing, Technological Educational Institute of Crete, Greece
kopanakis@e-bi.gr

Abstract. We propose a method for trajectory classification based on trajectory voting in Moving Object Databases (MOD). Trajectory voting is performed based on local trajectory similarity. This is a relatively new topic in the spatial and spatiotemporal database literature with a variety of applications like trajectory summarization, classification, searching and retrieval. In this work, we have used moving object databases in space, acquiring spatiotemporal 3-D trajectories, consisting of the 2-D geographic location and the 1-D time information. Each trajectory is modelled by sequential 3-D line segments. The global voting method is applied for each segment of the trajectory, forming a local trajectory descriptor. By the analysis of this descriptor the representative paths of the trajectory can be detected, that can be used to visualize a MOD. Our experimental results verify that the proposed method efficiently classifies trajectories and their sub-trajectories based on a robust voting method.

1 Introduction

Nowadays, there is a tremendous increase of moving objects databases due to location-acquisition technologies like GPS and GSM networks [1], and to computer vision based tracking techniques [2]. This explosion of information combines an increasing interest in the area of trajectory data mining and more generally the knowledge discovery from movement-aware data [3]. All these technological achievements require new services, software methods and tools for understanding, searching, retrieving and browsing spatiotemporal trajectories content.

A MOD consists of spatiotemporal trajectories of moving objects (e.g. humans, vehicles, animals, etc.). In general case, these trajectories encode the 2-D (two dimensional) or 3-D geographic location and the 1-D time information. Many of the existing approaches are interested in the trajectory shape analysis considering that the trajectory consists of sequential 2-D or 3-D spatial sampling positions ignoring the temporal dimension [4], [5]. In [6], a trajectory clustering

N. Adams et al. (Eds.): IDA 2009, LNCS 5772, pp. 131–142, 2009.
© Springer-Verlag Berlin Heidelberg 2009

algorithm is proposed that partitions a 2-D trajectory into a set of line segments, and then, groups similar line segments together into a cluster, while the notion of the *representative trajectory* of a cluster is defined. The algorithm is based on geometrical distances between line segments taking into account position and orientation. These methods can be applied on trajectory segmentation, classifications, searching and retrieval problems using shape based descriptors. Based on the idea of partial trajectories, Lee et al. [7] proposed an algorithm for trajectory classification showing that it is necessary and important to mine interesting knowledge on partial trajectories rather than on the whole. However, both of these algorithms cannot be applied with complex time-aware trajectories considering the whole route of the moving objects.

In addition, temporal dimension is ignored by almost all computer vision based methods, that are interested in human action and activity recognition. Many of them use 2-D trajectories from specific human points that are tracked in video sequences with constant frame rate [8], [9]. Another class of methods use temporally annotated sequences [10], [1], performing mining tasks. In these methods, as temporal dimension is used the transition time between sequentially points of the trajectory. Therefore, a trajectory of $n + 1$ points $S = (s_0, s_1, ..., s_n)$, is stored as $T = (S, \Delta t_1, \Delta t_2, ..., \Delta t_n)$, where Δt_i, denotes the transition time between the points s_{i-1} and s_i. The use of transition time takes into account that the sampling rate could be varied, providing information about speed. However, the format in temporal dimension changes and important information for some real world applications is missing. In real world, there are applications where the temporal dimension should be used unchanged. These applications concern traffic monitoring, security applications (e.g. identifying "illegal" trajectories under shape and space-time requirements), searching using space-time constraints, and so on.

In [11], distance-based criteria have been proposed for segmentation of object trajectories using spatiotemporal information. First, Minimum Bounding Rectangles (MBRs) is used in order to simplify the trajectories, taking advantage their tight integration with existing multidimensional indexes in commercial database management systems (such as R-trees). The use of R-trees reduces the computation cost of trajectory searching to $O(log(n))$, where n denotes the number of trajectories. The distance between two trajectories is defined used MBRs representation. Finally, the segmentation problem is given as a solution of a maximization problem, that attempts to create MBRs in such a way, that the original pairwise distances between all trajectories are minimized. In [12], a framework consisting of a set of distance operators based on primitive (space and time) as well as derived parameters of trajectories (speed and direction) has been introduced. They assume linear interpolation between sampled locations, so that a trajectory consists of a sequence of 3-D line segments, where each line segment represents the continuous development of the moving object during sampled locations. In [13], representative motion paths (frequently traveled trails of numerous moving objects) are detected in a distributed system under the assumption that the moving objects can communicate with a central unit (coordinator) and all processing must be performed in a single pass over the stream. The location measurements of each object is modeled with some uncertainty tolerance ϵ and a one-pass greedy algorithm, termed RayTrace, is running on each object independently. They have proposed a one-pass greedy algorithm, termed RayTrace, running on each object independently. The coordinator utilizes a lightweight index structure, termed MotionPath, which stores the representative motion paths. The goal of this work is in the same context with the aim of our research. However, they ignore segments' orientation taking into account only the points of the trajectories. Moreover, they propose to the use of a step function to formulate the closeness of two points. On the other hand the use of a continuous decision function derived robust and smooth results (in our approach).

Most of the above mentioned approaches propose different similarity metrics which they utilize either for introducing indexing structures for vast trajectory retrieval, or for clustering purposes, focusing either on space criteria, ignoring temporal variation, minimizing predefined metric criteria on feature domain, simplifying the given trajectories or applying simple clustering-based techniques. We argue that all of the above approaches, as well as those which are dealing with vast volumes of trajectory datasets would benefit if they would be applied in a representative subset (consisting of the representative trajectories) that best describes the whole dataset. Consider for example the domain of visual analytics on movement data [14] in which it is meaningless to visualize datasets over a certain small size, as the human eye cannot distinguish any movement pattern due to the immense size of the data. On the contrary, in this paper, we don't simplify the given trajectories, as we use the original data unchanged. Moreover, the temporal information is taken into account.

We are proposing a global voting method that is applied for each segment of trajectory without any simplification. Then, we analyze the voting descriptor in order to detect the representative paths of the trajectory, that followed by many objects at almost the same time and space. Moreover, we classify the trajectories and the trajectory segments. The results of classification have been used to visualize a MOD. The proposed methodology can be applied under different distance metrics (e.g. non Euclidean) and higher trajectory dimensions.

The rest of the paper is organized as follows: Section 2 gives the problem formulation describing the proposed modelling. Section 3 presents the proposed method for trajectory voting and classification. The experimental results are given in Section 4. Finally, conclusions and discussion are provided in Section 5.

2 Problem Formulation

In this section the problem formulation is given. Let us assume a MOD $D = \{T_1, T_2, \cdots, T_n\}$, of n trajectories, where T_k denotes the k-trajectory of the dataset, $k \in \{1, 2, ..., n\}$. We assume that the objects are moving in the xy plane. Let $p_k(i) = (x_k(i), y_k(i), t_k(i))$, be the i-point, $i \in \{1, 2, ..., L_k\}$ of k-trajectory, where L_k denotes the number of points of k-trajectory. $x_k(i), y_k(i)$ and $t_k(i)$) denote the 2-D location and the time coordinate of point $p_k(i)$, respectively.

Similar to the work of [12], [6], we consider linear interpolation between successive sampled points $p_k(i)$, $p_k(i + 1)$, so that each trajectory consists of a sequence of 3-D line segments $e_k(i) = p_k(i)p_k(i + 1)$, where each line segment represents the continuous moving of the object during sampled points. The goal of this work is to detect representative paths[1] and trajectories, that followed by many objects at almost the same time and space. A method to detect them is to apply a voting process for each segment $e_k(i)$ of the given trajectory T_k. This means that $e_k(i)$ will be voted by the trajectories of MOD, according to the distance of $e_k(i)$ to each trajectory. The sum of these votes is related to the number of trajectories that are close to $e_k(i)$. If this number is high, means that the segments is representative, followed by many objects at almost the same time and space. Thus, the voting results will be used to detect the representative paths and trajectories. First, we have to determine the distance $d(e_k(i), T_m)$ between $e_k(i)$ and a trajectory T_m of the dataset that consists of line segments. $d(e_k(i), T_m)$ is defined as the distance between $e_k(i)$ and the closest line segment of T_m to $e_k(i)$:

$$d(e_k(i), T_m) = min_j d(e_k(i), e_m(j)) \tag{1}$$

So, we have to compute distances between 3-D line segments $(d(e_k(i), e_m(j)))$. In this framework, the meaning of $(d(e_k(i), e_m(j)))$ is equal to the minimum energy of transportation of line segment $e_k(i)$ to $e_m(j)$, or line segment $e_m(j)$ to $e_k(i)$. Between these two choices, the transportation of minimum energy is selected. This idea has been introduced on Earth Movers Distance (EMD) framework [15] and it has been successfully applied on pattern recognition and computer vision applications [16]. In our case, this energy can be defined by the sum of two energies:

- translation energy $d_\perp(e_k(i), e_m(j))$ and
- rotation energy $d_\angle(e_k(i), e_m(j))$,

that depend on the Euclidean distance and on angle between the line segments, respectively. Therefore, taking into account the orientation of line segments, we have added an expression $d_\angle(e_k(i), e_m(j))$ to the distance formula related to the angle θ between the line segments,

$$d(e_k(i), e_m(j)) = d_\perp(e_k(i), e_m(j)) + d_\angle(e_k(i), e_m(j)) \tag{2}$$

$$d_\angle(e_k(i), e_m(j)) = min(|e_k(i)|, |e_m(j)|) \cdot sin(\theta) \tag{3}$$

where $d_\perp(e_k(i), e_m(j))$ denotes the Euclidean distance between the 3-D line segments and $|e_k(i)|$ the Euclidean norm (length) of 3-D line segment $e_k(i)$. In order to minimize the energy according to EMD definition, we select to rotate the line segment of minimum length, see Equation 3. Moreover, $d_\angle(e_k(i), e_m(j))$ has been expressed in "Euclidean distance" units, measuring the maximum distance that a point of line segment of minimum length will cover during rotation. If $d_\angle(e_k(i), e_m(j))$ was expressed in rads, we should introduce a weight to make $d_\perp(e_k(i), e_m(j))$ and $d_\angle(e_k(i), e_m(j))$ comparable (similar with Equation 4).

---

[1] In this framework, "path" is used for a trajectory part.

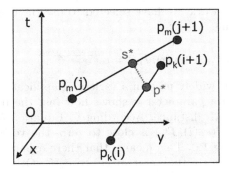

Fig. 1. The (closest) points p^* and s^* of 3-D line segments $p_k(i)p_k(i + 1)$ and $p_m(j)p_m(j + 1)$ define their distance

Fig. 1 illustrates the Euclidean distance (red dotted line) between the 3-D line segments $p_k(i)p_k(i + 1)$ and $p_m(j)p_m(j + 1)$.

The distance $d(e_k(i), e_m(j))$ cannot be expressed by a single formula, but it can be estimated in $O(1)$ [17]. This computation cost is not affected by the $d_\angle(e_k(i), e_m(j))$, since it a constant value for (each pair of points of the) two line segments. In order to estimate the Euclidean distance $d_\perp(p, s)$ between two points $p = (x, y, t)$ and $p' = (x', y', t')$, weights (w_1, w_2) can be used (Equation 4), making comparable location and time differences.

$$d_\perp(p, p') = \sqrt{w_1 \cdot (x - x')^2 + w_1 \cdot (y - y')^2 + w_2 \cdot (t - t')^2} \qquad (4)$$

The weights can be defined by the user. The ratio $\frac{w_2}{w_1}$ determines the spatial difference (e.g. how many meters) that "is equivalent" with one unit time difference (e.g. one second). This ratio can be estimated my the mean speed.

3   Global Voting and Classification

3.1   Voting Method

This section describes the proposed algorithm, the Global Voting Algorithm (GVA). The input of the algorithm is a MOD $D = \{T_1, T_2, \cdots, T_n\}$, a trajectory $T_k \in D$ and an intrinsic parameter σ of the method. The output of the method is the vector V_k of $L_k - 1$ components that can be considered as a trajectory descriptor along the T_k line segments. Each component of the vector $V_k(i)$ corresponds to the number of votes (representativeness) of $e_k(i), i \in \{1, 2, ..., L_k - 1\}$ of T_k.

According to the problem formulation, the algorithm for each line segment $e_k(i)$ of T_k and $T_m \in D, m \neq k$ computes the distance $d(e_k(i), T_m)$. This distance will be used to define the voting function $V(e_k(i), T_m)$. In literature, a lot of voting functions have been proposed, like the step functions or continuous functions [18]. In this work, we have selected to use the continuous function of gaussian kernel getting,

$$V(e_k(i), T_m) = e^{-\frac{d^2(e_k(i), T_m)}{2 \cdot \sigma^2}} \tag{5}$$

The gaussian kernel is widely used in a variety of applications of pattern recognition [19]. The control parameter σ shows how fast the function ("voting influence") decreases with distance. According to Equation 5, it holds that $0 \leq V(e_k(i), T_m) \leq 1$. If $d(e_k(i), T_m)$ is close to zero, the voting function gets its maximum value, giving 1.0. This means, that there exists a line segment of T_m that is being very close (in time and space) to $e_k(i)$. Otherwise, if $d(e_k(i), T_m)$ is high, e.g. greater than $5 \cdot \sigma$, the voting function gets almost 0, meaning that T_m is very far away (in time or space) from $e_k(i)$.

The use of a continuous voting function, like the gaussian kernel, gives smooth results for small changes on parameters (σ), and the possibility to get decimal values as results of voting process increasing the robustness of the method. Finally, $V_k(i)$ is estimated by getting the sum of votes for all of trajectories $T_m \in D, m \neq k$. Given the above discussion, a nice property that holds is that the proposed local trajectory descriptor V_k changes continuously over the trajectory segments. The pseudo-code of the above procedure is depicted at the end of the section (see Algorithm 1). The next subsection discusses the using of local trajectory descriptor V_k to classify trajectories and to detect representative paths.

---

input : The moving objects database $D = \{T_1, T_2, \cdots, T_n\}$ and a
         trajectory $T_k \in D$, parameter σ for voting.
output: The voting vector of T_k, V_k.

for $i = 1$ **to** $L_k - 1$ **do**
     $V_k(i) = 0$
     for $m = 1$ **to** n **do**
         if $m \neq k$ **then**
             $V_k(i) = V_k(i) + e^{-\frac{d^2(e_k(i), T_m)}{2 \cdot \sigma^2}}$
         end
     end
end

Algorithm 1. Global Voting Algorithm (GVA)

3.2   Trajectory Classification

In this section, we describe the analysis of local trajectory descriptor V_k in order to detect the representative paths of the trajectory and to classify the trajectories. Representative paths or representative trajectories are followed by many objects at almost the same time and space. In order to identify the representative trajectories, we will introduce E_k that is defined by the mean value of V_k over the line segments of T_k.

$$E_k = \frac{1}{L_k - 1} \sum_{i=1}^{L_k - 1} V_k(i) \tag{6}$$

This value is a measurement of trajectory representativeness. Therefore, a classification of the trajectories can be done using this value. Another trajectory feature is the maximum value of V_k, $M_k = max_i V_k(i)$. By the analysis of M_k, the representative line segments can be detected.

The trajectory classification results can be used for an efficient visualization and sampling of large datasets. The visualization of a large MOD suffers from the problem that the space-time density of the trajectories is extremely high (see Fig. 2(b)). A solution on this problem is given by an efficient sampling of the MOD, that can be provided by the classification results, using the detected representative trajectories (see Fig. 2(c)). In Section 4, we present experimental results concerning trajectory classification and visualization. The next subsection discusses the computational complexity issues of the proposed algorithm.

3.3   Computational Complexity Issues

Concerning the Global Voting Algorithm (GVA) complexity, the computational cost for each line segment $e_k(i)$, of T_k is $O(n)$. The computation cost of GVA (estimation of V_k) is $O(L_k \cdot n)$. If we perform GVA for each trajectory of the database, then the total computation cost is $O(\bar{L} \cdot n^2)$, where $\bar{L}$ denotes the mean number trajectory points (samples). Therefore the polynomial cost of the algorithm makes the algorithm efficient for large databases (i.e. more than 1000 trajectories needed few seconds).

However, it is possible to reduce this computation cost, in order to be able to execute the algorithm in even larger databases. MBRs can be used as initialization step, and the indexing of the line segments to MBRs should be stored. We have proposed the using MBRs because of their tight integration with multidimensional indexes (R-trees). Then, the cost of searching step of voting algorithm will be reduced in an MBR (or some MBRs). Thus, the use of R-trees will reduce the cost of GVA execution to $O(log(\bar{L} \cdot n))$, and the total cost to $O(n \cdot log(\bar{L} \cdot n))$.

4   Experimental Results

The method has been implemented using Matlab without any code optimization or using of R-trees structures. For our experiments, we used a Core 2 duo CPU at 1.5 GHz. A typical processing time of GVA execution, when $n = 1000$ and $\bar{L} = 100$, is about 3 seconds.

We have tested the proposed algorithm on the 'Athens trucks' MOD containing 1100 trajectories. The dataset is available online on [20]. In most of the figures we have depicted a subset (10% or 20%) of the trajectories of our dataset, due to visualization issues. Fig. 2 illustrates the trajectories of our dataset projected in 2-D spatial space ignoring time dimension (Fig. 2(a)) and in spatiotemporal 3-D

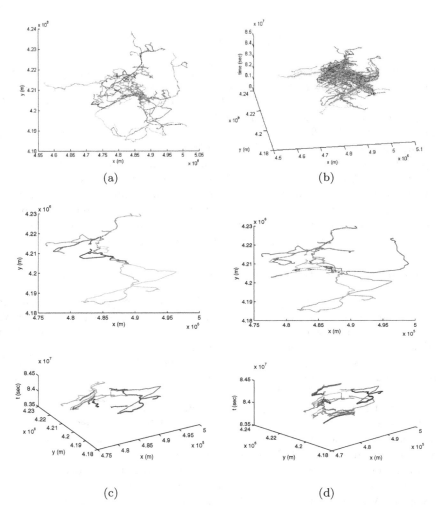

Fig. 2. The trajectories of our dataset (1100 traj.) projected in **(a)** 2-D spatial space ignoring time dimension and **(b)** spatiotemporal 3-D space. **(c)**, **(d)** The 20 and 50 most representative trajectories of the dataset projected in 2-D spatial space (up) and in 3-D spatiotemporal space (down), respectively.

space (Fig. 2(b)). The provided information of Figs. 2(a) and 2(b) can not be visualized efficiently, due to the large number of projected trajectories in almost the same time and space. In order to solve this problem, we have used the results of classification to sample the dataset. Figs. 2(c) and 2(d) illustrate an efficient sampling/visualazation of the dataset using the 20 and 50 most representative trajectories according to E_k criterion, respectively. According to the proposed method, the estimated representative trajectories have the property to be close to many other trajectories of the dataset and can be used efficiently to visualize

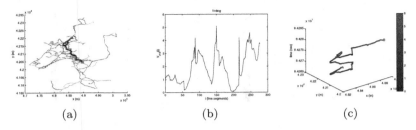

(a)   (b)   (c)

Fig. 3. Results of GVA for 219th trajectory of the dataset. **(a)** The 219th trajectory (bold black color) and some trajectories of our dataset projected in 2-D spatial space. **(b)** The voting descriptor V_{219}. **(c)** The 219th trajectory in 3-D space. The used colors correspond to the values of V_{219}, (red color for high values, blue color for low values).

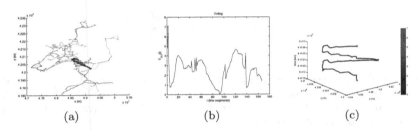

(a)   (b)   (c)

Fig. 4. Results of GVA for 253th trajectory of the dataset. **(a)** The 253th trajectory (bold black color) and some trajectories of our dataset projected in 2-D spatial space. **(b)** The voting descriptor V_{253}. **(c)** The 253th trajectory in 3-D space. The used colors correspond to the values of V_{253}, (red color for high values, blue color for low values).

them. In our framework we have used the weights $w_1 = 1/1000$, $w_2 = 1/30$ (see Equation 4) and $\sigma = 2.5$ (see Equation 5).

Figs. 3 and 4 show the results of GVA for the trajectories 219 and 253 of the dataset, respectively. Figs. 3(a) and 4(a) show with bold black color the trajectories 219 and 253 and some of the trajectories of the dataset projected in 2-D spatial space. The estimated voting descriptors V_{219} and V_{253} are illustrated in Figs. 3(b) and 4(b). As it was mentioned before, the estimated voting descriptors change continuously over the trajectory segments. Figs. 3(c) and 4(c) illustrate the trajectories 219 and 253 in 3-D using a blue-to-red color map according to the corresponding to segments voting values (red color for high values, blue color for low values). By the analysis of these figures, it can be observed that the most representative paths of the trajectory 219 are found at the middle and at the end of the trajectory, while the most representative path of the trajectory 253 is found at the start of the trajectory. Moreover, the maximum values of V_{219} and V_{253} descriptors shows how many trajectories are close to the most representative paths of 219 and 253 trajectories, respectively.

Fig. 5(a) illustrates the classification results for 220 trajectories of our dataset projected in 2-D spatial space using E_k descriptor. The used colors correspond

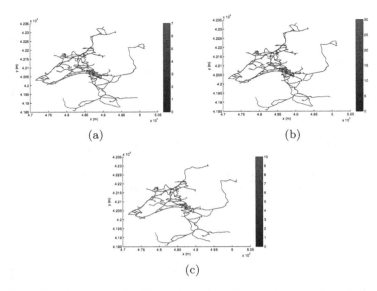

(a)   (b)

(c)

Fig. 5. Classification results for 220 trajectories of our dataset projected in 2-D spatial space using **(a)** E_k descriptor **(b)** using M_k descriptor, respectively. The used colors correspond to the class of the trajectory (red color for representative trajectories). **(c)** Classification results for trajectories line segments.

to the class of the trajectory (red color for representative trajectories). The most representative trajectory of the dataset is illustrated with red bold line. Similar results are obtained using M_k descriptor (see Fig. 5(b)). The most representative trajectories of the dataset are detected close to the center of the dataset, where most of the trajectories are crossed. Fig. 5(c) illustrates the classification results for trajectories line segments of 110 trajectories of the dataset projected in 2-D spatial space. The line segments with $V_k(i)$ greater than 10 are illustrated with red colors. The voting descriptor of the most representative line segment of the dataset has the value of 67.4. These figures are very useful for traffic monitoring, since they efficiently the trajectories and the segments, where the traffic is high.

5   Conclusions

In this paper, we have discussed the trajectory voting and classification problems in real spatiotemporal MOD. We have proposed an algorithm for trajectory voting and classification based on local trajectory similarity. Finally, a local trajectory descriptor per trajectory segment is estimated, that changes continuously over the trajectory segments. By the analysis of this descriptor the representative paths of the trajectory can be detected, that followed by many objects at almost the same time and at the same place. These results have been used to visualize a MOD. We have tested the proposed method under real databases and the experimental results shows that the method provides an efficient local (per segment) and global (per trajectory) classification of the dataset.

As future work, we plan to apply the voting results for trajectory segmentation, sampling, searching and retrieval. Segmentation and clustering algorithms can be applied on trajectory descriptor V_k providing a trajectory segmentation and a clustering of the dataset. Moreover, we plan to associate the voting results with an error function in order to measure the performance of the proposed sampling and to make comparisons with other works.

References

1. Giannotti, F., Nanni, M., Pinelli, F., Pedreschi, D.: Trajectory pattern mining. In: KDD 2007: Proceedings of the 13th ACM SIGKDD international conference on Knowledge discovery and data mining, pp. 330–339 (2007)
2. Wang, L., Hu, W., Tan, T.: Recent developments in human motion analysis. Pattern Recognition 36(3), 585–601 (2003)
3. Giannotti, F., Pedreschi, D.: Geography, mobility, and privacy: a knowledge discovery vision. Springer, Heidelberg (2007)
4. Vlachos, M., Gunopulos, D., Das, G.: Rotation invariant distance measures for trajectories. In: KDD 2004: Proceedings of the tenth ACM SIGKDD international conference on Knowledge discovery and data mining, pp. 707–712 (2004)
5. Chen, L., Özsu, M.T., Oria, V.: Robust and fast similarity search for moving object trajectories. In: SIGMOD 2005: Proc. of the 2005 ACM SIGMOD int. conf. on Management of data, pp. 491–502 (2005)
6. Lee, J.G., Han, J., Whang, K.Y.: Trajectory clustering: a partition-and-group framework. In: SIGMOD 2007: Proceedings of the 2007 ACM SIGMOD international conference on Management of data, pp. 593–604 (2007)
7. Lee, J.G., Han, J., Li, X., Gonzalez, H.: Traclass: trajectory classification using hierarchical region-based and trajectory-based clustering. In: PVLDB
8. Panagiotakis, C., Ramasso, E., Tziritas, G., Rombaut, M., Pellerin, D.: Shape-based individual/group detection for sport videos categorization. IJPRAI 22(6), 1187–1213 (2008)
9. Panagiotakis, C., Ramasso, E., Tziritas, G., Rombaut, M., Pellerin, D.: Shape-motion based athlete tracking for multilevel action recognition. In: Perales, F.J., Fisher, R.B. (eds.) AMDO 2006. LNCS, vol. 4069, pp. 385–394. Springer, Heidelberg (2006)
10. Giannotti, F., Nanni, M., Pedreschi, D.: Efficient mining of sequences with temporal annotations. In: Proc. SIAM Conference on Data Mining, pp. 346–357 (2006)
11. Anagnostopoulos, A., Vlachos, M., Hadjieleftheriou, M., Keogh, E., Yu, P.S.: Global distance-based segmentation of trajectories. In: KDD 2006: Proc. of the 12th ACM SIGKDD int. conf. on Knowledge discovery and data mining, pp. 34–43 (2006)
12. Pelekis, N., Kopanakis, I., Marketos, G., Ntoutsi, I., Andrienko, G., Theodoridis, Y.: Similarity search in trajectory databases. In: TIME 2007: Proc. of the 14th Int. Symposium on Temporal Representation and Reasoning, pp. 129–140 (2007)
13. Sacharidis, D., Patroumpas, K., Terrovitis, M., Kantere, V., Potamias, M., Mouratidis, K., Sellis, T.: On-line discovery of hot motion paths. In: EDBT 2008: Proc. of the 11th int. conf. on Extending database technology, pp. 392–403 (2008)

14. Andrienko, G., Andrienko, N., Wrobel, S.: Visual analytics tools for analysis of movement data. SIGKDD Explor. Newsl. 9(2), 38–46 (2007)
15. Rubner, Y., Tomasi, C., Guibas, L.J.: A metric for distributions with applications to image databases. In: ICCV 1998: Proceedings of the Sixth International Conference on Computer Vision (1998)
16. Shishibori, M., Tsuge, S., Le, Z., Sasaki, M., Uemura, Y., Kita, K.: A fast retrieval algorithm for the earth mover's distance using emd lower bounds. In: IRI, pp. 445–450 (2008)
17. Lumelsky, V.J.: On fast computation of distance between line segments 21, 55–61 (1985)
18. Patterson, D.: Artificial Neural Networks. Prentice-Hall, Englewood Cliffs (1996)
19. Yuan, J., Bo, L., Wang, K., Yu, T.: Adaptive spherical gaussian kernel in sparse bayesian learning framework for nonlinear regression. Expert Syst. Appl. 36(2), 3982–3989 (2009)
20. http://infolab.cs.unipi.gr/pubs/tkde2009/

Leveraging Call Center Logs for Customer Behavior Prediction

Anju G. Parvathy, Bintu G. Vasudevan, Abhishek Kumar,
and Rajesh Balakrishnan

Software Engineering Technology Labs
Infosys Technologies Limited
Bangalore 560100
{anjug_parvathy,bintu_vasudevan,abhishek_kumar25,rajeshb}@infosys.com

Abstract. Most major businesses use business process outsourcing for performing a process or a part of a process including financial services like mortgage processing, loan origination, finance and accounting and transaction processing. Call centers are used for the purpose of receiving and transmitting a large volume of requests through outbound and inbound calls to customers on behalf of a business. In this paper we deal specifically with the call centers notes from banks. Banks as financial institutions provide loans to non-financial businesses and individuals. Their call centers act as the nuclei of their client service operations and log the transactions between the customer and the bank. This crucial conversation or information can be exploited for predicting a customer's behavior which will in turn help these businesses to decide on the next action to be taken. Thus the banks save considerable time and effort in tracking delinquent customers to ensure minimum subsequent defaulters. Majority of the time the call center notes are very concise and brief and often the notes are misspelled and use many domain specific acronyms. In this paper we introduce a novel domain specific spelling correction algorithm which corrects the misspelled words in the call center logs to meaningful ones. We also discuss a procedure that builds the behavioral history sequences for the customers by categorizing the logs into one of the predefined behavioral states. We then describe a pattern based predictive algorithm that uses temporal behavioral patterns mined from these sequences to predict the customer's next behavioral state.

1 Introduction

Business Process Outsourcing (BPO) is a form of outsourcing which involves the contracting of the operations and responsibilities of a specific business function to a third-party service provider. Some businesses service internal functions through their own call centers to optimize their business activity. Call centers are used for the purpose of receiving and transmitting a large volume of requests through outbound and inbound calls to customers on behalf of a business or for a client to interact with their customers.

N. Adams et al. (Eds.): IDA 2009, LNCS 5772, pp. 143–154, 2009.
© Springer-Verlag Berlin Heidelberg 2009

A bank's call center acts as the nuclei of its client service operations and maintains logs of all customer transactions. This is later used by the bank to follow up the customers who have availed loans. This inturn enhances efficiency and helps the bank maintain better customer relations. In this paper we deal specifically with the call center notes from banks. We propose a technique to build a customer's behavioral history from these notes in order to predict the next behavioral state he is likely to exhibit. This proves beneficial to the bank as delinquent customers can easily be identified and appropriate action can be initiated.

The call center logs for a particular customer are first aggregated on a monthly basis. Then they are spell corrected using a domain specific spelling correction algorithm. An n-gram keywords based categorizer then categorizes the comments into one of the predefined behavioral states identified by a domain expert. The chain of behavioral categories across months constitute the Behavioral History Sequence (BHS) of a customer. Next we build the global and local behavioral pattern histories by mining behavioral patterns from the BHSs of different customers and the particular customer, respectively. We have developed a pattern based predictive algorithm which predicts a customer's behavioral state for the next month from his most recent behavioral pattern window, using the patterns mined.

The paper is organized as follows. In Sec.2 we examine related works. In Sec.3 we explain the methodology we have adopted. In Sec.4, we present our spelling correction algorithm. Sec.5 elaborates the procedure we use to build a customer's BHS and the categorization algorithm. Sec.6 describes the pattern based predictive algorithm from BHS wherein the deductive and non deductive modes are discussed in detail. In Sec.7 we discuss the results and in Sec.8 we bring in our conclusions.

2   Related Works

In the past various statistical techniques have been proposed for spelling correction. The edit distance method [1] devised by Levenshtein is a widely used metric for measuring syntactic string similarity. A variant of this is Damerau - Levenshtein distance [2] which measures the similarity between two strings by counting the minimum number of operations needed to transform one string into the other, where an operation is defined as an insertion, deletion, or substitution of a single character, or a transposition of two characters. The word in the dictionary that has the minimum edit distance with respect to the misspelled word is then chosen as the correction. Phonetic techniques like soundex [3] and double metaphone [4] generate codes for a given word and the word(s) with which the erroneous word shares the code, can be chosen as an appropriate correction. Some innovative methods for automatic spelling correction using trigrams [5] and for spelling correction in scientific and scholarly domain [6] have also been explored. A systematic approach to constructing a spelling corrector is detailed in [7]. Most of the approaches listed above are suitable for generic spelling correction involving human validation. Our domain specific spelling correction algorithm is designed to make spelling corrections automatically prior to categorization.

Techniques for customer behavior prediction using the transaction history sequence built from events have been presented by a number of researchers. One such method with application to detecting system failures in computer networks, has been proposed for predicting rare events in event sequences in [8]. In [9], a genetic algorithm based approach is used for predicting rare events. This is another rule-based method that defines predictive patterns, as sequences of events with temporal constraints between connected events. The rules for classification are constructed from the diverse set of predictive patterns identified. Another method, reported in [10], learns multiple sequence generation rules, like disjunctive normal form model, periodic rule model, etc., to infer properties (in the form of rules) about future events in the sequence. Frequent episode discovery [11][12] is a popular framework for mining temporal patterns in event streams. Sequence prediction based on categorical event streams and an algorithm for predicting target events based on estimating a generative model for event sequences in the form of a mixture of specialized hidden markov models has been presented in [13]. The algorithm for predicting a customer's behavioral state that we present is predominantly rule-based.

3   Methodology

Figure 1 shows the process flow for customer behavior prediction. The call logs that are sequential in nature by virtue of their timestamps are aggregated on a monthly basis for every customer.Then they are spell corrected using a domain specific algorithm that corrects the misspelled words in the logs to the most matching word in the domain dictionary (which consists of keywords that drive the categorizer). A customer's monthly interaction call comment is subsequently categorized into one of the eight predefined categories (Paid, Cooperative Compliance, Not Paid, Defaulter, Negotiation Failed, Fraud, Other and Not Available) using an n-gram keyword based categorizer. We then construct the BHS for a customer by episodically arranging the behavioral categories identified for the different months. Then we build a global behavioral pattern history by mining behavioral patterns from the BHSs of different customers. Each such

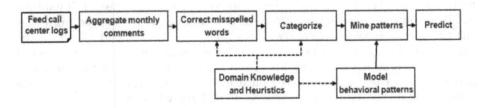

Fig. 1. Methodology

pattern is further described by its length, frequency, temporal information and the list of customers whose BHS contained them.

The local behavioral pattern history for a given customer is build by mining only his BHS for behavioral patterns. Our pattern based predictive algorithm then predicts a customer's behavioral state for the forthcoming month using his most recent behavioral pattern window.

4 Spelling Correction

In this section we introduce our domain specific spelling correction algorithm used to correct the call center logs. We also present a comparative analysis of our approach against other approaches like soundex and double metaphone (both of which are phonetic spelling correction algorithms) and the algorithm based on levenshtein distance (LD). The Domain Specific Acronym list (DSA), Domain Dictionary (DD), the English Dictionary (ED) and threshold parameters can be configured. The DD is built by a domain expert and consists of all keywords used for categorization. Correcting the misspelled words in the logs is a critical prerequisite for categorization as the categorizer is essentially keyword based. For example, if '*payments received*' is a phrase in the category '*Cooperative*' and the comment contained the phrase '*pymts rcvd*'; it is obvious that the customer was cooperative during the audit. However without spelling correction, the comment would not be correctly classified.

4.1 Spelling Correction - Our Approach

Every word or token is searched for in the ED using a tree based algorithm. If the token is found in the dictionary, it is retained in the comment; else the algorithm flags it as a misspelled word. Next, the word is checked against the seed DSA that contains acronyms like '*PFP*' ('*Promise for payment*'). If no correction exists, the algorithm proceeds to the next step which is construction of positional patterns with priority. For example, the positional patterns for the misspelled word '*cust*' in decreasing order of priority are '*cust*', '*cust.**', '*cus.*t*' '*cu.*s.*t*' and so on. Every word in the relevant subtree(the one representing all words that begin with the same alphabet as that of the token)in DD is matched against the patterns in order of priority. If a word that conforms to a pattern at a particular priority level is found, the remaining patterns are ignored. If multiple words in the subtree match a pattern at the same priority level; the word that has the least levenshtein distance (lesser than an upper bound 'β') from the token is chosen as the correction. If a valid correction is found at the end of this step, a series of checks are done to affirm its correctness. If the token is a prefix of the correction, we believe that it is a valid correction. Else, we perform a check based on Consonant Density Ratio (CDR), which is defined as:

$$CDR(word_1, word_2) = \frac{Number\ of\ consonants\ in\ word_1}{Number\ of\ consonants\ in\ word_2}$$

If $CDR(token, corrn) >$ threshold 'θ', we validate the correction and perform no further checks. Else, CDR($token_{stem}, corrn_{stem}$) is computed. If this is less than θ, the correction is invalidated. However, we check if there exists a word in the subtree that is relevant anagram of the token and if it exists, we return the anagram for a correction, else we check if the consonant character sets are the same for the token and the correction. If they are different, the correction is invalidated following which a check for relevant anagram is done. The token is retained as such or replaced by the relevant anagram as the case maybe and the higher process flow continued. If no correction could be found at the end of the above mentioned steps, the word in the subtree that has the same character set as that of the token and for which the levenshtein distance is minimum, is chosen as a correction. If the algorithm could successfully find a correction that passed the different validation checks, all occurrences of the token in the comment are replaced by the correction. The token is added as an acronym, with the correction as its expansion to the DSA. The process flow is detailed in algorithm 1.

Algorithm 1. Spelling Correction Algorithm

Input: *token*, DSA, DD, ED

Output: *correction* (*corrn*)

1: Construct positional patterns with priority for the *token*
2: **if** *token* is in ED **then**   do nothing
3: **else if** *token* is an acronym in DSA **then**
4:     *corrn* ← expansion of token in DSA
5: **else**
6: Assign *corrn* as a word from the DD matching the positional pattern at the highest priority with min LD $< \beta$
7:     **if** *corrn* exists **then**
8:         **if** *token* is a prefix of *corrn* **then**
9:             do nothing
10:         **else if** CDR($token, corrn$) $> \theta$ **then**
11:             do nothing
12:         **else if** CDR($token_{stem}, corrn_{stem}$)$\leq \theta$ **then**
13:             Invalidate current *corrn*
14:             Assign *corrn* to a Relevant anagram of the *token* (if it exists)
15:         **else if** Char_Set(*token*) = Char_Set(*corrn*) **then**
16:             do nothing
17:         **else**
18:             Assign *corrn* to a Relevant anagram of the *token* (if it exists)
19:         **end if**
20:     **else**
21: Assign *corrn* to a word in the concerned subtree in DD which shares the char_Set with the token and at min LD
22:     **end if**
23:     Update DSA with *token* to *corrn* entry
24: **end if**

4.2 Comparative Analysis

Here we compare our algorithm with other popular techniques for spelling correction like soundex and double metaphone (phonetic techniques) and the conventional levenshtein distance based approach. Table 1 shows the corrections provided by the algorithms for a few commonly encountered misspelled words.

A set of 534 misspelt words compiled from the log comments were corrected with respect to a DD that contained 141 words. The precision and recall for all the algorithms were calculated without considering the corrections made using DSA list. The upper section in Table 1 shows a set of misspelled words corrected using different algorithms and the lower section contains their respective precisions and recalls.

Table 1. Spelling correction comparison along with precision and recall

Misspelled word	Soundex	Double Metaphone	Levenshtein Distance	Customized Spell Checker
repymt	-	-	result	repayment
bowr	-	-	owner	borrower
appvl	-	-	april	approval
progess	process	-	process	progress
cmpltd	completed	completed	called	completed
Precision(%)	33.81	32.66	46.82	98.75
Recall(%)	38.44	35.84	99.71	99.42

5 Behavioral History Sequence (BHS)

In this section, we describe how we build a customer's BHS. The spell corrected call logs that are sequential in nature are aggregated on a monthly basis for every customer and categorized into one of the predefined categories. The BHS is then constructed by episodically arranging the behavioral categories identified for the different months.

5.1 Categorization

A set of predefined categories along with their characteristic keywords (unigrams, bigrams or trigrams) are identified by a domain expert. The expert also assigns global weights (between 0 and 1) for the words or phrases based on their relative importance(usually the highest for trigrams, the next for bigrams and the lowest for unigrams). However these weights can be overridden by specific weights at the discretion of the domain expert, say in some case, if a unigram is highly characteristic of a given category, then it can be assigned a weight greater than the default weight for unigrams. The probability for a log comment to belong to a category is calculated using a naive Bayes classification algorithm. The comment is classified into that category to which it has the maximum probability to belong to. The classifier also supports multi category classification. It assigns separate probabilities to the comment to belong to each of the different categories based on keyword hits, their frequencies and weights. For our domain, the eight different behavioral categories as identified by an expert are Paid, Cooperative Compliance, Not Paid, Defaulter, Negotiation Failed, Fraud, Other and Not Available.

5.2 Constructing the BHS

All comments that are logged for a customer in the same month are grouped together and categorized thus building BHS as a sequence of behavioral categories/states. A unique identification number is assigned to every delinquent customer and the comments for this customer are logged against this number. Let $U = \{u_1, u_2, ...u_n\}$ be the set of such unique IDs for n delinquent customers. A BHS is a time-ordered sequence of behavioral states and for a given customer's logs it is generated as $seq(u) = \{Ct_1, Ct_2, ..., Ct_m\}$; where $C = \{c_1, c_2, ...c_k\}$ is the set of predefined categories and t_1 to t_m ($t_1 \leq t \leq t_m$), the different months. The sequence length is uniform across all customers and it is based on the minimum and maximum dates in the timestamps of comments pulled from the database for analysis. If no comment has been logged for some month for a customer, the corresponding behavioral category is assigned '$Not\,Available$'. The temporally ordered BHSs thus built are then used for customer behavior prediction.

6   Customer Behavior Prediction

This section deals with building a model for customer behavior prediction. We describe a pattern based predictive algorithm that uses behavioral patterns mined to predict the next customer behavioral state. The algorithm first preprocesses the BHSs based on certain domain specific heuristics, then discovers the frequent and rare patterns from the BHSs in order to define the prediction space and then makes a prediction either deductively or non deductively using the most recent behavioral pattern window.

6.1   Domain Heuristics

The BHS for the customers are preprocessed based on a set of domain heuristics. If for a given customer there are no comments logged for a particular audit month, we interpret the corresponding missing behavioral state as '$Paid$', '$Not\,Paid$' or '$Not\,Available$'. The missing states at the two extreme ends of the BHS are understood as '$Not\,Available$'. If the comments are missing for exactly one month in between, this is also treated as a '$Not\,Available$' state as no conclusion can be drawn according to heuristics. According to domain experts, a customer is said to have turned delinquent if comments have been logged for more than two consecutive months. Hence, if in the BHS there is a contiguous sequence of missing states surrounded by other categories, the two most recent states in this sequence are considered as '*Not Paid*' and the rest of the states as '*Paid*'. The assumption is that since the comments were not logged, the customer might have paid.

6.2   Behavioral Pattern Selection

The behavioral patterns are learned from the BHS. A behavioral pattern (BP) is a sequence of behavioral states which inturn is a sub-sequence of the BHS. Each BP is characterized by ranks of its temporal positions (recency), frequency of occurrence and the list of unique customers whose $BHSs$ contained it. Thus two BPs which are constitutionally the same but have different temporal positions are considered different. The global BP history is built by mining patterns from the $BHSs$ of all available customers, keeping track of its frequency across the sequences. The local BP history for a given customer is build by mining only his BHS for BPs. The Behavioral Pattern Window (BPW), is the most recent BP in a BHS. Typically a prediction can be made if a BPW of length 'w' (the search window) is the trailing pattern of a BP of length '$w + 1$'where 'w' can vary from 2 to user defined value value 'k'. In our experiment BP lengths range from 2 to 5. Many a times the $BHSs$ are sparse (contain few non NA categories), and hence a BPW of length 'w' is built from the BHS by gathering together the 'w' most recent non NA states in the BHS in order. Figure 2 shows an example of how a BPW of length 3 is matched against a BP of length 4, to enable prediction.

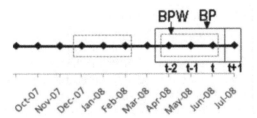

Fig. 2. An illustration for BP and BPW

6.3   Deductive Approach

In the deductive approach we use the most recent $BPW(s)$ of the BHS of a given customer and make the prediction by deductively assessing the length, recency and frequency of the BPs. The different factors jointly attribute a weight to the prediction which is indicative of its believability. We denote the predicted state of a customer's behavior by PS. The $BPWs$ of length 'w', where $w = 1, 2, ...k$ (k - upper bound for window length) are mined from the BHS. They are used in decreasing order of lengths, for prediction. For a given BPW of length 'w', we refer to the local BP history first and then the global BP history.

Within the active history we first identify the BPs of length '$w + 1$' and order them according to their recency of occurrence. For every BP in this ordered set, it is checked if BPW is its trailing pattern. If yes, $BP[w + 1]$ becomes a *candidate PS* and BP its parent pattern. In order to choose the PS from the set of *candidate PSs*, we compute the similarity between the customer's BHS and each of the $BHSs$ which contains its parent BP. The *candidate PS* linked to the highest similarity index is chosen as the PS. We believe this helps us bias our prediction towards the behavior of a customer with whom he/she has the most behavioral resemblance. The weight of prediction is computed as the ratio of frequency of parent BP of the chosen PS to the sum of frequencies of all parent BPs of the *candidate PSs*. The detailed algorithm is given below.

Algorithm 2. The deductive approach for customer behavior prediction

---

Input: BHS, Local and Global BP Histories, BPW
Output: PS for $t + 1^{th}$ month
1: Order Local and Global BP Histories according to length and recency of occurrence
2: Mine $BPWs$ of lengths $w = k, k - 1, ...1$ 'k' - upper bound for window length
3: $PS \leftarrow NA$
4: weight of prediction $\leftarrow 0$
5: *candidate PSs* $\leftarrow \{\}$
6: Active BP History $\leftarrow$ Ordered Local BP History
7: **for all** $BPWs$ **do**
8:      **for all** BPs in Active BP History **do**
9:          **if** BPW of length 'w' is a trailing pattern of BP of length '$w + 1$' **then**
10:             Add $BP[w + 1]$ to *candidate PSs* with BP as its parent pattern
11:         **end if**
12:     **end for**
13: **end for**
14: If *candidate PSs* is empty, repeat step 7 through 10 with Active BP History as the global one
15: Select PS as the *candidate PS* with whose BHS, the current BHS is most similar to
16: weight of prediction$\leftarrow \dfrac{Frequency\,of\,parent\,pattern\,of\,PS}{Sum\,of\,frequencies\,of\,parent\,patterns\,of\,candidate\,PSs}$

6.4   Non Deductive Approach

In the non deductive approach, we do not restrict ourselves to one scheme of action for prediction. We predict the customer's next behavioral state from $BPWs$ of lengths $w = 1, 2..., k$ using both local and global histories considering frequency of patterns in the history and their recency. Hence for a given customer, we have multiple predictions (along with weights) for the next state. The highest weighed prediction is chosen as the customer's next behavioral state.

7   Discussions

In this section, we discuss the results of our experiments on the call center log feed obtained from the bank. The dataset consisted of comments logged from

Jul 05 t0 *Jun* 08 for 750 different delinquent customers. Hence the *BHS* of every customer was a sequence of length 36 of which each state was one of the eight predefined states identified by the domain expert. The behavioral patterns mined from these *BHSs* were used for predicting the customer's behavioral state in the next month, using the deductive and non deductive approaches to prediction.

Fig 3. shows the partial *BHS* of a particular customer illustrated as a date vs category graph. The x axis timelines the 36 months and the y axis represents the different categorical states[1]. For validation the known state of the customer's behavior for the tth month was predicted using the states during the timeperiod $(t-1, t-2, ..., t-k)$, where k is the window size. This was validated against the known t^{th} state. Next, prediction for customer's behavioral state for the $(t+1)^{th}$ month $(Jul\,08)$ was made using the most recent behavioral pattern window $(t, t-1, ..., t-k)$.

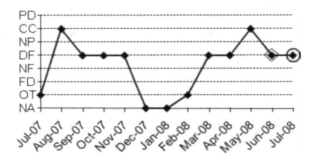

Fig. 3. Prediction for t^{th} state shown in diamond and the actual prediction $(t+1)^{th}$ state months shown in circle

According to this particular customer's *BHS*, he started showing a delinquent behavior from *Jun* 07. However he became cooperative in the month of *Aug* 07 only to become a defaulter from *Sep* 07 *to Nov* 07. The log comments for the subsequent months of *Dec* 07 *to Feb* 08 did not provide any conclusive pointer to the customer's behavior. He/she then stayed a defaulter till *Apr* 08 before becoming cooperative again in the month of *May* 08. This was only to turn delinquent in the next month(*Jun* 08). In order to test the accuracy of our prediction we predict the customer behavior for *Jun* 08 by using the *BPs* of the previous months. Our prediction for this case (using the deductive approach)is the same as the actual behavior the customer exhibited in *Jun* 08. This is because even in the past the customer has become a defaulter after being cooperative. We now predict that he will be a defaulter in the month of *Jul* 08 too. Our prediction is supported by the fact that from *Aug* 07 to *Oct* 07 the *BHS* of the customer's behavior showed a similar trend.

---

[1] PD=Paid,   CC=Cooperative-Compliance,   NP=Not-Paid,   DF=Defaulter
NF=Negotiation-Failed, FD=Fraud, OT=Other and NA=Not-Available

The precision and recall measures for gauging our prediction were calculated post the validation phase. The recall measure is equal to the fraction of customers for whom our mechanism yielded a prediction. The precision of the approaches was calculated by comparing the predictions for the most recent months against the actual customer behavioral states in the respective months. Predictions were made for the $(t)^{th}$, $(t-1)^{th}$, and $(t-2)^{th}$ months using the "*corresponding*" most recent $BPWs$ $((t)^{th}$ predicted using the states at $(t-1, t-2, ..., t-k)$, $(t-1)^{th}$ using the states at $(t-2, t-3, ..., t-k)$ and $(t-2)^{th}$ state using those at $(t-3, t-4, ..., t-k))$ respectively. Table 2 contains the precisions and recalls for both the deductive and non deductive approaches when tested on the validation set.

Table 2. Precision and recall, deductive and non deductive approaches

Prediction for months	Deductive approach		Non deductive approach	
	Precision(%)	Recall(%)	Precision(%)	Recall(%)
$(t)^{th}$	53.33	72.63	62.66	72.63
$(t-1)^{th}$	50.95	79.57	59.38	79.57
$(t-2)^{th}$	52.10	85.61	64.28	85.61

The deductive approach works based on predefined heuristics wherein the different factors influencing prediction are prioritized and ordered. In the non-deductive approach, different combinations (ordering) of these factors give different predictions, of which the best is chosen. Though this approach is time consuming, the precision values are marginally better than those for the deductive approach. During a separate training phase, each of the schemes can be judged based on the precision and recall for the past predictions to zero in on the best plan.

8   Conclusion

In this paper, we have proposed a new technique for predicting a customer's behavior based on a bank's call center logs. A domain specific spelling correction algorithm, an n-gram keyword based categorization algorithm and a behavior pattern based predictive algorithm have been used in a phased manner. Our algorithm for spelling correction can be employed to process unstructured documents in the telecom, healthcare and other domains which have well defined domain dictionaries. The precision for this algorithm shows that it performs well in a given domain. The categorization algorithm also works well with the set of keywords identified by a domain expert and can be fine tuned through supervised training. Though the accuracy of the predictive algorithm can be bettered it seems to be very promising for predicting a customer's behavioral state in the current scenario. Careful evaluation by domain experts can be used to refine the algorithm. Important areas of future work include the evaluation of the proposed temporal pattern discovery framework for efficient search using machine learning techniques and employing genetic algorithm to arrive at the optimal prediction heuristic.

References

1. Levenshtein, V.I.: Binary codes capable of correcting deletions, insertions, and reversals. Soviet Physics Doklady 10, 707–710 (1966)
2. Damerau, F.J.: A technique for computer detection and correction of spelling errors. Communications of the ACM 7(3), 171–176 (1964)
3. Knuth, D.E.: The Art of Computer Programming, Volume 3: Sorting and Searching. Addison-Wesley Publishing Company, Reading (1998)
4. Philips, L.: The double metaphone search algorithm. C/C++ Users J. (June 2000)
5. Angell, R., Freund, G., Willet, P.: Automatic spelling correction using a trigram similarity measure. Information Processing and Management 19(4), 305–316 (1983)
6. Pollock, J.J., Zamora, A.: Automatic spelling correction in scientific and scholarly text. Communications of the ACM 27(4), 358–368 (1984)
7. Peterson, J.L.: Computer programs for detecting and correcting spelling errors. Communications of the ACM 23(12), 676–687 (1980)
8. Vilalta, R., Ma, S.: Predicting rare events in temporal domains. In: Proc. of the 2002 IEEE International Conference on Data Mining, ICDM 2002, pp. 474–481 (2002)
9. Weiss, G.M., Hirsh, H.: Learning to predict rare events in event sequences. In: Proc. of KDD 1998, pp. 359–363 (1998)
10. Dietterich, T.G., Michalski, R.S.: Discovering patterns in sequences of events. Artificial Intelligence 25(2), 187–232 (1985)
11. Mannila, H., Toivonen, H., Verkamo, A.I.: Discovery of frequent episodes in event sequences. Data Min. Knowl. Discov. 1(3), 259–289 (1997)
12. Laxman, S., Unnikrishnan, K.P., Sastry, P.S.: Discovering frequent episodes and learning hidden markov models: A formal connection. IEEE Trans. on Knowl. and Data Eng. 17(11), 1505–1517 (2005)
13. Laxman, S., Tankasali, V., White, R.W.: Stream prediction using a generative model based on frequent episodes in event sequences. In: Proc. of the 14th ACM SIGKDD Int'l. Conf. on Knowledge discovery and data mining, pp. 453–461 (2008)

Condensed Representation of Sequential Patterns According to Frequency-Based Measures

Marc Plantevit and Bruno Crémilleux

GREYC-CNRS-UMR 6072
Université de Caen Basse-Normandie
Campus Côte de Nacre,
14032 Caen Cedex, France

Abstract. Condensed representations of patterns are at the core of many data mining works and there are a lot of contributions handling data described by items. In this paper, we tackle sequential data and we define an exact condensed representation for sequential patterns according to the frequency-based measures. These measures are often used, typically in order to evaluate classification rules. Furthermore, we show how to infer the best patterns according to these measures, i.e., the patterns which maximize them. These patterns are immediately obtained from the condensed representation so that this approach is easily usable in practice. Experiments conducted on various datasets demonstrate the feasibility and the interest of our approach.

1 Introduction

It is well-known that the "pattern flooding which follows data flooding" is unfortunate consequence in exploratory Knowledge Discovery in Databases (KDD) processes. There is a large range of methods to discover the patterns of a potential user's interest but the most significant patterns are lost among too much trivial, noisy and redundant information. Many works propose methods to reduce the collection of patterns, such as the constraint-based paradigm [15], the pattern set discovery approach [4,11], the so-called condensed representations [3,27] as well as the compression of the dataset by exploiting the Minimum Description Length Principle [19]. In practice, these methods often tackle data described by items (i.e., itemsets) and/or specific contexts, such as the largely studied frequent patterns extraction issue (a pattern X is said *frequent* if the number of examples in the database supporting X exceeds a given threshold). Many applications (e.g., security network, bioinformatics) require sequence mining. Oddly enough, even more than in the item domain, sequence mining suffers from the massive output of the KDD processes. However, little works focused on this aspect mainly because the difficult formalization required for sequential patterns. For instance, although there are many condensed representations of frequent itemsets, only closed sequential patterns have been proposed as a exact condensed representation for all the frequent sequential patterns [27]. Moreover,

N. Adams et al. (Eds.): IDA 2009, LNCS 5772, pp. 155–166, 2009.
© Springer-Verlag Berlin Heidelberg 2009

some concise representations of itemset patterns cannot be used in order to condense frequent sequential patterns [17]. This illustrates the intrinsic difficulty to extend such works from itemsets to sequential patterns.

This paper addresses the issue of condensed representations of sequential patterns. The idea is to compute a representation $\mathcal{R}$ of the extracted patterns which is lossless: the whole collection of patterns can be efficiently derived from $\mathcal{R}$. This approach has been mainly developed in the context of frequency [3,27] and there are very few works addressing other measures [8,21,22]. In this paper, we investigate exact condensed representations of sequential patterns based on many interestingness measures, the so-called *frequency-based measures* (see Section 3). These measures (e.g., frequency, confidence, lift, growth rate, information gain) are precious in real-world applications to evaluate the interestingness of patterns and the quality of classification rules [20]. For instance, the emerging measure is very useful to characterize classes and classify them. Initially introduced in [5], emerging patterns (EPs) are patterns whose frequency strongly varies between two datasets (i.e., two classes). An EP can be seen as a classification rule and EPs are at the origin of various works such as powerful classifiers [13]. From an applicative point of view, we can quote many works on the characterization of biochemical properties or medical data [14]. A condensed representation of itemsets according to frequency-based measures has already been proposed [22], but it is only limited to the item domain.

The contribution of this paper is twofold. First, we define an exact condensed representation of sequential patterns according to the frequency-based measures. *Exact* means that we are able to infer not only the patterns, but also the measure values associated to the patterns without accessing the data. This is useful because the user is mainly interested in these values. For that purpose, the key idea is to show that the value of a frequency-based measure of any sequential pattern can be deduced from one of its closed sequential patterns. This idea has already been used in the item domain [22], but not in sequential data. Contrary to itemsets, a sequential pattern may have several closed sequential patterns, our method overcomes this difficulty. As this condensed representation is based on the closed sequential patterns and there are efficient algorithms to extract these patterns, these algorithms are also efficient to mine such a condensed representation. Second, we define the notion of strong sequential patterns (SPs) according to frequency-based measures. Given a frequency-based measure, these patterns maximize it. This is interesting because it highlights the best patterns with respect to the measure and moreover it reduces the output. On the other hand, the SPs are immediately obtained from the condensed representation. Finally, experiments conducted on various datatsets demonstrate the feasibility of our approach and quantify the interests of SPs.

This paper is organized as follows. Section 2 provides the preliminaries which are needed for the rest of the paper. In Section 3, we propose a condensed representation of sequential patterns according to the frequency-based measures. Section 4 defines the strong frequency-based measures and the SPs. Section 5 provides in depth experimental results and we review related work in Section 6.

2   Preliminary Concepts and Definitions

Let $\mathcal{I} = \{i_1, i_2, \ldots, i_n\}$ be a finite set of items. An *itemset* I is a subset of $\mathcal{I}$. A sequence $s = \langle I_1, I_2, \ldots, I_n \rangle$ is an ordered list of itemsets. A sequence $s_\alpha = \langle A_1, A_2, \ldots, A_n \rangle$ is said to be contained in another sequence $s_\beta = \langle B_1, B_2, \ldots, B_m \rangle$ if there exist integer $1 \leq i_1 < i_2 < \ldots < i_n \leq m$ such that $A_1 \subseteq B_{i_1}, A_2 \subseteq B_{i_2}, \ldots, A_n \subseteq B_{i_n}$ (denoted by $s_\alpha \sqsubseteq s_\beta$). If the sequence s_α is contained in the sequence s_β, s_α is called a *subsequence* of s_β and s_β a *supersequence* of s_α.

Input data in sequential pattern mining consists in a collection of sequences. As previously highlighted in the introduction section, frequency based-measures are mainly used to assess the quality of classification rules and a class identifier is associated to each data sequence. Therefore, the input database $\mathcal{D}$ consists in a collection of tuples (sid, s, c) where sid is a sequence identifier, s is sequence and c is a class identifier (see the example given in Tab. 1). $\mathcal{D}$ corresponds to a partition of i subsets $\mathcal{D}_i$ where each $\mathcal{D}_i$ contains all tuples (sid, s, c_i) in $\mathcal{D}$. Each sequence belongs to a single subset $\mathcal{D}_i$. A tuple (sid, s, c) is said to *contain* a sequence s_α if $s_\alpha \sqsubseteq s$. The *intersection* of a set of sequences $S = \{s_1, s_2, \ldots, s_n\}$, denoted $\bigcap s_i \in S$, is the set of all maximal subsequences contained into all the s_i. For example, the intersection of $s = \langle c, b, c, a \rangle$ and $s' = \langle c, b, a, c, c, c \rangle$ is $\{\langle c, b, a, \rangle, \langle c, b, c \rangle\}$.

Table 1. Toy database $\mathcal{D}$ with class values

Seq_id	Sequence	Class
s_1	$\langle c, b, c, a \rangle$	c_1
s_2	$\langle c, b, a, c, c, c \rangle$	c_1
s_3	$\langle a, a, a, c, c, a, a \rangle$	c_2
s_4	$\langle a, a, b, a, c, c \rangle$	c_2

Frequency-based measures are linked to the notions of *support*. The *absolute support* of a sequence s_α in $\mathcal{D}$ is the number of tuples in $\mathcal{D}$ that contain s_α, denoted by $support(s_\alpha, \mathcal{D})$. The *relative support* of s_α is the percentage of tuples in $\mathcal{D}$ that contain s_α. For instance, $support(\langle c, a \rangle, \mathcal{D}) = 3$. Unless otherwise stated, we use the absolute support all along this paper.

Let $minsup$ be a minimum support threshold. A sequence s_α is a frequent sequence on $\mathcal{D}$ if $support(s_\alpha, \mathcal{D}) \geq minsup$. A frequent sequence s_α is a closed frequent sequence if there does not exist a sequence s_β such that $support(s_\alpha, \mathcal{D}) = support(s_\beta, \mathcal{D})$ and $s_\alpha \sqsubset s_\beta$. Then, given $\mathcal{D}$ and $minsup$, the problem of mining frequent closed sequential patterns is to find the complete set of frequent closed sequences. Function $Closed(x, \mathcal{D})$ from Definition 1 return the set of closed sequential patterns in sequence database $\mathcal{D}$ which contains a sequence s.

Definition 1 ($Closed(x, \mathcal{D})$). *Let x be a sequential pattern and $\mathcal{D}$ be a sequence database.*

$$Closed(x, \mathcal{D}) = \bigcap \{s \in \mathcal{D} | x \sqsubseteq s\}$$

Following our example in Table 1, we get: $Closed(\langle c, b \rangle, \mathcal{D}) = \{\langle c, b, a \rangle, \langle c, b, c \rangle\}$. These two sequences are closed in $\mathcal{D}$. Finally, we recall the notion of classification sequential rule.

Definition 2 (Classification sequential rules). *Let $\mathcal{C} = \{c_1, c_2, \ldots, c_m\}$ be a set of class values, a classification sequential rule is a rule $R = s \rightarrow c_i$ where s is a sequential pattern and $c_i \in \mathcal{C}$.*

3   Exact Condensed Representation of Sequential Pattern According to Frequency Based Measures

Various measures [7] are used to evaluate the quality of classification rules. Many measures are based on the frequency of the sequential patterns s and the concatenation of s and c_i, i.e. $\langle s, \{c\} \rangle$. These measures are called frequency-based measures and are defined as follows:

Definition 3 (Frequency-Based Measure). *Let $\mathcal{D}$ be a sequence database partitioned into k subsets denoted $\mathcal{D}_1, \mathcal{D}_2, \ldots, \mathcal{D}_k$, a frequency-based measure M_i to characterize $\mathcal{D}_i$ is a function F of supports: $support(s, \mathcal{D}_1)$, $support(s, \mathcal{D}_2), \ldots, support(s, \mathcal{D}_k)$, i.e. $M_i(s) = F(support(s, \mathcal{D}_1), support(s, \mathcal{D}_2), \ldots, support(s, \mathcal{D}_k))$.*

With the notation M_i, the subscript i denotes the dataset $\mathcal{D}_i$ which is characterized according to the measure M. A frequency-based measure consists of a finite combination of supports of a pattern s on several sequence data sets $\mathcal{D}_i$. More precisely, a frequency-based measure cannot contain other parameters. Table 2 lists some well-known frequency-based measures that are commonly used in the literature. These measures are given here by using the absolute support whereas the literature often writes them in term of conditional probabilities [7] ($P(X|c_i)$) corresponds to $\frac{support(X, \mathcal{D}_i)}{|\mathcal{D}_i|}$ where X is a (sequential) pattern. Note that some frequency-based measures (e.g., J-Measure, confidence, lift, growth rate) are expressed with supports that are not restricted to sets $\mathcal{D}_1, \mathcal{D}_2, \ldots, \mathcal{D}_k$. However, these measures respect Definition 3 because these supports can be computed from $support(s, \mathcal{D}_1), support(s, \mathcal{D}_2), \ldots support(s, , \mathcal{D}_k)$. For instance, $support(s, \mathcal{D}) = \sum_{j=1}^{k} support(s, \mathcal{D}_j)$.

To compute the value of a frequency-based measure for a rule $s \rightarrow c_i$, computing the support of s in datasets $\mathcal{D}$ and $\mathcal{D}_i$ ($support(s, \mathcal{D})$ and $support(s, \mathcal{D}_i)$) is enough. An important result is that these frequencies can be obtained thanks to the set of closed sequential patterns in $\mathcal{D}$ and $\mathcal{D}_i$. Indeed, we have:

- $\forall e \in Closed(s, \mathcal{D})$ $support(s, \mathcal{D}) = support(e, \mathcal{D})$
- $\forall e \in Closed(s, \mathcal{D}_i)$ $support(s, \mathcal{D}_i) = support(e, \mathcal{D}_i)$

Consequently, the computation of $Closed(s, \mathcal{D})$ and $Closed(s, \mathcal{D}_i)$ are enough to compute $support(s, \mathcal{D})$ and $support(s, \mathcal{D}_i)$. Furthermore, the following property indicates that the computation of the function $Closed$ can be made only once:

Table 2. Examples of frequency-based measures characterizing $\mathcal{D}_i$

Frequency-based measure	Formula	Strong
J-Measure	$\frac{support(s,\mathcal{D}_i)}{|\mathcal{D}|} \times \log \frac{support(s,\mathcal{D}_i)\times|\mathcal{D}|}{|\mathcal{D}_i|\times support(s,\mathcal{D})}$	no
Relative support	$\frac{support(s,\mathcal{D}_i)}{|\mathcal{D}|}$	yes
Confidence	$\frac{support(s,\mathcal{D}_i)}{support(s,\mathcal{D})}$	yes
Sensitivity	$\frac{support(s,\mathcal{D}_i)}{|\mathcal{D}_i|}$	yes
Success rate	$\frac{support(s,\mathcal{D}_i)}{|\mathcal{D}|} + \frac{|\mathcal{D}\backslash\mathcal{D}_i|-support(s,\mathcal{D}\backslash\mathcal{D}_i)}{|\mathcal{D}|}$	yes
Specificity	$\frac{|\mathcal{D}\backslash\mathcal{D}_i|-support(s,\mathcal{D}\backslash\mathcal{D}_i)}{|\mathcal{D}|}$	yes
Piatetsky-Shapiro's (PS)	$\frac{support(s,\mathcal{D}_i)}{|\mathcal{D}|} - \frac{support(s,\mathcal{D})}{|\mathcal{D}|} \times \frac{|\mathcal{D}_i|}{|\mathcal{D}|}$	yes
Lift	$\frac{support(s,\mathcal{D}_i)\times|\mathcal{D}|}{support(s,\mathcal{D})\times|\mathcal{D}_i|}$	yes
Odd ratio (α)	$\frac{support(s,\mathcal{D}_i)\times(|\mathcal{D}\backslash\mathcal{D}_i|-support(s,\mathcal{D}\backslash\mathcal{D}_i))}{(support(s,\mathcal{D})-support(s,\mathcal{D}_i))\times(|\mathcal{D}_i|-support(s,\mathcal{D}_i))}$	yes
Growth rate (GR)	$\frac{|\mathcal{D}|-|\mathcal{D}_i|}{|\mathcal{D}|} \times \frac{support(s,\mathcal{D}_i)}{support(s,\mathcal{D})-support(s,\mathcal{D}_i)}$	yes
Information Gain	$\log \frac{support(s,\mathcal{D}_i)\times|\mathcal{D}|}{support(s,\mathcal{D})\times|\mathcal{D}_i|}$	yes

Property 1. Let s be a sequential pattern and $\mathcal{D}_i$ a subset of $\mathcal{D}$, $\forall e \in Closed(s,\mathcal{D})$, $support(s,\mathcal{D}_i) = support(e,\mathcal{D}_i)$

Proof. According to Definition 1, sequence e from $Closed(s,\mathcal{D})$ is a super-sequence of s having the same support in $\mathcal{D}$. Since s is a subsequence of e, all sequences from $\mathcal{D}$ that contain e also contain s. Moreover, sequences s and e have the same support in $\mathcal{D}$. Thus, they are contained by the same sequences of $\mathcal{D}$. Since $\mathcal{D}_i$ is a subset of $\mathcal{D}$, sequences e and s are contained in the same sequences of $\mathcal{D}_i$. Thus, $support(s,\mathcal{D}_i) = support(e,\mathcal{D}_i)$.

As said in Section 2, a sequential pattern s may have several closed patterns. Theorem 1 shows that all closed patterns of s have the same value for a measure. Consequently, the value of a frequency-based measure for s can be deduced from any of its closed sequential patterns:

Theorem 1. *Let s be a sequential pattern, we have:*

$$\forall e \in Closed(s,\mathcal{D}), \ M_i(s) = M_i(e)$$

Proof. Let s be a sequential pattern. Since $\forall e \in Closed(s,\mathcal{D})$, $support(s,\mathcal{D}_i) = support(e,\mathcal{D}_i)$ (property 1), we can express $M_i(s) = F(support(s,\mathcal{D}_1), support(s,\mathcal{D}_2),\ldots, support(s,\mathcal{D}_k))$ by $M_i(s) = F(support(e,\mathcal{D}_1), support(e,\mathcal{D}_2),\ldots, support(e,\mathcal{D}_k)) = M_i(e)$ where $e \in Closed(s,\mathcal{D})$. Thus $M_i(s) = M_i(e)$.

For example, $Closed(\langle c,b\rangle,\mathcal{D}) = \{\langle c,b,a\rangle, \langle c,b,c\rangle\}$, and $Confidence_{c_1}(\langle c,b\rangle) = Confidence_{c_1}(\langle c,b,a\rangle) = 1$. The closed sequential patterns with their values of the measure M_i are enough to synthesize the set of sequential patterns according to M_i. As a consequence, the closed sequential patterns with their values of the measure M_i are an exact condensed representation of the whole set of sequential patterns according to M_i. In practice, the number of closed patterns is lower (and often much lower) than the complete set of sequential patterns. More generally, this condensed representation benefits from all the advantages of the condensed representation based on the closed sequential patterns [27,25].

4 Strong Sequential Patterns According to Frequency-Based Measures

In practice, the number of patterns satisfying a given threshold for a measure M_i can be very large and hampers their individual analysis. In this section, we show that our approach easily enables us to highlight the best patterns according to measures, that is to say the patterns which maximize such measures. To achieve this result, we have to consider a slightly different set of measures, the strong frequency-based measures:

Definition 4 (Strong Frequency-Based Measure). *A frequency-based measure M_i which decreases with $support(s, \mathcal{D})$ when $support(s, \mathcal{D}_i)$ remains unchanged, is a strong frequency-based measure.*

Most frequency-based measures are also strong frequency-based measures (in Table 2, only the J-measure is not a strong frequency-based measure). More generally, Definition 4 is less restrictive than the property P_3 of Piatetsky-Shapiro's framework [16] which defines three properties which have to be satisfied by an interestingness measure to be qualified as a "good" one.

Theorem 2 indicates that the closed sequential patterns satisfy an interesting property w.r.t. the strong frequency-based measures.

Theorem 2. *Let M_i be a strong frequency-based measure and s be a sequential pattern, we have $\forall e \in Closed(s, \mathcal{D}_i), M_i(s) \leq M_i(e)$. The elements from $Closed(s, \mathcal{D}_i)$ are called strong sequential patterns (SPs) in class i or dominant sequential patterns for M_i.*

Proof. Let M_i be a strong frequency-based measure and s be a sequential pattern. $\forall e \in Closed(s, \mathcal{D}_i)$, we have $support(s, \mathcal{D}_i) = support(e, \mathcal{D}_i)$ (see Definition 1). As $s \sqsubseteq e$, we obtain that $support(s, \mathcal{D}) \geq support(e, \mathcal{D})$ thanks to the anti-monotonicity of the support. By definition 4, we conclude that $M_i(s) \leq M_i(e)$.

The result given by Theorem 2 is important: it means that the closed sequential patterns in $\mathcal{D}_i$ maximize any strong frequency-based measure M_i. In other words, a sequential pattern that is not closed in $\mathcal{D}_i$ has a lower (or equal) value than one of its closed sequential patterns in $\mathcal{D}_i$ for any measure M_i.

However, Theorem 2 means that mining all closed sequential patterns in each $\mathcal{D}_i$ is needed, which indeed require a lot of computation. Lemma 1 links closed sequential patterns in $\mathcal{D}_i$ with closed sequential patterns in $\mathcal{D}$. For that, we first have to define the sequence concatenation. Given a sequence $s_\alpha = \langle A_1, A_2, \ldots, A_n \rangle$ and a class c, the concatenation of sequence s_α with $\langle c \rangle$, denoted $s_\alpha \bullet c$ is $\langle A_1, A_2, \ldots, A_n, \{c\} \rangle$. We then consider the sequence database $\mathcal{D}'$

from $\mathcal{D}$ where each data sequence contains a new item that represents their class value. For each tuple (sid, s, c) we add the tuple $(sid, s \bullet c, c)$ in $\mathcal{D}'$. Then, like $\mathcal{D}$, $\mathcal{D}'$ corresponds to a partition of i subsets $\mathcal{D}'_i$ where each $\mathcal{D}'_i$ contains all tuples $(sid, s \bullet c_i, c_i)$. Note that we have the relation $support(s, \mathcal{D}_i) = support(s \bullet c_i, \mathcal{D}'_i)$.

Lemma 1. *If the sequence $s \bullet c_i$ is a closed sequential pattern in $\mathcal{D}'_i$ then $s \bullet c_i$ is a closed sequential pattern in $\mathcal{D}'$.*

Proof. By construction of the subsets of $\mathcal{D}'$, a class value c_i is only contained in $\mathcal{D}'_i$ and not in the other datasets. So we have $support(s, \mathcal{D}'_i) = support(s \bullet c_i, \mathcal{D}')$.

Thanks to Lemma 1, we can give Property 2 which indicates that mining only the closed sequential patterns in $\mathcal{D}'$ is enough. In other words, only one extraction of closed patterns is needed.

Property 2 (SPs: computation of their frequency-based measure). If s is a strong sequential pattern in $\mathcal{D}_i$, then $M_i(s)$ can directly be computed with the supports of the condensed representation based on the closed sequential patterns of $\mathcal{D}'$.

Proof. Let s be a SP in $\mathcal{D}_i$. Thus, $s \bullet c_i$ is a closed sequential pattern in $\mathcal{D}'_i$. To compute $M_i(s)$, it is necessary to know $support(s, \mathcal{D}'_i)$ and $support(s, \mathcal{D}')$. By definition of $\mathcal{D}'_i$, $support(s, \mathcal{D}'_i) = support(s \bullet c_i, \mathcal{D}')$ and lemma 1 ensures that $s \bullet c_i$ is closed in $\mathcal{D}'$. As a consequence, its support is provided by the condensed representation of the closed sequential patterns of $\mathcal{D}'$. To compute $support(s, \mathcal{D}')$, two cases are possible: (i) if s is a closed sequential pattern in $\mathcal{D}'$, its support is directly available; (ii) if, s is not a closed sequential pattern in $\mathcal{D}'$, then $s \bullet c_i$ belongs to $Closed(s, \mathcal{D}')$ and $support(s, \mathcal{D}') = support(s \bullet c_i, \mathcal{D}')$.

We have defined a theoretical framework for SPs in database $\mathcal{D}$ and its subset $\mathcal{D}_i$. In practice, these patterns can be discovered in $\mathcal{D}'$ and their frequencies can also be computed in $\mathcal{D}'$ thanks to any closed sequential pattern mining algorithm. Indeed, if s is a strong sequential pattern in $\mathcal{D}_i$, then $s \bullet c_i$ is a closed sequential pattern in $\mathcal{D}'$, $support(s, \mathcal{D}_i) = support(s \bullet c_i, \mathcal{D}')$ and $support(s, \mathcal{D}) = support(s, \mathcal{D}')$.

Example 1. Following our example in Table 1, with $minsup = 2$, we have 11 closed frequent sequential patterns. In particular, $\langle c, b, a \rangle$ and $\langle c, b, c \rangle$ are SPs for class c_1, $\langle a, a, a, c, c \rangle$ is a SP for class c_2. Thus these sequences maximize any frequency-based measure M_i.

Let M_2 be the confidence measure, $\langle a, a, a, c, c \rangle \bullet c_2$ is a closed sequential pattern in $\mathcal{D}'$. To compute its confidence, we need to know $support(\langle a, a, a, c, c \rangle, \mathcal{D}')$. Since sequence $\langle a, a, a, c, c \rangle$ is not a closed sequential pattern in $\mathcal{D}'$, then $support(\langle a, a, a, c, c \rangle, \mathcal{D}') = support(\langle a, a, a, c, c \rangle \bullet c_2, \mathcal{D}') = 2$. Thus the confidence of SP $\langle a, a, a, c, c \rangle$ for class c_2 is 1.

5   Experiments

Experiments have been carried out on real datasets by considering the emerging measure (Growth Rate) [5]. The emerging measure is very useful to characterize classes and classify them. Emerging patterns (EPs) are patterns whose frequency strongly varies between two datasets (i.e., two classes). Note that any frequency-based measure can be used. However, due to the space limitation, we only report experiments on strong emerging frequent sequential patterns (SESPs). To mine closed sequential patterns, we have implemented Bide algorithm [25] in Java language (JVM 1.5). Furthermore, we do not report results about the runtime of the discovery of SEPSs. However, it is important to note that the computation of SP growth rates is negligible compared to the step of frequent closed sequential pattern mining. We can conclude that the scalability issue of SPs' discovery is Bide-dependent and Bide is known as being a scalable and robust algorithm. Consequently, the discovery of SESPs is then scalable.

In these experiments, we consider the following real datasets:

- *E.Coli Promoters dataset:* The *E. Coli* Promoters data set [23] is available on the UCI machine learning repository [1]. The data set is divided into two classes: 53 *E. Coli* promoter instances and 53 non-promoter instances. We consider pairs of monomers (*e.g.*, aa, ac, etc.) as items.

- *PSORTdb v.2.0 cytoplasmic dataset:* The cytoplasmic data set was obtained from PSORTdb v.2.0 [6]. The data set contains 278 cytoplasmic Gram-negative sequences and 194 Gram-positive sequences. We consider items in the same way as in the previous dataset.

- *Greenberg's Unix dataset:* We transform the original Unix dataset [9] into a new data set that contains 18681 data sequences where a data sequence contains a session of a Unix command shell user. These sequences are divided into 4 classes: 7751 sequences about navigation of computer scientists, 3859 sequences for *experienced-programmers*, 1906 sequences for *non-programmers* and 5165 sequences about *novice-programmers*.

- *Entree Chicago Recommendation Dataset:* We use the data set underlying the Entree system [2]. This data set is also available on the UCI machine learning repository [1]. For each restaurant, a sequence of features is associated. We consider 8 classes (Atlanta, Boston, Chicago, Los Angeles, New Orleans, New York, San Francisco and Washington DC) that respectively contain $267, 438, 676, 447, 327, 1200, 414$ and 391 sequences.

These experiments aim at studying several quantitative results of the discovery of strong sequential patterns satisfying both a growth rate threshold and a support threshold.

Figures from Fig. 1 report the number of frequent closed, strong and emerging sequential patterns according to the minimum support threshold. Obviously, the number of patterns decreases when the support threshold increases. We note that the number of SESPs is much lower than the number of SPs which is itself significantly lower than the number of sequential patterns (the figure uses a logarithmic scale). It indicates a high condensation of patterns reducing the output and highlighting the most valuable patterns according to the measures. Note that there is no SESP (and no SP in E.coli and Entree datasets) when *minsup* is high because no pattern can satisfy the growth rate measure.

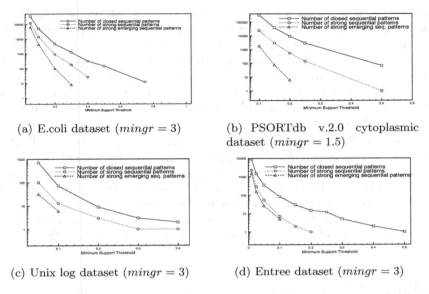

(a) E.coli dataset ($mingr = 3$)

(b) PSORTdb v.2.0 cytoplasmic dataset ($mingr = 1.5$)

(c) Unix log dataset ($mingr = 3$)

(d) Entree dataset ($mingr = 3$)

Fig. 1. Numbers of closed, strong and emerging sequential patterns according to the minimum support threshold

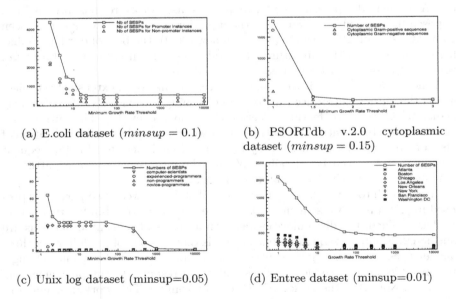

(a) E.coli dataset ($minsup = 0.1$)

(b) PSORTdb v.2.0 cytoplasmic dataset ($minsup = 0.15$)

(c) Unix log dataset (minsup=0.05)

(d) Entree dataset (minsup=0.01)

Fig. 2. Number of strong emerging sequential patterns and their repartition according to the growth rate threshold

Figures from Fig. 2 report the number of SESPs and their distribution among the different class values according to the growth rate threshold. The number of SESPs decreases when the growth rate threshold increases. However, this number does not always tend to zero. Indeed, some SESPs with an infinite growth rate can appear (see Fig. 2(a) and Fig. 2(d)). These particular SESPs are called jumping SESPs (JSESPs). They are sequential patterns that appear for only one class value, and never appear for other class values. It should be noticed that the repartition of SESPs among the class values is not necessary uniform. For instance, class value *novice-programmers* for Fig. 2(c) and class value *Washington DC* for Fig. 2(d) contain a significantly greater number of SESPs than the others.

Discovered SESPs: Sequential pattern ⟨(aa)(at)(ta)(gc)⟩ is a JSESP for promoter sequences in *E. coli* dataset while ⟨(tg)(cg)(ac)(tg)⟩ is a JS-ESP for non-promoter sequences. According to *Entree* dataset, the sequence ⟨(Week-end Dining)(Parking-Valet)⟩ is a SESP for class *Washington DC* with a growth rate $gr = 10.02$. Sequential pattern ⟨pix, umacs, pix, umacs⟩ is a SESP for class *novice-programmers* with a growth rate $gr = 450$. Let us recall that to the best of our knowledge, our method is the unique method to discover such patterns.

6   Related Work

Main works on condensed representations have been outlined in the introduction. A condensed representation of frequency-based measures has already been proposed in [22], but it is limited to the item domain and our work can be seen as an extension of [22] to the sequence framework. To the best of our knowledge, there is no work in the literature that addresses condensed representations of sequential patterns w.r.t. any frequency-based measure.

In the literature, classification on sequence data has been extensively studied. In [26], the authors introduce the problem of mining sequence classifiers for early prediction. Criteria for feature selection are proposed in [12]. The authors use the confidence to quantify the features. Our work can lead to a generalization of this work by allowing the use of any frequency-based measure. In [18] frequent subsequences are used for classification but the interestingness of a pattern is valued according to the only confidence measure.

An approach to detect sequential pattern changes between two periods is proposed in [24]. First, two sequential pattern sets are discovered in the two-period databases. Then, the dissimilarities between all pairs of sequential patterns are considered. Finally, a sequential pattern is classified as one of the following three change-types: an emerging sequential pattern, an unexpected sequence change, and an added/perish sequence. These latter correspond to jumping emerging sequences. Note that the notion of EPs differs from [5]. This work does not consider condensed representations. Moreover, two databases have to be mined and then similarities between each pair of sequences have to be computed whereas our framework needs only one database mining and no computation of sequence similarities.

7   Conclusion

In this paper, we have investigated condensed representations of sequential patterns according to many interestingness measures and we have proposed an exact condensed representation of sequential patterns according to the frequency-based measures. Then, we have defined the strong sequential patterns which are the best patterns according to the measures. These patterns are straightforwardly obtained from the condensed representation so that this approach can be easily used in practice. Experiments show the feasibility and the interest of the approach.

We think that condensed representations of patterns have a lot of applications and their use is not limited to obtain more efficiently patterns associated to their interestingness measures. As they can be used as cache mechanisms, they make interactive KDD processes more easily and are a key concept of inductive databases. Moreover, their properties are useful for high-level KDD tasks such as classification or clustering. Finally, the behavior of interestingness measures has been studied in [10] and the next step is to determine lower bounds for weighted combinations of frequency-based measures in order to ensure a global quality according to a set of measures.

Acknowledgments. The authors would like to thank Arnaud Soulet (Université François Rabelais de Tours, Fr) for very fruitful comments and invaluable discussions. This work is partly supported by the ANR (French National Research Agency) funded project Bingo2 ANR-07-MDCO-014.

References

1. Asuncion, A., Newman, D.: UCI machine learning repository (2007), http://www.ics.uci.edu/~mlearn/MLRepository.html
2. Burke, R.D.: The wasabi personal shopper: A case-based recommender system. In: AAAI/IAAI, pp. 844–849 (1999)
3. Calders, T., Rigotti, C., Boulicaut, J.-F.: A survey on condensed representations for frequent sets. In: Constraint-Based Mining and Inductive Databases, pp. 64–80 (2004)
4. De Raedt, L., Zimmermann, A.: Constraint-based pattern set mining. In: SDM (2007)
5. Dong, G., Li, J.: Efficient mining of emerging patterns: discovering trends and differences. In: ACM SIGKDD 1999, San Diego, CA, pp. 43–52. ACM Press, New York (1999)
6. Gardy, J.L., Spencer, C., Wang, K., Ester, M., Tusnady, G.E., Simon, I., Hua, S.: PSORT-B: improving protein subcellular localization prediction for Gram-negative bacteria. Nucl. Acids Res. 31(13), 3613–3617 (2003)
7. Geng, L., Hamilton, H.J.: Interestingness measures for data mining: A survey. ACM Comput. Surv. 38(3) (2006)
8. Giacometti, A., Laurent, D., Diop, C.T.: Condensed representations for sets of mining queries. In: Knowledge Discovery in Inductive Databases, 1st International Workshop, KDID 2002 (2002)

9. Greenberg, S.: Using Unix: Collected traces of 168 users. Research Report, 88/333/45, Department of Computer Science, University of Calgary, Calgary, Canada (1988), http://grouplab.cpsc.ucalgary.ca/papers/

10. Hébert, C., Crémilleux, B.: A unified view of objective interestingness measures. In: Perner, P. (ed.) MLDM 2007. LNCS (LNAI), vol. 4571, pp. 533–547. Springer, Heidelberg (2007)

11. Knobbe, A.J., Ho, E.K.Y.: Pattern teams. In: Fürnkranz, J., Scheffer, T., Spiliopoulou, M. (eds.) PKDD 2006. LNCS (LNAI), vol. 4213, pp. 577–584. Springer, Heidelberg (2006)

12. Lesh, N., Zaki, M.J., Ogihara, M.: Mining features for sequence classification. In: KDD, pp. 342–346 (1999)

13. Li, J., Dong, G., Ramamohanarao, K.: Making use of the most expressive jumping emerging patterns for classification. Knowledge and Information Systems 3(2), 131–145 (2001)

14. Li, J., Wong, L.: Emerging patterns and gene expression data. Genome Informatics 12, 3–13 (2001)

15. Ng, R.T., Lakshmanan, L.V.S., Han, J., Pang, A.: Exploratory mining and pruning optimizations of constrained associations rules. In: ACM SIGMOD 1998, pp. 13–24. ACM Press, New York (1998)

16. Piatetsky-Shapiro, G.: Discovery, analysis, and presentation of strong rules. In: Knowledge Discovery in Databases, pp. 229–248. AAAI/MIT Press (1991)

17. Raïssi, C., Calders, T., Poncelet, P.: Mining conjunctive sequential patterns. Data Min. Knowl. Discov. 17(1), 77–93 (2008)

18. She, R., Chen, F., Wang, K., Ester, M., Gardy, J.L., Brinkman, F.S.L.: Frequent-subsequence-based prediction of outer membrane proteins. In: Getoor, L., Senator, T.E., Domingos, P., Faloutsos, C. (eds.) KDD, pp. 436–445. ACM, New York (2003)

19. Siebes, A., Vreeken, J., van Leeuwen, M.: Item sets that compress. In: Proceedings of the Sixth SIAM International Conference on Data Mining, Bethesda, MD, USA. SIAM, Philadelphia (2006)

20. Smyth, P., Goodman, R.M.: Rule induction using information theory. In: Knowledge Discovery in Databases, pp. 159–176. AAAI Press, Menlo Park (1991)

21. Soulet, A., Crémilleux, B.: Adequate condensed representations of patterns. Data Min. Knowl. Discov. 17(1), 94–110 (2008)

22. Soulet, A., Crémilleux, B., Rioult, F.: Condensed representation of eps and patterns quantified by frequency-based measures. In: KDID 2004, Revised Selected and Invited Paperss, pp. 173–190 (2004)

23. Towell, G.G., Shavlik, J.W., Noordewier, M.O.: Refinement of approximate domain theories by knowledge-based neural networks. In: AAAI, pp. 861–866 (1990)

24. Tsai, C.-Y., Shieh, Y.-C.: A change detection method for sequential patterns. Decis. Support Syst. 46(2), 501–511 (2009)

25. Wang, J., Han, J., Li, C.: Frequent closed sequence mining without candidate maintenance. IEEE Trans. Knowl. Data Eng. 19(8), 1042–1056 (2007)

26. Xing, Z., Pei, J., Dong, G., Yu, P.S.: Mining sequence classifiers for early prediction. In: SDM, pp. 644–655 (2008)

27. Yan, X., Han, J., Afshar, R.: Clospan: Mining closed sequential patterns in large databases. In: SDM (2003)

ART-Based Neural Networks for Multi-label Classification

Elena P. Sapozhnikova

Department of Computer and Information Science, Box M712,
University of Konstanz, 78457 Konstanz, Germany
elena.sapozhnikova@uni-konstanz.de

Abstract. Multi-label classification is an active and rapidly developing research area of data analysis. It becomes increasingly important in such fields as gene function prediction, text classification or web mining. This task corresponds to classification of instances labeled by multiple classes rather than just one. Traditionally, it was solved by learning independent binary classifiers for each class and combining their outputs to obtain multi-label predictions. Alternatively, a classifier can be directly trained to predict a label set of an unknown size for each unseen instance. Recently, several direct multi-label machine learning algorithms have been proposed. This paper presents a novel approach based on ART (Adaptive Resonance Theory) neural networks. The Fuzzy ARTMAP and ARAM algorithms were modified in order to improve their multi-label classification performance and were evaluated on benchmark datasets. Comparison of experimental results with the results of other multi-label classifiers shows the effectiveness of the proposed approach.

1 Introduction

In the past decades, machine learning and neural network classifiers have been extensively studied in the one-class-per-instance setting. However, many real world problems produce more complex data sets, for example, with classes that are not necessarily mutually exclusive. So, a gene can have multiple biological functions or a text document can be assigned to multiple topics. For this reason, Multi-label Classification (MC) when an instance could belong to more than one class becomes increasingly important particularly in the fields of web or text mining and bioinformatics.

Generally, a MC task is more difficult to solve than a single-label classification task. The main problem is a large number of possible class label combinations and the corresponding sparseness of available data. In addition, standard classifiers cannot be directly applied to a MC problem for two reasons. First, most standard algorithms assume mutually exclusive class labels, and second, standard performance measures are not suitable for evaluation of classifiers in a MC setting.

A traditional approach to MC is to learn multiple independent binary classifiers to separate one class from the others and then to combine their outputs. However, in such a case the labels are treated as independent that can significantly reduce

N. Adams et al. (Eds.): IDA 2009, LNCS 5772, pp. 167–177, 2009.
© Springer-Verlag Berlin Heidelberg 2009

classification performance [1], [2]. Recently, several approaches [1-8] which can handle multi-label data directly have been specially developed for solving a MC task. Some of them are based on "black-box" machine learning techniques like Support Vector Machine (SVM) [2-4], k-Nearest Neighbor (kNN) classifier [5-7] or neural networks [8]. Alternatively, interpretable decision tree algorithms have been applied to MC in bioinformatics [9-11] where knowledge extraction from a trained classifier is preferable for testing the results by biologists in order to determine their plausibility. Two other generalizations of decision trees to MC are reported in [12] and [13].

In this paper, another type of interpretable multi-label classifiers is studied. These are multi-label extensions of two neuro-fuzzy models based on the Adaptive Resonance Theory (ART): Fuzzy ARTMAP (FAM) [14] and fuzzy ARAM [15] which allow for fuzzy rule extraction. Besides their interpretability, these networks possess also the advantage of fast learning i.e. learning only for one epoch, a valuable property of the ART family. The Multi-Label FAM (ML-FAM) and Multi-Label ARAM (ML-ARAM) algorithms are proposed and compared with their standard counterparts as well as with some other multi-label classifiers. To date, the has been no attempt to apply an ART network to MC, except for [16], where the authors, however, were interested in obtaining hierarchical relationships between output classes, not in solving a MC task.

The paper is organized as follows. Section 2 presents a brief description of the FAM and ARAM base algorithms pointing to the differences between them. In Section 3, their multi-label extensions are introduced. Section 4 describes experiments and compares obtained results with the results of other multi-label classifiers taking into account a large set of different performance measures. And finally, Section 5 concludes the paper.

2   FAM and ARAM Neural Networks

The reader is supposed to be familiar with the FAM and ARAM algorithms. Due to space constraints only the basic concepts are discussed below. A FAM system generally consists of two self-organizing two-layer Fuzzy ART modules, ART_a and ART_b, which process inputs A and targets B, respectively, by building prototype categories in their second layers F_2 (Fig. 1a). In classification tasks, the target vectors usually represent class labels, for example, in the binary form.

Each neuron of F_2 stores in its weight connections a prototype, i.e. a set of relevant features, describing a cluster of inputs in the feature space (inputs belonging to the same cluster share common characteristics). The F_2 fields are linked by the Map Field – an associative memory, which contains associations between ART_a and ART_b prototype categories. This enables FAM to learn mapping between input and target pairs during training.

Initially, all weight vectors are set to unity and prototype nodes are said to be uncommitted. Vectors A and B represent in the complement coded form [14] an input pattern X and its class label vector Y. After presentation of an input vector A to F_1

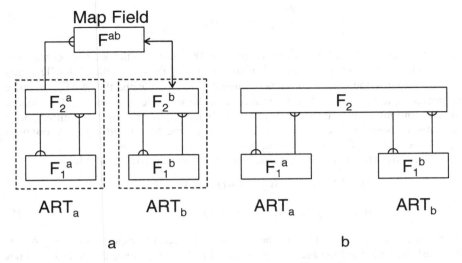

Fig. 1. FAM (a) and ARAM (b) neural networks

field of ART_a, the activation function T_j (1) is calculated for each node in F_2 and the winner is then chosen by the Winner-Take-All (WTA) rule (2):

$$T_j(A) = \frac{|A \wedge W_j|}{\alpha + |W_j|} \tag{1}$$

where "$\wedge$" denotes the fuzzy AND, element-wise min operator, and $\alpha > 0$ is called the choice parameter.

$$T_J = \max\{T_j : j = 1,...,N\} \tag{2}$$

The network category choice J must be confirmed by checking the match criterion:

$$\frac{|A \wedge W_J|}{|A|} \geq \rho_a \tag{3}$$

where $\rho_a \in [0,1]$ is the user-defined vigilance parameter which defines the minimum required similarity between an input vector and the prototype of the node it can be associated with. If (3) fails, the system inhibits the winning node J and enables another category to be selected. When no existing prototype provides the satisfactory similarity, then the network will chose a new (uncommitted) neuron. Thus, the network has a dynamically self-organizing structure within which the number of prototypes grows to adapt to the environment.

The operations at ART_b are identical to those at ART_a, except for checking the Map Field vigilance criterion (4) that tests the correctness of the association between the ART_a node J coding an input vector A and the corresponding ART_b node K coding a target vector B:

$$\frac{|\,U^{\,b} \wedge W_{J}^{\,ab}\,|}{|\,U^{\,b}\,|} \ge \rho_{ab} \tag{4}$$

where U^{b} denotes the ART_{b} output vector and W_{J}^{ab} denotes the weights connecting the Jth ART_{a} node and the Map Field. If the inequality fails, then the Match Tracking process initiates the choice of a new category by increasing the ART_{a} vigilance parameter ρ_{a} to the value slightly greater than the left-side term of (3). This search process continues until the input is either assigned to an existing (committed) node that satisfies both the ART_{a} match criterion (3) and the Map Field vigilance criterion (4) or to activation of an uncommitted neuron.

A successful end of search starts the learning process at ART_{a} and ART_{b}:

$$W_{J}^{(new)} = \beta(A \wedge W_{J}^{(old)}) + (1 - \beta)W_{J}^{(old)}$$

$$W_{J}^{(new)} = \beta(B \wedge W_{J}^{(old)}) + (1 - \beta)W_{J}^{(old)} \tag{5}$$

where $\beta \in [0,1]$ is the learning rate. The fast learning mode corresponds to setting $\beta = 1$. With fast learning, the Map Field weights are set to $U^{b} \wedge W_{J}^{ab}$ once the node J becomes committed and this association is permanent.

Although fuzzy ARAM may be tuned to be functionally equivalent to FAM, it significantly differs in the architecture (Fig. 1b). ARAM contains no Map Field because its F_{2} nodes are linked directly to F_{1} fields of both ART_{a} and ART_{b} modules. In its simplified version, ARAM first computes the activation function (1) and selects the winning node according to (2). Then the winner choice should be confirmed by the pair of match criteria simultaneously:

$$\frac{|\,A \wedge W_{J}^{a}\,|}{|\,A\,|} \ge \rho_{a} \qquad \frac{|\,B \wedge W_{J}^{b}\,|}{|\,B\,|} \ge \rho_{b} \cdot \tag{6}$$

If any of the inequalities is violated, reset occurs and the match tracking process can be started if needed. When each match criterion is satisfied in the respective module, resonance occurs and the learning process follows. During learning the selected node J learns to encode the input and target vectors by adjusting its weight vectors W_{J}^{a} and W_{J}^{b} similarly to (5).

3 ML-FAM and ML-ARAM

By encoding associations between input and target vectors FAM and ARAM are generally able to learn multi-label predictions in the case when the class label vector Y contains not a single label but a set of labels (usually represented as a binary vector). However, the direct use of FAM and ARAM for solving a MC task can be ineffective due to the WTA choice rule (2). It allows only the most highly activated category to be selected. Though justified for mutually exclusive classes, in the multi-label setting, this can lead to poor performance because only one set of labels can be predicted for an instance no matter how close to each other may be different label combinations. Thus, it would be advantageous to utilize distributed prototype activation during the classification stage in order to extract more information about dependencies and correlations between different label sets.

Distributed activation has been already shown to be effective in classification by ART networks [16], [17] and considered in default ARTMAP [18]. A better prediction can be made by joining the class information of those prototype categories at F_2 which are about equally activated. In MC, this can be achieved by combining the multi-label predictions of several most activated nodes. The proposed method is implemented in both ML-FAM and ML-ARAM as follows.

First, a set of N best categories with the largest activation values T_j is chosen according to the following rule: a category j is included in the set, if the relative difference $(T_{max}-T_j)/T_{max}$ lies below a predefined threshold t. Then the activations are normalized as

$$u_j^a = T_j \bigg/ \sum_{n=1}^{N} T_n .$$

(7)

The idea is to take into account only those categories which have a small relative decrease of activation in respect to the maximum value and to use the labels associated with them for obtaining the resulting multi-label prediction. This method differs from the CAM power rule [18] and is computationally less expensive.

The resulting distributed output pattern P is then a weighted sum of N individual predictions p_j

$$P = \sum_{n=1}^{N} u_j^a p_j$$

(8)

where $p_j = W_k^b$ with k such that $W_{jk}^{ab} = 1$ for FAM and $p_j = W_j^b$ for ARAM. Thus, P contains a score for each label which is proportional to the frequency of predicting this label among N best categories.

Since fast learning of ART-networks leads to varying performance with the input presentation order, voting across several networks trained with different orderings of a training set is usually utilized. It typically improves classification performance and reduces its variability. Voting also presents a useful option for averaging multi-label predictions. So, multiplying the number of output categories N in (8) by the number of voters V enables building the sum over all voters and producing a collective distributed output pattern P, which may be then used to determine the predicted classes.

Finally, a post processing filter method [16] outputs only those labels from P for which the following holds: if P_q signals corresponding to q output classes are arranged from largest to smallest, their P_q values lie up to the point of maximum decrease in the signal size from one class to the next. An advantage of this instance-based filtering is that a distributed prediction is thresholded by means of no a priori fixed parameter.

An addition modification is made in the ART_b weights which now count label frequencies for each ART_b category during learning. After commitment they are increased by 1 each time the corresponding label occurs. The weight matrix W^b then contains information about the frequency with which each label was coded by a node k.

It should be noted that this modification causes some changes in the ART_b activation function of FAM which is now computed as $T_j = |\boldsymbol{B} \wedge \boldsymbol{W}_j^b|$.

Another difference of ML-FAM and ML-ARAM as compared to the standard algorithms is that they do not use match tracking with raising vigilance because it leads to category proliferation [19]. The winner node is simply inhibited and a new search is started when the chosen ART_a category does not code the proper label set of a training instance.

It should be noted that despite a relatively large number of potentially adjustable parameters in the ART networks the most of them have only little impact on the classification performance and are usually set to their default values (α close to 0, $\beta = 1$ for fast learning, $V = 5$) [18]. Typically, only tuning of the vigilance parameter is essential for classification because the network size as well as its generalization ability depends on the value of ρ: with high values, the network tends to create a large number of small clusters to represent the underlying class distributions while with low ρ the clusters are large. Therefore one possible default setting accepted in the literature is $\rho = 0$ for the minimum network size. However this leads oft to relatively poor classification performance. An alternative default value of ρ is 0.8 which is usually set for achieving a compromise between the code compression and the classification performance. In the multi-label extensions, the threshold t plays also an important role. The lower it is, the fewer categories are taken into account during building a multi-label prediction. From extensive experiments on different datasets the value of 5% has emerged to be a good choice for this parameter.

4 Experimental Results

4.1 Datasets and Performance Measures

The effectiveness of the standard FAM and ARAM with the WTA choice rule was compared to that of the modified algorithms in the multi-label setting. The performance was evaluated on two datasets from different application areas. The first one was the well-known *Yeast* dataset [2] describing the 2417 genes of *Saccharomyces cerevisiae*. Each gene is characterized by 103 features derived from the micro-array expression data and phylogenetic profiles. 14 possible classes in the top level of the functional hierarchy are considered with the average value of 4.24 labels per gene. The large average number of labels associated with each instance makes this dataset difficult to classify. The second dataset *Emotions* [20] concerns classification of music into 6 mood classes (e.g. happy-pleased, quiet-still, sad-lonely etc.). It consists of 593 songs described by 72 rhythmic and timbre features. The average number of labels per song is 1.87.

Due to a more complex nature of MC applications, there exist a large set of possible evaluation criteria for multi-label classifiers as opposite to single-label ones. The most common example-based performance measures for MC which are also used in this paper include *Hamming Loss (HL)*, *Accuracy (A)* and *F-measure (F)* [21]. Given a set $S = \{(\mathbf{X}_1, \mathbf{Y}_1), \ldots, (\mathbf{X}_n, \mathbf{Y}_n)\}$ of n test examples where $\mathbf{Y}_i$ is the proper label set for an instance $\mathbf{X}_i$, let $\mathbf{Z}_i$ be the set of predicted labels for $\mathbf{X}_i$ and $\mathbf{L}$ the finite set of possible labels. Then *HL* counts prediction errors when a false label is predicted as well as missing errors when a true label is not predicted:

$$HL = \frac{1}{n} \sum_{i=1}^{n} \frac{|\mathbf{Y}_i \Delta \mathbf{Z}_i|}{|\mathbf{L}|} \tag{9}$$

where Δ denotes the symmetric difference of two sets and corresponds to the XOR operation in Boolean logic. The smaller is the *HL* value, the better is the MC performance.

Accuracy and *F-measure* are defined as follows:

$$A = \frac{1}{n} \sum_{i=1}^{n} \frac{|\mathbf{Y}_i \cap \mathbf{Z}_i|}{|\mathbf{Y}_i \cup \mathbf{Z}_i|} \tag{10}$$

$$F = \frac{1}{n} \sum_{i=1}^{n} \frac{2|\mathbf{Y}_i \cap \mathbf{Z}_i|}{|\mathbf{Z}_i| + |\mathbf{Y}_i|} . \tag{11}$$

The performance becomes perfect as they approach unity.

A limitation of these performance measures is that they are based on the predicted labels and therefore are influenced by the chosen threshold value. Alternatively, ranking-based evaluation criteria as, for example, *One-Error* (*OE*) can be used. Given a real-valued function f which assigns larger values to labels in $\mathbf{Y}_i$ than those not in $\mathbf{Y}_i$, the *OE* metric can be defined as:

$$OE = \frac{1}{n} \sum_{i=1}^{n} \delta\left([\arg \max f(\mathbf{x}_i, y)] \notin \mathbf{Y}_i\right) \tag{12}$$

where δ is a function that outputs 1 if its argument is true and 0 otherwise. The performance is perfect when *OE* equals to 0.

Ranking-based performance measures stem from the information retrieval field and are better suited for document classification. Additionally to *OE*, *Coverage* (*C*), *Ranking Loss* (*RL*), and *Average Precision* (*AP*) are often used [6]. *Ranking Loss* is defined as the average fraction of pairs of labels that are ordered incorrectly. *Coverage* evaluates how far we need, on average, to go down the list of labels in order to cover all the proper labels of the instance. The smaller is its value, the better is the performance. *Average Precision* evaluates the average fraction of labels ranked above a particular label y in Y which actually are in Y. If it is equal to 1, the performance is perfect.

Besides the above discussed example-based performance measures, label-based ones are also often utilized [21]. They are calculated using binary evaluation measures for each label: the counts for true positives (*tp*), true negatives (*tn*), false positives (*fp*), and false negatives (*fn*). The most popular of them are recall, precision and their harmonic mean *F1*-measure.

$$F1 = \frac{2tp}{2tp + fn + fp} \tag{13}$$

In this paper, micro-averaged precision and recall are used for computing the *F1*-measure. Micro-averaged precision corresponds to the proportion of predicted labels in the whole test set that are correct and recall to the proportion of labels that are correctly predicted. The closer the $F1$ value is to 1, the better is the performance.

4.2 Performance Comparison

4.2.1 Experiment Settings
In this paper, four ART-based algorithms FAM, ML-FAM, ARAM and ML-ARAM were tested in the following setting. The fast learning mode with $\beta = 1$ was used. The choice parameter α was set to 0.0001. The number of voters V was five. The parameter t of ML-FAM and ML-ARAM was chosen to be 0.05. In FAM and ARAM, predicted multi-labels were simply averaged over the voters as described in Section 3. The vigilance parameter of ART_a was set to 0.8, the vigilance parameter of ART_b was set to its default value of 1.

The experiments on both datasets were made by 10-fold cross-validation. For comparison, a multi-label k-nearest neighbor classifier (ML-kNN) was selected as a state-of-the-art high-performance classification algorithm. The implementations for all ART-networks and ML-kNN were written in MatLab. ML-kNN classifier was tested with the number of neighbors equal to 10.

4.2.2 *Yeast* Dataset
Table 1 reports the average classification performance of FAM, ML-FAM, ARAM, ML-ARAM, and ML-kNN evaluated by all measures discussed in Section 4.1. The best result on each metric is shown with bold typeface. The last line contains the number of times an algorithm achieves the best result.

As expected, FAM and ARAM were outperformed by their multi-label extensions as well as by ML-kNN on all nine performance measures. Due to match tracking, the network sizes of FAM and ARAM with the same vigilance parameters were twice as large as those of the modified algorithms (about 1635 categories against 778 on average). This is a clear sign of category proliferation. ML-ARAM performed slightly better than ML-FAM and both networks were superior to ML-kNN. Taking into account that the latter algorithm memorizes all 2175 training examples and therefore requires a longer training phase, it is clear that ML-FAM and ML-ARAM with much fewer prototype categories showed a significant advantage in the classification performance.

It is interesting to note that the relative performance differences reflected by various evaluation metrics vary from small to large, i.e. they have different discriminative power. In this experiment, the largest performance differences between FAM (ARAM) and multi-label classifiers were indicated by RL.

For comparison, Table 2 presents results obtained on the same dataset by other multi-label classifiers: BoosTexter, Alternating Decision Tree (ADT), Rank-SVM, and multi-label neural network BP-MLL as reported in [8]. For these experiments, only a reduced set of five performance measures including HL, OE, C, RL, and AP was used (RL values were not reported for ADT). The results show that the multi-label extensions of FAM and ARAM are superior to all other multi-label classifiers.

Table 1. Classification results (mean, std) of FAM, ARAM, ML-FAM, ML-ARAM, and ML-kNN on the *Yeast* data

	FAM		ARAM		ML-FAM		ML-ARAM		ML-kNN	
	mean	std	mean	std	mean	std	mean	std	mean	std
HL	0.198	0.009	0.197	0.011	0.194	0.006	**0.193**	0.007	**0.193**	0.007
A	0.515	0.015	0.519	0.022	**0.539**	0.016	0.536	0.019	0.519	0.019
F	0.610	0.015	0.614	0.022	**0.635**	0.018	0.633	0.019	0.622	0.018
OE	0.272	0.028	0.276	0.033	0.225	0.029	**0.218**	0.027	0.230	0.028
C	6.039	0.220	6.037	0.242	**5.967**	0.260	6.005	0.237	6.220	0.280
RL	0.267	0.017	0.270	0.021	0.168	0.015	**0.165**	0.014	**0.165**	0.015
AP	0.762	0.017	0.761	0.018	0.773	0.019	**0.779**	0.017	0.767	0.019
F1-mic	0.641	0.014	0.645	0.019	0.654	0.015	**0.662**	0.017	0.648	0.016
wins	0		0		3		6		2	

Table 2. Classification results (mean, std) of multi-label classifiers on the *Yeast* data from [8]

	ADT		BoosTexter		BP-MLL		Rank-SVM	
	mean	std	mean	std	mean	std	mean	std
HL	0.207	0.010	0.220	0.011	0.206	0.011	0.207	0.013
OE	0.244	0.035	0.278	0.034	0.233	0.034	0.243	0.039
C	6.390	0.203	6.550	0.243	6.421	0.237	7.090	0.503
RL	-	-	0.186	0.015	0.171	0.015	0.195	0.021
AP	0.744	0.025	0.737	0.022	0.756	0.021	0.749	0.026

4.2.3 *Emotions* Dataset

Table 3 summarizes the classification performance of FAM, ML-FAM, ARAM, ML-ARAM, and ML-kNN obtained on the *Emotions* dataset. While the average number of created categories in this experiment was about 355 for FAM and ARAM, it was about 218 for their multi-label extensions. On this dataset, the latter algorithms again outperformed their single-label counterparts on all performance measures. This time,

Table 3. Classification results (mean, std) of FAM, ARAM, ML-FAM, ML-ARAM, and ML-kNN on the *Emotions* data

	FAM		ARAM		ML-FAM		ML-ARAM		ML-kNN	
	mean	std	mean	std	mean	std	mean	std	mean	std
HL	0.204	0.015	0.204	0.017	**0.196**	0.016	0.200	0.015	0.206	0.023
A	0.563	0.036	0.559	0.046	**0.596**	0.031	0.577	0.034	0.527	0.035
F	0.652	0.035	0.643	0.042	**0.683**	0.030	0.666	0.036	0.614	0.041
OE	0.286	0.048	0.309	0.047	0.273	0.056	**0.271**	0.066	0.302	0.047
C	1.895	0.224	1.919	0.243	**1.703**	0.221	1.734	0.188	1.868	0.235
RL	0.263	0.032	0.268	0.033	**0.158**	0.022	0.160	0.020	0.175	0.023
AP	0.786	0.023	0.782	0.023	**0.809**	0.030	0.805	0.027	0.783	0.028
F1-mic	0.671	0.030	0.665	0.035	**0.697**	0.039	0.687	0.037	0.653	0.039
wins	0		0		7		2		0	

however, ML-FAM dominated ML-ARAM in almost all metrics. Surprisingly, results achieved on this dataset by the single-label ART-networks in terms of five measures (HL, A, F, OE, and $F1$) were better than those of ML-kNN.

5 Conclusions

In this paper, two multi-label extensions ML-FAM and ML-ARAM of ART-based fuzzy networks are presented. They can be successfully used for solving a MC task. The experimental results evaluated by a large set of performance measures show that the performance of both proposed extensions is superior to that of the single-label algorithms. They also outperform several state-of-the-art multi-label classifiers. This work should be continued by evaluating ML-FAM and ML-ARAM on other multi-label datasets in order to ensure a more reliable performance comparison.

References

1. Schapire, R.E., Singer, Y.: BoosTexter: a Boosting-based System for Text Categorization. Machine Learning 39, 135–168 (2000)
2. Elisseeff, A., Weston, J.: A Kernel Method for Multi-labelled Classification. In: Advances in Neural Information Processing Systems, pp. 681–687 (2001)
3. Godbole, S., Sarawagi, S.: Discriminative methods for multi-labeled classification. In: Dai, H., Srikant, R., Zhang, C. (eds.) PAKDD 2004. LNCS (LNAI), vol. 3056, pp. 22–30. Springer, Heidelberg (2004)
4. Boutell, M.R., Shen, X., Luo, J., Brown, C.: Learning Multi-label Semantic Scene Classification. Pattern Recognition 37, 1757–1771 (2004)
5. Zhang, M.-L., Zhou, Z.-H.: A k-Nearest Neighbor Based Algorithm for Multi-label Classification. In: International Conference on Granular Computing, pp. 718–721 (2005)
6. Zhang, M.-L., Zhou, Z.-H.: ML-kNN: A Lazy Learning Approach to Multi-label Learning. Pattern Recognition 40, 2038–3048 (2007)
7. Younes, Z., Abdallah, F., Denoeux, T.: Multi-label Classification Algorithm Derived from k-Nearest Neighbor Rule with Label Dependencies. In: 16th European Signal Processing Conference (2008)
8. Zhang, M.-L., Zhou, Z.-H.: Multi-label Neural Networks with Applications to Functional Genomics and Text Categorization. IEEE transactions on Knowledge and Data Engineering 18, 1338–1351 (2006)
9. Clare, A., King, R.D.: Knowledge Discovery in Multi-label Phenotype Data. In: 5th European Conference on Principles of Data Mining and Knowledge Discovery, pp. 42–53 (2001)
10. Blockeel, H., Bruynooghe, M., Džeroski, S., Ramon, J., Struyf, J.: Hierarchical Multi-classification. In: First International Workshop on Multi-Relational Data Mining, pp. 21–35 (2002)
11. Vens, C., Struyf, J., Schietgat, L., Džeroski, S., Blockeel, H.: Decision Trees for Hierarchical Multi-label Classification. Machine Learning 73, 185–214 (2008)
12. Suzuki, E., Gotoh, M., Choki, Y.: Bloomy Decision Tree for Multi-objective Classification. In: 5th European Conference on Principles of Data Mining and Knowledge Discovery, pp. 436–447 (2001)

13. Comité, F.D., Gilleron, R., Tommasi, M.: Learning Multi-label Alternating Decision Trees from Texts and Data. In: Perner, P., Rosenfeld, A. (eds.) MLDM 2003. LNCS, vol. 2734, pp. 251–274. Springer, Heidelberg (2003)
14. Carpenter, G., et al.: Fuzzy ARTMAP: A Neural Network Architecture for Incremental Supervised Learning of Analog Multidimensional Maps. IEEE transactions on Neural Networks 3, 698–713 (1992)
15. Tan, A.-H.: Adaptive Resonance Associative Map. Neural Networks 8, 437–446 (1995)
16. Carpenter, G., Martens, S., Ogas, O.: Self-organizing Information Fusion and Hierarchical Knowledge Discovery: a New Framework Using ARTMAP Neural Networks. Neural Networks 18, 287–295 (2005)
17. Carpenter, G., Ross, W.D.: ART-EMAP: A Neural Network Architecture for Object Recognition by Evidence Accumulation. IEEE Transactions on Neural Networks 6, 805–818 (1995)
18. Carpenter, G.A.: Default ARTMAP. In: International Joint Conference on Neural Networks (IJCNN), pp. 1396–1401 (2003)
19. Sapojnikova, E.: ART-based Fuzzy Classifiers: ART Fuzzy Networks for Automatic Classification. Cuvillier Verlag, Goettingen (2004)
20. Trohidis, K., Tsoumakas, G., Kalliris, G., Vlahavas, I.: Multilabel Classification of Music into Emotions. In: 9th International Conference on Music Information Retrieval, ISMIR (2008)
21. Tsoumakas, G., Vlahavas, I.P.: Random k-labelsets: An ensemble method for multilabel classification. In: Kok, J.N., Koronacki, J., Lopez de Mantaras, R., Matwin, S., Mladenič, D., Skowron, A. (eds.) ECML 2007. LNCS (LNAI), vol. 4701, pp. 406–417. Springer, Heidelberg (2007)

Two-Way Grouping by One-Way Topic Models

Eerika Savia, Kai Puolamäki, and Samuel Kaski

Helsinki Institute for Information Technology HIIT
Department of Information and Computer Science
Helsinki University of Technology
P.O. Box 5400, FI-02015 TKK, Finland
{forename.surname}@tkk.fi

Abstract. We tackle the problem of new users or documents in collaborative filtering. Generalization over users by grouping them into user groups is beneficial when a rating is to be predicted for a relatively new document having only few observed ratings. The same applies for documents in the case of new users. We have shown earlier that if there are both new users and new documents, two-way generalization becomes necessary, and introduced a probabilistic Two-Way Model for the task. The task of finding a two-way grouping is a non-trivial combinatorial problem, which makes it computationally difficult. We suggest approximating the Two-Way Model with two URP models; one that groups users and one that groups documents. Their two predictions are combined using a product of experts model. This combination of two one-way models achieves even better prediction performance than the original Two-Way Model.

1 Introduction

This paper considers models for the task of predicting relevance values for user–item pairs based on a set of observed ratings of users for the items. In particular, we concentrate on the task of predicting relevance when very few ratings are known for each user or item.[1]

In so-called collaborative filtering methods the predictions are based on the opinions of similar-minded users. Collaborative filtering is needed when the task is to make personalized predictions but there is not enough data available for each user individually. The early collaborative filtering methods were memory-based (see, e.g., [1,2]). Model-based approaches are justified by the poor scaling of the memory-based techniques. Recent work includes probabilistic and information-theoretic models, see for instance [3,4,5,6].

A family of models most related to our work are the latent topic models, which have been successfully used in document modeling but also in collaborative filtering [7,8,9,10,11,12,13,14,15,16,17]. The closest related models include probabilistic Latent Semantic Analysis (pLSA; [3]), Latent Dirichlet Allocation

---

[1] The models we discuss are generally applicable, but since our prototype application area has been information retrieval we will refer to the items as documents.

N. Adams et al. (Eds.): IDA 2009, LNCS 5772, pp. 178–189, 2009.
© Springer-Verlag Berlin Heidelberg 2009

(LDA; [18,19]), and User Rating Profile model (URP; [20]), which all assume a one-way grouping. In addition, there is a two-way grouping model, called Flexible Mixture Model (FMM; [21]). We have discussed the main differences between our Two-Way Model and these related models in [22].

1.1   Cold-Start Problem

Since a collaborative filtering system has to rely on the past experiences of the users, it will have problems when assessing new documents that have not yet been seen by most of the users. Making the collaborative filtering scheme item-based, that is, grouping items or documents instead of users, would in turn imply the problem where new users that have only few ratings will get poor predictions. This problem of unseen or almost unseen users and documents is generally referred to as the *cold-start problem* in recommender system literature, see for instance [23]. The Two-Way Model was proposed to tackle this problem of either new users or new documents [22,24].

1.2   Approximating Two-Way Model with Two One-Way Models

It has been shown for hard biclustering of binary data matrices, that clustering the marginals independently to produce a check-board-like biclustering is guaranteed to achieve fairly good results compared to the NP-hard optimal solution. An approximation ratio for the crossing of two one-way clusterings has been proven [25,26]. Inspired by this theoretical guarantee, we suggest approximating the Two-Way Model with two User Rating Profile models (URP, [20]); one that groups users and one that groups documents. The combination of the two Gibbs-sampled probabilistic predictions is made using a product of experts model [27].

We have followed the experimental setups of our earlier study [22] in order to be able to compare the results in a straightforward manner. We briefly describe the experimental scenarios, the performance measures and the baseline models in Sect. 3. In Sect. 4.1 we demonstrate with clearly clustered toy data how the product of two URP models improves the relevance predictions of the corresponding one-way models. Finally, in Sect. 4.2 we show in a real-world case study from our earlier paper that the proposed method works as expected also in practice.

We expected the proposed method to have the advantage of giving better predictions than the individual one-way models with the computational complexity of the one-way model. The one-way grouping models are faster and more reliable in their convergence than the Two-Way Model, basically because of the difference in the intrinsic complexity of the tasks they are solving.

2   Method

Originally, User Rating Profile model was suggested to be estimated by variational approximation (variational URP, [20]), but we have introduced also Gibbs-sampled variants of the model in [22,24] (Gibbs URP and Gibbs URP-GEN).

The difference between a one-way model and the Two-Way Model is whether to cluster only users (documents) or to cluster both users and documents. Another difference between URP and the Two-Way Model is whether the users and documents are assumed to be generated by the model or treated as covariates of the model. In our earlier study [22] it was found that unless the data marginals are especially misleading about the full data, it is always useful to design the model to be fully generative, in contrast to seeing users and documents as given covariates of the model. Therefore, we have only included the generative variants of Gibbs URP models is this study (Gibbs URP-GEN).

2.1   One-Way Grouping Models

In Fig. 1 we show graphical representations of the generative Gibbs URP model introduced in [22] (User Gibbs URP-GEN), and the corresponding document-grouping variant (Doc Gibbs URP-GEN). They are the one-way grouping models used as the basis of our suggested method. Our main notations are summarized in Table 1.

2.2   Two-Way Grouping Model

In Fig. 2 we show a graphical representation of the Two-Way Model that our suggested method approximates. The Two-Way Model generalizes the generative user-grouping URP by grouping both users and documents. It has been shown to predict relevance more accurately than one-way models when the target consists of both new documents and new users. The reason is that generalization over documents becomes beneficial for new documents and at the same time generalization over users is needed for new users. Finally, Table 2 summarizes the differences between the models.

2.3   Approximation of Two-Way Model by Product of Experts

We propose a model where we estimate predictive Bernoulli distributions separately with user-based URP and document-based URP and combine their results with a product of experts model [27]. To be exact, we took the product of the Bernoulli relevance probabilities given by the user-based URP ($P_U(r = 1|u, d)$) and the document-based URP ($P_D(r = 1|u, d)$) and normalized the product distributions, as follows:

$$P_{PoE}(r = 1|u, d) = \frac{P_U(r = 1|u, d) \, P_D(r = 1|u, d)}{\sum_{r=0,1} P_U(r|u, d) \, P_D(r|u, d)} \quad . \tag{1}$$

2.4   Baseline Models

We compared our results to two simple baseline models. These models mainly serve as an estimate of the lower bound of performance by making an assumption that the data comes from one cluster only. The *Document Frequency Model* does not take into account differences between users or user groups at all. It simply

Table 1. Notation

SYMBOL	DESCRIPTION
u	user index
d	document index
r	binary relevance (relevant $= 1$, irrelevant $= 0$)
u^*	user group index (attitude in URP)
d^*	document cluster index
N_U	number of users
N_D	number of documents
N	number of triplets (u, d, r)
K_U	number of user groups
K_D	number of document clusters

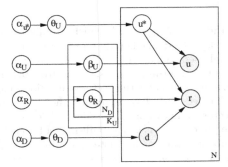
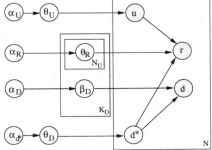

(a) User Gibbs URP-GEN groups only users and assumes that the relevance depends solely on the user group and the document.

(b) Doc Gibbs URP-GEN groups only documents and assumes that the relevance depends solely on the document cluster and the user.

Fig. 1. Graphical model representations of the generative Gibbs URP models with user grouping (User Gibbs URP-GEN) and with document grouping (Doc Gibbs URP-GEN). The grey circles indicate observed values. The boxes are "plates" representing replicates; the value in a corner of each plate is the number of replicates. The rightmost plate represents the repeated choice of N (user, document, rating) triplets. The plate labeled with K_U (or K_D) represents the different user groups (or document clusters), and β_U (or β_D) denotes the vector of multinomial parameters for each user group (or document cluster). The plate labeled with N_D (or N_U) represents the documents (or users). In the intersection of these plates there is a Bernoulli-model for each of the $K_U \times N_D$ (or $K_D \times N_U$) combinations of user group and document (or document cluster and user). Since α_D and θ_D (or α_U and θ_U) are conditionally independent of all other parameters given document d (or user u), they have no effect on the predictions of relevance $P(r \mid u, d)$ in these models. They only describe how documents d (or users u) are assumed to be generated. A table listing distributions of all the random variables can be found in the Appendix.

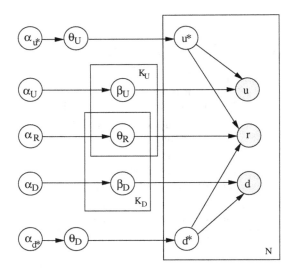

Fig. 2. Graphical model representation of the Two-Way Model, which groups both users and documents and assumes that the relevance depends only on the user group and the document cluster instead of individual users/documents. The rightmost plate represents the repeated choice of N (user, document, rating) triplets. The plate labeled with K_U represents the different user groups, and β_U denotes the vector of multinomial parameters for each user group. The plate labeled with K_D represents the different document clusters, and β_D denotes the vector of multinomial parameters for each document cluster. In the intersection of these plates there is a Bernoulli-model for each of the $K_U \times K_D$ combinations of user group and document cluster. A table listing distributions of all the random variables can be found in the Appendix.

Table 2. Summary of the models (**u**=user, **d**=document). The column "**Gibbs**" indicates which of the models are estimates by Gibbs sampling, in contrast to variational approximation. Prefix "2-way" stands for combination of two one-way models.

Model Abbreviation	Generates u,d	Gibbs	Groups u	Groups d
Two-Way Model	•	•	•	•
2-way Gibbs URP-GEN	•	•	•	•
2-way Gibbs URP	–	•	•	•
2-way Variational URP	–	–	•	•
1-way User Gibbs URP-GEN	•	•	•	–
1-way User Gibbs URP	–	•	•	–
1-way User Var URP	–	–	•	–
1-way Doc Gibbs URP-GEN	•	•	–	•
1-way Doc Gibbs URP	–	•	–	•
1-way Doc Var URP	–	–	–	•

models the probability of a document being relevant as the frequency of $r = 1$ in the training data for the document:

$$P(r = 1 \mid d) = \frac{\sum_u \#(u, d, r = 1)}{\sum_{u,r} \#(u, d, r)} \quad . \tag{2}$$

The *User Frequency Model*, on the other hand, does not take into account differences between documents or document groups. It is the analogue of Document Frequency Model, where the roles of users and documents have been interchanged.

3 Experiments

3.1 Experimental Scenarios

In this section we describe the different types of experimental scenarios that were studied with both data sets. The training and test sets were taken from the earlier study [22]. The scenarios have various levels of difficulty for models that group only users, only documents, or that group both.

- **Only "New" Documents.** This scenario had been constructed to correspond to prediction of relevances for new documents in information retrieval. It had been taken care that each of the randomly selected test documents had only 3 ratings in the training data. The rest of the ratings for these documents had been left to the test set. For the rest of the documents, all the ratings were included in the training set. Hence, the models were able to use "older" documents (for which users' opinions are already known) for training the user groups and document clusters. This scenario favors models that cluster documents.
- **Only "New" Users.** The experimental setting for new users had been constructed in exactly the same way as the setting for new documents but with the roles of users and documents reversed. This scenario favors models that cluster users.
- **Either User or Document is "New".** In an even more general scenario either the users or the documents can be "new." In this setting the test set consisted of user-document pairs where either the user is "new" and the document is "old" or vice versa. This scenario brings out the need for two-way generalization.
- **Both User and Document are "New".** In this setting all the users and documents appearing in the test set were "new," having only 3 ratings in the training set. This case is similar to the previous setting but much harder, even for the two-way grouping models.

3.2 Measures of Performance

For all the models, we used log-likelihood of the test data set as a measure of performance, written in the form of perplexity,

$$\text{perplexity} = e^{-\frac{\mathcal{L}}{N}} \ , \text{ where } \mathcal{L} = \sum_{i=1}^{N} \log P(r_i \mid u_i, d_i, \mathcal{D}) \quad . \tag{3}$$

Here $\mathcal{D}$ denotes the training set data, and N is the size of the test set. Gibbs sampling gives an estimate for the table of relevance probabilities over all (u, d) pairs, $P(r \mid u, d, \mathcal{D})$, from which the likelihood of each test pair (u_i, d_i) can be estimated as $P(r_i \mid u_i, d_i, \mathcal{D})$.[2]

We further computed the accuracy, that is, the fraction of the triplets in the test data set for which the prediction was correct. We took the predicted relevance to be $\arg\max_{r \in \{0,1\}} P(r \mid u, d, \mathcal{D})$, where $P(r \mid u, d, \mathcal{D})$ is the probability of relevance given by the model. Statistical significance was tested with the Wilcoxon signed rank test.

3.3   Demonstration with Artificial Data

The artificial data sets were taken from the earlier study [22]. The experimental setting is described in detail in the technical report [28]. In brief, the data was designed such that it contained bicluster structure with $K_U = K_D = 3$. There were 10 artificial data sets of size 18,000, that all followed the same bicluster structure.

All the models were trained with the known true numbers of clusters. For each of the 10 data sets the models were trained with a training set and tested with a separate test set, and the final result was the mean of the 10 test set perplexities.

The generative Gibbs URP models were combined as a product of experts model. According to our earlier studies, the variational URP generally seems to produce extreme predictions, near either 0 or 1. Therefore, the variational URP models (User Var URP and Doc Var URP) were combined as a hard biclustering model, as follows. The MAP estimates for cluster belongings from the distributions of the one-way variational URP models were used to divide all the users and documents into bins to produce a hard check-board-like biclustering. In each bicluster the $P(r = 1 | u, d)$ was set to the mean of the training data points that lay in the bicluster.

3.4   Experiments with Parliament Data

We selected the cluster numbers using a validation set described in [28]. The validated cluster numbers (Two-Way Model $K_U = 4$ and $K_D = 2$, User Gibbs URP-GEN $K_U = 2$, Doc Gibbs URP-GEN $K_D = 2$) were used in all experimental scenarios. The choices from which the cluster numbers were selected were $K_U \in \{1, 2, 3, 4, 5, 10, 20, 50\}$ for the user groups and $K_D \in \{1, 2, 3, 4, 5, 10, 20\}$ for the document clusters.

4   Results

4.1   Results of Experiments with Artificial Data

The results of the experiment with artificial data are shown in Table 3. The proposed product of two generative Gibbs URP models outperformed even the

---

[2] Theoretically, perplexity can grow without a limit if the model predicts zero probability for some element in the test data set, so in practice, we clipped the probabilities to the range $[e^{-10}, 1]$.

Two-Way Model in all the scenarios, being the best in all but the "both new" case, where the hard clustering of MAP estimates of variational URP models was the best. The hard biclustering model worked very well for the variational URP (See Table 3), in contrast to the product of experts -combination, which did not perform well for variational URP[3]. The prediction accuracy of the best model varied between 83–84%, while the prediction accuracy of the best baseline model varied between 50–52%. The full results with all the accuracy values can be found in the technical report [28].

Table 3. Perplexity of the various models in experiments with artificial data. In each column, the best model (underlined) differs statistically significantly from the second-best one (P-value $\leq$ 0.01). Small perplexity is better; 2.0 corresponds to binary random guessing and 1.0 to perfect prediction.

Method	New Doc	New User	Either New	Both New
Two-Way Model	1.52	1.54	1.53	1.70
2-way Gibbs URP-GEN	1.46	1.47	1.45	1.70
2-way Var URP	1.55	1.57	1.54	1.52
User Gibbs URP-GEN	1.68	1.57	1.62	1.83
User Var URP	7.03	2.07	3.45	9.27
Doc Gibbs URP-GEN	1.56	1.69	1.62	1.81
Doc Var URP	1.86	5.99	3.08	6.90
User Freq.	2.02	5.65	3.25	4.99
Document Freq.	5.29	2.01	3.21	5.92

4.2 Results of Experiments with Parliament Data

The product of two generative Gibbs URP models outperformed even the Two-Way Model in all the scenarios, being the best in all cases (see Table 4). The prediction accuracy of the best model varied between 93–97%, while the prediction accuracy of the best baseline model varied between 64–71%. The full results with all the accuracy values can be found in the technical report [28].

5 Discussion

We have tackled the problem of new users or documents in collaborative filtering. We have shown in our previous work that if there are both new users and new documents, two-way generalization becomes necessary, and introduced a probabilistic Two-Way Model for the task in [22].

In this paper we suggest an approximation for the Two-Way Model with two User Rating Profile models — one that groups users and one that groups

---

[3] We only show the performance of Variational URP for the artificial data since our implementation is too inefficient for larger data sets.

Table 4. Parliament Data. Comparison between the models by perplexity over the test set. In each column, the best model (underlined) differs statistically significantly from the second-best one (P-value $\leq$ 0.01). Small perplexity is better; 2.0 corresponds to binary random guessing and 1.0 to perfect prediction.

Method	New Doc	New User	Either New	Both New
Two-Way Model	1.37	1.40	1.38	1.62
2-way Gibbs URP-GEN	<u>1.19</u>	<u>1.22</u>	<u>1.20</u>	<u>1.45</u>
User Gibbs URP-GEN	1.47	1.34	1.41	1.64
Doc Gibbs URP-GEN	**1.34**	1.54	1.43	1.68
User Freq.	2.00	5.68	3.32	4.78
Document Freq.	5.36	1.76	3.12	5.85

documents — which are combined as a product of experts (PoE). We show with two data sets from the earlier study [22], that the PoE model achieves the performance level of the more principled Two-Way Model and even outperforms it.

The task of finding such a two-way grouping that best predicts the relevance is a difficult combinatorial problem, which makes convergence of the sampling hard to achieve. This work was motivated by the finding that hard biclustering of binary data can be approximated using two one-way clusterings with a proven approximation ratio.

The main advantage of the proposed method, compared to earlier works, is the ability to make at least as good predictions as the Two-Way Model but with the computational complexity of the one-way model. We assume that the reason why the product of experts combination outperformed the Two-Way Model lies in the less reliable and slower convergence of the Two-Way Model compared to the one-way grouping models. This is basically due to the difference in the intrinsic complexity of the tasks they are solving.

Acknowledgments

This work was supported in part Network of Excellence of the EC. This publication only reflects the authors' views. Access rights to the data sets and are restricted due to other commitments. This work was done in the Adaptive Informatics Research Centre, a Centre of Excellence of the Academy of Finland.

References

1. Konstan, J., Miller, B., Maltz, D., Herlocker, J.: GroupLens: Applying collaborative filtering to usenet news. Communications of the ACM 40(3), 77–87 (1997)
2. Shardanand, U., Maes, P.: Social information filtering: Algorithms for automating 'word of mouth'. In: Proceedings of the ACM CHI 1995 Human Factors in Computing Systems Conference, pp. 210–217 (1995)

3. Hofmann, T.: Latent semantic models for collaborative filtering. ACM Trans. Inf. Syst. 22(1), 89–115 (2004)
4. Jin, R., Si, L.: A Bayesian approach towards active learning for collaborative filtering. In: Proceedings of the Twentieth Conference on Uncertainty in Artificial Intelligence, UAI 2004, pp. 278–285. AUAI Press (2004)
5. Wettig, H., Lahtinen, J., Lepola, T., Myllymäki, P., Tirri, H.: Bayesian analysis of online newspaper log data. In: Proc. of the 2003 Symposium on Applications and the Internet Workshops (SAINT 2003), pp. 282–287. IEEE Computer Society, Los Alamitos (2003)
6. Zitnick, C., Kanade, T.: Maximum entropy for collaborative filtering. In: Proceedings of the 20th Conference on Uncertainty in Artificial Intelligence, UAI 2004, pp. 636–643. AUAI Press (2004)
7. Blei, D.M., Jordan, M.I.: Modeling annotated data. In: Proceedings of the 26th Annual International ACM SIGIR Conference on Research and Development in Information Retrieval, pp. 127–134. ACM Press, New York (2003)
8. Buntine, W., Jakulin, A.: Discrete component analysis. In: Saunders, C., Grobelnik, M., Gunn, S., Shawe-Taylor, J. (eds.) SLSFS 2005. LNCS, vol. 3940, pp. 1–33. Springer, Heidelberg (2006)
9. Erosheva, E., Fienberg, S., Lafferty, J.: Mixed membership models of scientific publications. Proc. of the National Academy of Sciences 101, 5220–5227 (2004)
10. Keller, M., Bengio, S.: Theme topic mixture model: A graphical model for document representation. In: PASCAL Workshop on Text Mining and Understanding (2004)
11. Marlin, B., Zemel, R.S.: The multiple multiplicative factor model for collaborative filtering. In: ICML 2004: Proceedings of the 21th International Conference on Machine Learning, p. 73. ACM Press, New York (2004)
12. McCallum, A., Corrada-Emmanuel, A., Wang, X.: The author-recipient-topic model for topic and role discovery in social networks: Experiments with Enron and Academic Email. Technical report, University of Massachusetts (2004)
13. Popescul, A., Ungar, L., Pennock, D., Lawrence, S.: Probabilistic models for unified collaborative and content-based recommendation in sparse-data environments. In: Proceedings of UAI 2001, pp. 437–444. Morgan Kaufmann, San Francisco (2001)
14. Pritchard, J.K., Stephens, M., Donnelly, P.: Inference of population structure using multilocus genotype data. Genetics 155, 945–959 (2000)
15. Rosen-Zvi, M., Griffiths, T., Steyvers, M., Smyth, P.: The author-topic model for authors and documents. In: Proceedings of the 20th Conference on Uncertainty in Artificial Intelligence, UAI 2004, pp. 487–494. AUAI Press (2004)
16. Yu, K., Yu, S., Tresp, V.: Dirichlet enhanced latent semantic analysis. In: Cowell, R.G., Ghahramani, Z. (eds.) Proceedings of the Tenth International Workshop on Artificial Intelligence and Statistics, AISTATS 2005, pp. 437–444. Society for Artificial Intelligence and Statistics (2005)
17. Yu, S., Yu, K., Tresp, V., Kriegel, H.-P.: A probabilistic clustering-projection model for discrete data. In: Jorge, A.M., Torgo, L., Brazdil, P.B., Camacho, R., Gama, J. (eds.) PKDD 2005. LNCS (LNAI), vol. 3721, pp. 417–428. Springer, Heidelberg (2005)
18. Blei, D., Ng, A.Y., Jordan, M.I.: Latent Dirichlet allocation. Journal of Machine Learning Research 3, 993–1022 (2003)
19. Buntine, W.: Variational extensions to EM and multinomial PCA. In: Elomaa, T., Mannila, H., Toivonen, H. (eds.) ECML 2002. LNCS (LNAI), vol. 2430, pp. 23–34. Springer, Heidelberg (2002)

20. Marlin, B.: Modeling user rating profiles for collaborative filtering. In: Advances in Neural Information Processing Systems 16, pp. 627–634. MIT Press, Cambridge (2004)
21. Si, L., Jin, R.: Flexible mixture model for collaborative filtering. In: Fawcett, T., Mishra, N. (eds.) Proceedings of the Twentieth International Conference on Machine Learning, ICML 2003, pp. 704–711. AAAI Press, Menlo Park (2003)
22. Savia, E., Puolamäki, K., Kaski, S.: Latent grouping models for user preference prediction. Machine Learning 74(1), 75–109 (2009)
23. Lam, X.N., Vu, T., Le, T.D., Duong, A.D.: Addressing cold-start problem in recommendation systems. In: ICUIMC 2008: Proceedings of the 2nd international conference on Ubiquitous information management and communication, pp. 208–211. ACM, New York (2008)
24. Savia, E., Puolamäki, K., Sinkkonen, J., Kaski, S.: Two-way latent grouping model for user preference prediction. In: Bacchus, F., Jaakkola, T. (eds.) Uncertainty in Artificial Intelligence 21, pp. 518–525. AUAI Press, Corvallis (2005)
25. Puolamäki, K., Hanhijärvi, S., Garriga, G.C.: An approximation ratio for biclustering. Information Processing Letters 108, 45–49 (2008)
26. Anagnostopoulos, A., Dasgupta, A., Kumar, R.: Approximation algorithms for co-clustering. In: Proceedings of the Twenty-Seventh ACM SIGMOD-SIGACT-SIGART symposium on Principles of database systems, pp. 201–210. ACM, New York (2008)
27. Hinton, G.E.: Training Products of Experts by Minimizing Contrastive Divergence. Neural Computation 14(8), 1771–1800 (2002)
28. Savia, E., Puolamäki, K., Kaski, S.: On two-way grouping by one-way topic models. Technical Report TKK-ICS-R15, Helsinki University of Technology, Department of Information and Computer Science, Espoo, Finland (May 2009)

A   Distributions in the Models

Table 5. Summary of the distributions in User Gibbs URP-GEN

SYMBOL	DESCRIPTION
$\beta_U(u^*)$	Vector of multinomial parameters defining the probabilities of certain user group u^* to contain each user
θ_U	Multinomial probabilities of user groups u^* to occur
θ_D	Multinomial probabilities of documents d to occur (needed only for the generative process)
$\theta_R(u^*, d)$	Vector of Bernoulli parameters defining the probabilities of certain user group u^* to consider document d relevant or irrelevant
α_U	Dirichlet prior parameters for all β_U
α_{u^*}	Dirichlet prior parameters for θ_U
α_D	Dirichlet prior parameters for θ_D (needed only for the generative process)
α_R	Dirichlet prior parameters for all θ_R

Table 6. Summary of the distributions in the Two-Way Model

SYMBOL	DESCRIPTION
θ_U	Multinomial probabilities of user groups u^* to occur
$\beta_U(u^*)$	Vector of multinomial parameters defining the probabilities of certain user group u^* to contain each user
θ_D	Multinomial probabilities of document clusters d^* to occur
$\beta_D(d^*)$	Vector of multinomial parameters defining the probabilities of certain document cluster d^* to contain each document
$\theta_R(u^*, d^*)$	Vector of Bernoulli parameters defining the probabilities of certain user group u^* to consider document cluster d^* relevant or irrelevant
α_U	Dirichlet prior parameters for all β_U
α_{u^*}	Dirichlet prior parameters for θ_U
α_D	Dirichlet prior parameters for all β_D
α_{d^*}	Dirichlet prior parameters for θ_D
α_R	Dirichlet prior parameters for all θ_R

Selecting and Weighting Data for Building Consensus Gene Regulatory Networks

Emma Steele and Allan Tucker

School of Information Systems Computing and Maths,
Brunel University, Uxbridge UB8 3PH, UK
{emma.steele,allan.tucker}@brunel.ac.uk

Abstract. Microarrays are the major source of data for gene expression activity, allowing the expression of thousands of genes to be measured simultaneously. Gene regulatory networks (GRNs) describe how the expression level of genes affect the expression of the other genes. Modelling GRNs from expression data is a topic of great interest in current bioinformatics research. Previously, we took advantage of publicly available gene expression datasets generated by similar biological studies by drawing together a richer and/or broader collection of data in order to produce GRN models that are more robust, have greater confidence and place less reliance on a single dataset. In this paper a new approach, Weighted Consensus Bayesian Networks, introduces the use of weights in order to place more influence on certain input networks or remove the least reliable networks from the input with encouraging results on both synthetic data and real world yeast microarray datasets.

Keywords: Consensus, Microarray, Networks, Weighting.

1 Introduction

Microarrays are the major source of data for gene expression activity, allowing the expression of thousands of genes to be measured simultaneously. Gene regulatory networks (GRNs) describe how the expression level of genes affect the expression of the other genes. Modelling GRNs from expression data is a topic of great interest in current bioinformatics research [12,5]. An ongoing issue concerns the problem of data quality and the variation of GRNs generated from microarray studies in different laboratories. One potential solution to this problem lies in the integration of multiple datasets into a single unifying GRN. Previously, we took advantage of publicly available gene expression datasets generated by similar biological studies by drawing together a richer and/or broader collection of data in order to produce GRN models that are more robust, have greater confidence and place less reliance on a single dataset [14]. This was achieved by using a method we called Consensus Bayesian Networks (CBNs) where consistencies across a set of network-edges learnt from all, or a certain proportion of the input datasets are then identified.

N. Adams et al. (Eds.): IDA 2009, LNCS 5772, pp. 190–201, 2009.
© Springer-Verlag Berlin Heidelberg 2009

In this paper we extend the CBN approach in order to address the fact that the reliability of networks varies across the datasets from which they are generated. A generalised consensus approach is presented, which we call Weighted Consensus Bayesian Networks, that can act on input networks with statistical confidences attached to each edge (generated using bootstrapping as in [14]), The new approach introduces the use of weights in order to place more influence on certain input networks or remove the least reliable networks from the input.

Whilst comparing and combining microarray expression datasets is a popular topic of research in bioinformatics [9,2], Wang et al.[10] are the first (to our knowledge) to address the issue with regards to modelling GRNs that combines the models for each dataset into an overall, consistent solution. In [14], we introduced CBNs for building consensus networks. However, these did not make use of weights to bias the input data as we do in this paper. Using weights to vary the influence of each dataset when learning or modelling from multiple data sources is obviously not new, most previous research has focused on combining classifiers, whereas we are concerned with learning and combining network structures. In particular, the idea of combining models has some similarities to ensemble learning, such as boosting [13], which aims to combine several weak classifiers into one strong classifier. In general the weak classifiers are combined according to some weighting that is related to their accuracy. Similarly, Bayesian Model Averaging (BMA) [6] is a technique that calculates a (weighted) average over the posterior distributions of a set of potential models. There has been research in using BMA with Bayesian networks for classification and prediction. However, BMA is not model combination. Instead it is designed to address uncertainty in model selection given a particular dataset. Both types of technique are designed to improve classification or prediction for a model across a single dataset. By contrast, we are concerned with combining models generated from multiple datasets, which have their own biases and levels of noise. This means that we may have a high quality dataset requiring a high weight, and a low quality dataset that requires a low weight. Ensemble learning and BMA have not been designed to deal with this type of input.

In the next section we describe the improvements to the original CBN algorithm before describing the experiments and the datasets used to test them in Section 3. In Section 4 we discuss the implications of the results and potential future work.

2   Methods

This section presents the Consensus Bayesian Network (CBN) approach before introducing the Weighted Consensus Bootstrapped Networks (WCBNs) that can act on input networks with statistical confidences attached to each edge. These confidences are generated using a bootstrapping approach on each input dataset. We explain how weighting can be incorporated into the method to allow each input network to have a different influence on the final consensus network.

Consensus Bayesian Networks: The CBN approach involves learning a network from each dataset using bootstrapping so that each edge contains a confidence measure. For this paper, we use a score and-search based method incorporating simulated annealing and Bayes Information Criterion. Uniform priors are assumed and the bootstrap involved repeatedly learning networks from each resampled dataset 100 times in order to calculate confidences in links between nodes (based upon the number of times they appeared in the different bootstraps). Each network with its confidences are then used as inputs to the consensus algorithm which selects links based upon a threshold. The threshold determines at what level of confidence the links appear in the final consensus network and the whole process is summarised in Figure 1 for a number of different thresholds showing how the process relates to graph union when the threshold is low and intersection when it is high. Undirected edges represent edges that have appeared in either direction in the input networks.

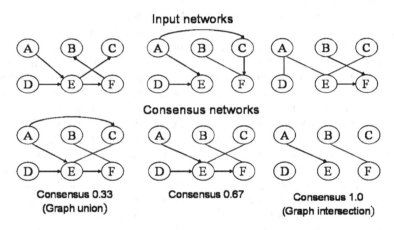

Fig. 1. The basic Consensus Bayesian Network approach for different consensus thresholds

Weighted Consensus Bootstrapped Networks: The WCBNs approach also takes as input a set of bootstrapped networks, from each input dataset, with a confidence attached to each edge and a consensus threshold t. As output it returns a consensus network (also with a confidence attached to each edge). In the input bootstrapped networks, the statistical confidence attached to each edge is referred to as the *input confidence*. In the consensus network we refer to the confidence attached to each edge as the *consensus confidence*. The WCBN approach updates the CBNs approach in that input networks with higher weights have more influence over each edge's consensus confidence than input networks with lower weights. In order to achieve this it exploits *predictive accuracy* for calculating weights in order to bias the effect of each input dataset: A dataset that is used to build a network model that can predict node values with high accuracy on other independent datasets can be said to be better at generalising to other data. This is a fairly intuitive measure as it shows that the network does not overfit the dataset from which it was generated, as it can perform well on other datasets.

To calculate prediction-based weights for a set of input datasets, the following method is used: for each dataset a network is learnt and the prediction accuracy is calculated for every node, over a random sample of observations equally distributed across the other datasets. Then a median prediction accuracy is calculated for each network, derived across all nodes (the use of the median is to account for outliers, nodes that perform unusually badly or well). We now explore two methods for using this weighting information: Prediction-based network weighting and Prediction-based network selection, before describing the WCBN algorithm in detail.

Prediction-based network weighting: The prediction-based reliability measures for a set of networks can be translated to network weights, compatible for use with the WCBN approach, described in detail shortly. Simply, each network weight is calculated by its median prediction accuracy as a percentage of the total sum of median prediction accuracies across the set of all networks:

$$w_i = \frac{a_i}{\sum_{k=1}^{N} a_k}$$

where N is the number of networks, w_i is the weight for network i and a_i is the median predictive accuracy for network i. This method of calculation ensures that all weights sum to 1, as required by the WCBN algorithm. Then each weight represents the proportional influence for that network in the consensus algorithm.

Prediction-based network selection: Previous results in [14] indicate that in some cases, using only a subset of available datasets/networks can produce a better consensus model than when using all available datasets. Therefore, as an alternative, the prediction-based reliability measures for a set of networks can also be used to select a subset of input networks, where each network has an equal influence on the consensus process. In this case, instead of weighting the influence of individual networks, the least reliable networks are simply discarded from the input. Using the prediction-based reliability measure, the input networks can be ranked from most reliable to least reliable and either the n networks with the highest median prediction accuracy are selected as input, or a threshold, x, is used to determine the cutoff for accuracy.

The Weighted Consensus Bootstrapped Network Algorithm: The WCBN algorithm proceeds as follows (also see Algorithm 1 for step-by-step details). Each edge is considered in turn. For each edge, a set C is created of input confidence-weight pairs, where the kth pair contains the input confidence for that edge in the kth input network, and the weight w_k of the kth input network. The pairs in C are ordered descendingly by input confidence. Then a subset $C_{max} \subseteq C$ of this ordered set is created, which contains the pairs with the highest input confidences. The size of the subset C_{max} depends on the network weights and the consensus threshold t: the sum of network weights of the pairs in subset C_{max} must equal or exceed the consensus threshold t. Then, the edge consensus confidence is the minimum input confidence in the subset C_{max}.

Input: Set of n bootstrapped networks, each with an attached weight w_i indicating its influence such that $\sum_i w_i = 1$ and a consensus threshold (between 0 and 1)
Output: Consensus network

for *each pair of nodes i,j* **do**
    1. Create set $C = \{(c_{ij_1}, w_1), ..., (c_{ij_n}, w_n)\}$ where c_{ij_k} is the edge confidence for the edge between nodes $i \to j$ and w_k is the weight for the kth input network
    2. Reorder and re-index the confidence-weight pairs in C from highest to lowest confidence where c_{ij_1} is the highest edge confidence and c_{ij_n} is the lowest
    3. Create subset $C_{max} \subseteq C$ such that $C_{max} = \{(c_{ij_1}, w_1), ..., (c_{ij_A}, w_A)\}$ where $\sum_{k=1:A} w_k \geq t$
    4. Define the edge consensus confidence as $Con_{ij} = \min_{conf}(C_{max})$

end

Algorithm 1. Weighted Consensus Bootstrapped Networks

3 Experiments and Results

Since the inputs and outputs of the original CBN in [14] and the WCBN algorithm are different, it is not straightforward (or appropriate) to make a direct performance comparison between them. Instead we compare three different variations of the WCBN: firstly a method that only involved using the bootstrap weights without any input data weightings. We refer to this method from now on as Equal Weightings and can be achieved by using Algorithm 1 but with each input weight set to $1/N$. Secondly we use the full WCBN with prediction-based network weighting and prediction-based network selection as described earlier. Additionally, we compare the different WCBNs approaches against the individual input networks and against a network generated from an aggregate dataset formed by combining the input datasets followed by scale normalisation.

The performance of the different consensus networks are evaluated by comparing them against the true network in terms of TP and FP interactions. (Also compared is the network generated from the aggregated normalised input datasets as a straw man base-line). These comparisons can be represented by single values by using ROC curve analysis to calculated the Area Under Curve (AUC). A higher AUC value (assuming values are above 0.5) indicates a better performing network. Note that a different network is generated for each consensus threshold from 1-100%. This means that the AUC value may vary across consensus thresholds for each different approach. Therefore in the comparison, for each approach we consider the maximum AUC value achieved and the corresponding consensus threshold.

All algorithms are tested on 18 different combinations of synthetic data (with varying levels of noise) as well as real-world yeast datasets, both of which we describe now.

3.1 Datasets and Experiment Design

In order to examine the relationship between the input network quality and the performance of the resulting consensus network, experiments were first carried out on synthetic microarray datasets to provide a controlled setting, before moving on to evaluation on a real data application. In this section, the different types of datasets that are used are described, and the experiments performed and evaluation is explained.

Synthetic datasets: The synthetic datasets are based on a synthetic network, which is generated based on a regulatory network structure of 13 genes. Four different time-series gene expression datasets were generated for the network using differential equations to mimic a transcriptional gene network. See Figure 2 for the network structure. The change of the expression value of each gene is determined by a function composed of three parts: activation by a single other gene, repression by a single other gene and decay. Each dataset has a varying number of samples ranging from 40-120. In order to investigate how network quality affects the consensus approach a number of further datasets were generated by adding Gaussian noise, with various variances, to each dataset. Each

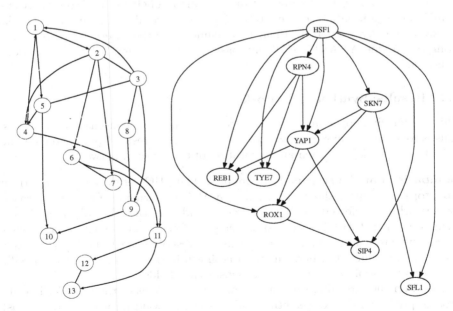

Fig. 2. Structures of the Synthetic and Yeast networks

Table 1. Summary of yeast datasets

Dataset [reference]	Description	No. Samples
Beissbarth [1]	Heat-shock response	12
Eisen [3]	Cold-shock and heat-shock response	14
Gasch [4]	Environmental changes inc heat-shock response	173
Grigull [11]	Heat-shock response	27
Spellman [8]	Cell-cycle	73

dataset was discretised using an equal frequency methods with three bins. Input networks with edge confidences were then generated from each dataset using bootstrapping. Each consensus approach was run on 18 different combinations of input networks. Each set of input networks contains 4 networks, where each network is generated from a version of one of the original four datasets (no set contains more than one version of each dataset). The sets of networks are ordered by their *collective level of reliability*, which is measured as the median of median predictive accuracies for the input networks generated from each dataset in the set. Therefore synthetic dataset 1 is considered the most reliable combination of input data, followed by set 2 down to set 18 (which contains the noisiest datasets as expected).

Yeast datasets: We also use microarray data from well-documented yeast studies. We focus on a sub-network of 9 regulatory genes that are related to heat-shock response. The microarray datasets are publicly available on the Yeast-BASE expression database - see Table 1 for more details. The learnt networks are evaluated by comparing them to documented gene interactions, obtained from the online YEASTRACT database [7]. See Figure 2 for the network structure.

3.2   Results: Synthetic Networks

This section first looks at a comparison of the different consensus approaches. This is followed by an exploration of the consensus threshold parameter for the weighted networks and then an exploration of the selected networks.

Comparison of Consensus Approaches: The AUC performance comparison plot (top) in Figure 3 shows the maximum AUC performance for each approach by network set (referred to as sets 1-18). In addition, the AUC of the best performing input network is recorded on the plot. Recall that a consensus network of each type is generated for each consensus threshold from 1-100%. This means that the maximum AUC performance (as shown in the plot) relates to a specific consensus threshold or interval of consensus thresholds.

Based on these results, we can see that both versions of the new WCBN approach, prediction-based weighting and selection, always improve on or at least equal the performance of the Equal Weightings approach. In particular, both

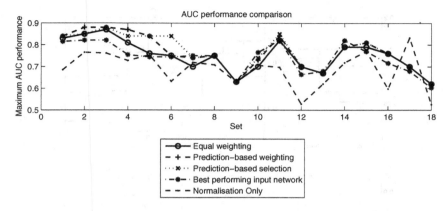

Fig. 3. Synthetic results: across all network sets, a comparison of the maximum AUC and the length of consensus threshold interval for which this is achieved

WCBN approaches are able to outperform the best performing input network and the Normalisation only network in more cases than the Equal Weightings approach. In order to obtain a p-value indicating the statistical significance of these findings, a paired t-test was used. All consensus-based approaches outperform the Normalisation only network with $p \leq 0.003$. The weighting and selection approaches also both outperform the best input network with statistical significance (p=0.016 and 0.006 respectively), whilst the Equal Weightings approach only obtains a p-value of 0.27 for outperforming the best input network. The weighting and selection approaches also outperform the Equal Weightings approach with statistical significance (p=0.008 and 0.009 respectively), but there is no significant difference between the weighting and selection approaches (p=0.63).

Recall that the network sets are ordered in terms of their collective reliability (networks in set 1 have a higher prediction accuracy than those in set 18). It can be seen in this plot that in general the maximum AUC decreases as the collective level of reliability decreases as expected. It also seems that the largest increases in performance for the new methods are found with the most reliable collections of datasets (e.g. sets 1-6). This implies that combining better quality data can produce even greater increases in performance.

Exploration of Consensus Threshold: The AUC in the previous discussion corresponds to a specific interval of consensus thresholds. An issue that is raised for both WCBN approaches is, how do we predict the exact consensus threshold where the maximum AUC value is to be found? In order to consider this in more detail, Fig 4 compares the AUC performance for the different approaches, across all consensus thresholds from 1-100%, for a selection of the network sets. First, we can see that the performance of the prediction-based weighting approach varies considerably by consensus threshold. The thresholds where the AUC changes correspond to 'weighting boundaries' - thresholds at which an additional network is able to influence the final consensus network. In general,

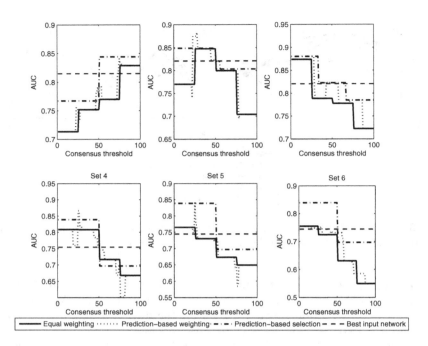

Fig. 4. Synthetic networks: approaches performance comparison for a selection of individual network sets

these weighting boundaries correspond to the areas around the standard equal weighted boundaries - (25%,50% and 75% thresholds for 4 datasets). This is because in these cases network weights are close to being equal. However, even in these cases, the use of weights can cause a significant improvement in AUC as in data set 5 at the 25% threshold, where the AUC value rises from around 0.77 (Equal Weightings) to 0.85 (prediction-based weighting). However, there are also intervals of consensus threshold where the weighting can cause a significant decrease in AUC such as in set 4 at the 25% threshold. It seems that selection of the optimum consensus threshold for prediction-based weighting is not trivial.

Prediction-based selection: Method Comparison: We now discuss the results of investigating the optimum number of selected input networks (n) and the highest median prediction accuracy of the networks that have not been selected (this is the threshold x). Firstly, we found that it is not possible to choose a prediction accuracy threshold x for network selection that is appropriate to all cases, since the highest median predictive accuracies of the unselected networks vary from 0.15 to 0.44. However, further analysis reveals that there is a relationship between n and x. Where fewer networks are selected, the selected networks have a higher threshold for x, the highest median predictive accuracy of the networks not selected. Together, these patterns imply that network sets with higher predictive accuracies (i.e. higher reliability) require a lower number of input networks, whilst network sets with lower predictive accuracies (i.e. lower reliability) require more input networks. This is intuitive - the higher the quality of data the less we need in order to build better performing models.

3.3   Results: Yeast Heat-Stress Network

In this section, the new consensus approaches are applied to a set of real yeast microarray datasets, that have been generated by different heat-stress microarray studies. Figure 5 shows the AUC for each consensus approach, across all consensus thresholds from 0 to 100%, as well the best performing individual input network AUC (generated from the Spellman dataset). The prediction-based weighting network achieves the highest AUC, making an improvement of almost 0.05 in AUC over the Equal Weightings approach. However, this is only at a single consensus threshold of 78%. Fig 5 shows that it is around the thresholds relating to equal weights (20%,40%,60%,80%) where the prediction-based weighting network attains the largest increases in AUC. This is because the prediction-based weights largely reflect an equal distribution (see Table 2). The prediction-based selection network also achieves a significant improvement in the maximum AUC when compared to the Equal Weightings or input networks. The best result is achieved when the top three input networks are selected ranked by the highest median prediction accuracy, giving the Gasch, Grigull and Spellman inputs (see Table 2). This network attains a maximum AUC of 0.613 between the consensus thresholds of 34% and 67%, showing that removing the two noisiest datasets significantly improves the performance of the consensus network. Although the prediction-based weighting network attains a slightly higher maximum AUC of

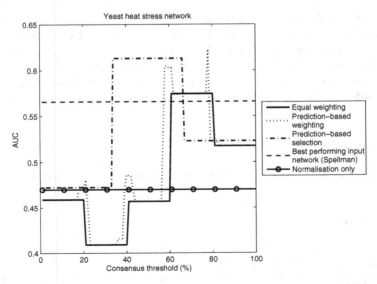

Fig. 5. Yeast heat stress network: comparison of approaches by AUC

Table 2. Yeast heat stress datasets: prediction-based reliability measures

Dataset	Median predictive accuracy	Prediction-based weight	Individual network AUC
Beissbarth	0.25	0.17	0.43
Eisen	0.26	0.18	0.43
Gasch	0.31	0.22	0.50
Grigull	0.30	0.21	0.49
Spellman	0.31	0.22	0.57

0.623 at a 78% threshold, this is at a single threshold point whilst prediction-based selection improves network performance more consistently across thresholds (as in the synthetic network results).

4 Conclusions

This paper has furthered the consensus approach to learning gene regulatory networks by exploring ways to weight or select the input datasets. A measure of reliability for sets of data, based on the prediction of node values has been introduced. A comparison of these different consensus approaches has been applied using synthetic and real microarray datasets with promising results. In all cases presented in the comparison, prediction-based weighting and selection was able to equal or improve the performance of both the standard consensus Bayesian network approach with equal weighting and a normalisation approach where data was aggregated prior to network learning. The performance of the consensus network, whichever approach is used, can vary considerably by consensus threshold. However, for prediction-based selection in particular, the experimental results indicated that there could be a relationship between the reliability of the individual input networks and the optimum consensus threshold. More reliable sets of input networks attained their maximum AUC values with higher consensus thresholds than noisier, less reliable sets of input networks. The experiments presented in this paper have been carried out on a limited set of data with small-scale networks. There is no reason why the algorithm should not scale and in order to draw firmer conclusions, a larger set of datasets, based on larger networks will be investigated.

Acknowledgements

This work was part funded by a Royal Society International Joint Project Grant (2006/R3).

References

1. Beissbarth, T., et al.: Processing and quality control of DNA array hybridization data. Bioinformatics 16(11) (2000)
2. Conlon, E.M., Song, J.J., Liu, J.S.: Bayesian models for pooling microarray studies with multiple sources of replications. BMC Bioinformatics 7(247) (2006)

3. Eisen, M.B., et al.: Cluster analysis and display of genome-wide expression patterns. PNAS 95(25), 14863–14868 (1998)
4. Gasch, A., et al.: Genomic expression program in the response of yeast cells to environmental changes. Mol. Cell 11, 4241–4257 (2000)
5. Pe'er, D., et al.: Inferring subnetworks from perturbed expression profiles. In: Proceedings of the Int. Conference on Intelligent Systems for Molecular Biology (2001)
6. Hoeting, J.A., et al.: Bayesian model averaging: a tutorial. Statistical Science 14(4), 382–417 (1999)
7. Teixeira, M., et al.: The YEASTRACT database: a tool for the analysis of transcription regulatory associations in Saccharomyces cerevisiae. Nucleic Acids Research 34, D446–D451 (2006)
8. Spellman, P., et al.: Comprehensive identification of cell cycle-regulated genes of the yeast saccharomyces cerevisiae by microarray hybridization. Mol. Cell 9, 3273–3297 (1998)
9. Ng, S.K., et al.: On combining multiple microarray studies for improved functional classification by whole-dataset feature selection. Genome Informatics (14), 44–53 (2003)
10. Wang, Y., et al.: Inferring gene regulatory networks from multiple microarray datasets. Bioinformatics 22(19), 2413–2420 (2006)
11. Grigull, J., et al.: Genome-wide analysis of mrna stability using transcription inhibitors and microarrays reveals post-transcriptional control of ribosome biogenesis factors. Mol. Cell 24(12), 5534–5547 (2004)
12. de Morais, S.R., Aussem, A.: A novel scalable and data efficient feature subset selection algorithm. In: Daelemans, W., Goethals, B., Morik, K. (eds.) ECML PKDD 2008, Part II. LNCS (LNAI), vol. 5212, pp. 298–312. Springer, Heidelberg (2008)
13. Schapire, R.E.: The boosting approach to machine learning: An overview. In: Denison, D.D., et al. (eds.) Nonlinear Estimation and Classification. Springer, Heidelberg (2003)
14. Steele, E., Tucker, A.: Consensus and meta-analysis regulatory networks for combining multiple microarray gene expression datasets. Jounral of Biomedical Informatics 41(6), 914–926 (2008)

Incremental Bayesian Network Learning for Scalable Feature Selection

Grégory Thibault[1], Alex Aussem[1], and Stéphane Bonnevay[2]

[1] University of Lyon, LIESP, F-69622 Villeurbanne Cedex, France
[2] University of Lyon, ERIC, F-69622 Villeurbanne Cedex, France

Abstract. Our aim is to solve the feature subset selection problem with thousands of variables using an incremental procedure. The procedure combines incrementally the outputs of non-scalable search-and-score Bayesian network structure learning methods that are run on much smaller sets of variables. We assess the scalability, the performance and the stability of the procedure through several experiments on synthetic and real databases scaling up to 139 351 variables. Our method is shown to be efficient in terms of both running time and accuracy.

1 Introduction

Feature subset selection (FSS for short) is an essential component of quantitative modeling, data-driven construction of decision support models or even computer-assisted discovery. No a priori information or selection of variables is required. Therefore, no previous knowledge premise will bias the final models. The FSS enables the classification model to achieve good or even better solutions with a restricted subset of features, and it helps the human expert to focus on a relevant subset of features. However, databases have increased many fold in recent years and most FSS algorithms do not scale to thousands of variables. Also, large-scale databases presents enormous opportunities and challenges for knowledge discovery and machine learning.

There have been a number of comparative studies for feature selection but few scale up to (say) 100 000 variables. Moreover, findings reported at low dimensions do not necessarily apply in high dimensions. While SVM are efficient and well suited for scalable feature selection [1] (e.g., SVM-RFE stand for SVM Recursive Feature Elimination), there is still much room for improvement. In microarray data analysis for instance, it is common to use statistical testing to control precision (often referred to as the false discovery rate) while maximizing recall, in order to obtain high quality gene (feature) sets. [1] show that none of the above SVM-based method provide such control. Moreover, not only model performance but also robustness of the feature selection process should be taken into account [2]. [3] show experimentally that SVM-RFE is highly sensitive to the "filter-out" factor and that the SVM-RFE is an unstable algorithm. [4,5] showed recently through extensive comparisons with high-dimensional genomic data that none of the considered feature-selection methods performs best across all scenarios. Thus, there is still room for work to be conducted in this area.

N. Adams et al. (Eds.): IDA 2009, LNCS 5772, pp. 202–212, 2009.
© Springer-Verlag Berlin Heidelberg 2009

In this paper, we report the use of a probabilistic FSS technique to identify "strongly" relevant features, among thousands of potentially irrelevant and redundant features. A principled solution to the FSS problem is to determine the *Markov boundary* (MB for short) of the class variable. A MB of a variable T is any minimal subset of U (the full set of variables) that renders the rest of U independent of T. If the probability distribution underlying the data can be faithfully represented by a Bayesian network, the MB of T is unique. In recent years, there have been a growing interest in inducing the MB automatically from data. Very powerful correct, scalable and data-efficient constraint-based (CB) algorithms have been proposed recently [6,7,8,9]. CB discovery methods search a database for conditional independence relations. In contrast to search-and-score methods, CB methods are able to construct the local MB structure without having to construct the whole BN first. Hence their ability to scale up to thousands of variables. This was, so far, a key advantage of CB methods over search-and-score methods.

Our specific aim is to solve the feature subset selection (FSS) problem with thousands of variables using an incremental procedure that combines the result of search-and-score methods run on small sets of variables. We assess the accuracy, the scalability and the robustness of the procedure through several experiments on synthetic and real-world databases scaling up to 139 351 variables.

2   Feature Selection

Feature selection techniques can be divided into three categories, depending on how they interact with the classifier. Filter methods directly operate on the dataset, and provide a feature weighting, ranking or subset as output. These methods have the advantage of being fast and independent of the classification model, but at the cost of inferior results. Wrapper methods perform a search in the space of feature subsets, guided by the outcome of the model (e.g. classification performance on a cross-validation of the training set). They often report better results than filter methods, but at the price of an increased computational cost. Finally, embedded methods use internal information of the classification model to perform feature selection (e.g. use of the weight vector in support vector machines). They often provide a good trade-off between performance and computational cost.

Finding the minimal set of features require an exhaustive search among all subsets of relevant variables, which is an NP-complete problem, and may not be unique. In this study, the FSS is achieved in the context of determining the Markov boundary of the class variable that we want to predict. Markov boundary (MB for short) learning techniques can be regarded as in between filter and embedded methods. They solve the feature subset selection (FSS) problem and, in the meantime, they build a local Bayesian network around the target variable that can be used afterwards as a probabilistic classifier.

3   Bayesian Networks

For the paper to be accessible to those outside the domain, we recall first the principle of Bayesian network. We denote a variable with an upper-case, X, and value of that variable by the same lower-case, x. We denote a set of variables by upper-case bold-face, $\mathbf{Z}$, and we use the corresponding lower-case bold-face, $\mathbf{z}$, to denote an assignment of value to each variable in the set. We denote the conditional independence of the variable X and Y given $\mathbf{Z}$, in some distribution P with $X \perp_P Y|\mathbf{Z}$. In this paper, we only deal with discrete random variables.

Formally, a BN is a tuple $< \mathcal{G}, P >$, where $\mathcal{G} =< \mathcal{V}, \mathcal{E} >$ is a directed acyclic graph (DAG) with nodes representing the random variables $\mathcal{V}$ and P a joint probability distribution on $\mathcal{V}$. In addition, $\mathcal{G}$ and P must satisfy the Markov condition: every variable, $X \in \mathcal{V}$, is independent of any subset of its non-descendant variables conditioned on the set of its parents, denoted by $\mathbf{Pa}_i^{\mathcal{G}}$.

A Markov blanket $\mathbf{M}_T$ of the T is any set of variables such that T is conditionally independent of all the remaining variables given $\mathbf{M}_T$. A Markov boundary, $\mathbf{MB}_T$, of T is any Markov blanket such that none of its proper subsets is a Markov blanket of T. We say that $< \mathcal{G}, P >$ satisfies the faithfulness condition when $\mathcal{G}$ entails all and only conditional independencies in P.

Theorem 1. *Suppose $< \mathcal{G}, P >$ satisfies the faithfulness condition. Then X and Y are not adjacent in $\mathcal{G}$ iff $\exists \mathbf{Z} \in \mathbf{V} \setminus \{X \cup Y\}$ such that $X \perp_P Y|\mathbf{Z}$. Moreover, for all X, the set of parents, children of X, and parents of children of X is the unique Markov boundary of X.*

A proof can be found for instance in [10]. We denote by $\mathbf{PC}_T$, the set of parents and children of T in $\mathcal{G}$, and by $\mathbf{SP}_T$, the set of *spouses* of T in $\mathcal{G}$. The *spouses* of T are the parents of the children of T. These sets are unique for all $\mathcal{G}$, such that $< \mathcal{G}, P >$ is faithful and so we will drop the superscript $\mathcal{G}$.

Two graphs are said *equivalent* iff they encode the same set of conditional independencies via the d-separation criterion. The equivalence class of a DAG $\mathcal{G}$ is a set of DAGs that are equivalent to $\mathcal{G}$. The next result showed by [11], establishes that equivalent graphs have the same undirected graph but might disagree on the direction of some of the arcs.

Theorem 2. *Two DAGs are equivalent iff they have the same underlying undirected graph and the same set of v-structures (i.e. converging edges into the same node, such as $X \to Y \leftarrow Z$.*

Moreover, an equivalence class of network structures can be uniquely represented by a completed partially directed DAG (CPDAG), also called a DAG pattern. The DAG pattern is defined as the graph that has the same links as the DAGs in the equivalence class and has oriented all and only the edges common to all the DAGs in the equivalence class.

4   Incremental MB Structure Learning for Scalable FSS

The key idea in this paper is that an incremental procedure could help in alleviating the complexity obstacle by aggregating the outputs of several feature

---

Algorithm 1. Generic Incremental FSS by MB Search

1: **function** IFSS($\mathcal{D}, target, selsize, Vars$)
2:     $MB \leftarrow \emptyset$
3:     **repeat**
4:         $Testvars \leftarrow \{target\} \cup MB$
5:         $Testvars \leftarrow Testvars \cup$ SELECTION($selsize$)
6:         $G \leftarrow$ BNLEARNING($\mathcal{D}, Testvars$)
7:         $MB \leftarrow$ EXTRACT_MB(G)   ▷ MB extraction
8:     **until** stop_criterion
9:     **return** MB ▷ features = variables in MB
10: **end function**

---

selectors working on much fewer variables. More specifically, a collection of single FSS models is run on small subsets of variables in incremental fashion. The output of one feature selector serves as input to the next. The feature selector used in our method is based on a BN structure identification algorithm. Algorithm 1 displays our incremental feature selection process based on Markov Boundary search. Input parameters are:

- $\mathcal{D}$: data used for supervised learning,
- $target$: the target variable,
- $selsize$: number of new variables at each iteration,
- $Vars$: set of variables except the $target$ variable.

Standard search-and-score $BNLearning$ methods do not scale to high-dimensional data sets of variables. The aim of the meta-procedure is to learn many small MBs (in regard to the whole set of variables) from many small subsets of variables. $BNLearning$ can be implemented by any BN structure algorithm. In this study, it is implemented with the GES scoring-based greedy search algorithm discussed by Chickering in [12].

At the beginning, the set of variables, $Testvars$, used to learn a Bayesian network, G, is chosen at random. A first Markov Boundary, MB, is extracted from G. At each iteration, variables in MB are kept into the set $Testvars$ and some other variables are added by a uniform random selection without replacement. The size of this selection, $selsize$, is adapted according to the size of the Markov Boundary, MB. Our variables selection process assumes that, in the first part of the algorithm, each variable of $Vars$ is selected once; then, when all variables have been selected once, the process restart with the whole set of variables, $Vars$. The algorithm stops when all variables have been selected twice. At the end, the selected features are returned. Under the faithfulness assumptions and assuming that the induction algorithm is correct, IFSS returns the correct Markov Boundary. This a sample limit property. In practice, our hope is to output the features that GES would have found on the complete database.

Indeed, after the first part of algorithm (when all variables have been selected once), MB contains all the parents and the children of the target, because by definition, the variable adjacent to the target cannot be d-separated from the target, given any other variable. During the second part of the algorithm (when all variables have been selected at least twice), the spouses of the *target* enter the candidate MB set.

5 Experiments

In this section, we assess the accuracy, the scalability and the robustness of IFSS through several empirical experiments on benchmark data sets. We use a state-of-the-art search-and-score BN structure learning algorithm called GES as our BN learner (*BNLearning*). First, we compare IFSS against GES in terms of accuracy on several synthetic data sets. Second, we assess the scalability of IFSS on a high-dimensional data sets that was provided at the KDD-Cup 2001. Third, we assess the IFSS's robustness.

5.1 Accuracy

We report here the results of our experiments on six common benchmarks: ASIA, ASIA8 (ASIA tilled 8 times), ALARM, INSULIN, INSURANCE and HAIL-FINDER, (see [8] and references therein). For ASIA8, the tiling is performed in a way that maintains the structural and probabilistic properties of the original network, ASIA, in the tiled network. Description of the benchmarks is showed in Table 1. For each benchmark, 10 databases with independent and identically distributed samples were generated by logic sampling. The amount of data was chosen large enough to avoid the bias due to a lack of data. The task is to learn the MB of the variable that appears in the third column in Table 1. The size of the MB varies from 5 to 18 variables as may be observed. We compare IFSS against GES in terms of true positive rate (TPR, i.e., the number of true positives variables in the output divided by the number of variables in the output), false positive rate (FPR, i.e., the number of false positives divided by the the number of variables in the output), the Kappa index (κ), the weighted accuracy (WAcc), computed as the average of the accuracy on true positives and the accuracy on true negatives and finally, the time in seconds. Kappa is a measure that assesses improvement over chance is appropriate. The following ranges of

Table 1. Description of the Bayesian networks used in these experiments to assess the comparative accuracy of IFSS and GES Markov boundary discovery on the target variable

Benchmark	# var	# edges	target	MB size	# samples
ASIA	8	8	OR	5	10 000
ASIA8	64	64	OR	5	10 000
ALARM	37	46	HR	8	30 000
INSULIN	35	52	IPA	18	50 000
INSURANCE	27	52	Accident	10	30 000
HAILFINDER	56	66	Scenario	17	50 000

agreement for the Kappa statistic suggested in the literature are: poor $K < 0.4$, good $0.4 < K < 0.75$ and excellent $K > 0.75$. In all our experiments, GES is trained to maximize the Bayesian Dirichlet scoring criterion defined as:

$$BD(\mathcal{B} \mid \mathcal{D}) = p(\mathcal{B}) \cdot \prod_{i=1}^{n} \prod_{j=1}^{q_i} \frac{\Gamma(\alpha_{ij})}{\Gamma(N_{ij} + \alpha_{ij})} \prod_{k=1}^{r_i} \frac{\Gamma(N_{ijk} + \alpha_{ijk})}{\Gamma(\alpha_{ijk})}$$

Note that no *a priori* information structure is used for tests on synthetic data (*i.e.*, $p(\mathcal{B})$ is uniform). Moreover, the prior on parameters is set so as to be non-informative, that is, an equivalent uniform Dirichlet prior with an equivalent sample size (ESS) equal to the greatest variable modality (see [10] for details). Table 2 summarizes the average performance indexes over 10 runs for each benchmark. As may be observed, IFSS performs as well as GES on all benchmarks, except on INSURANCE where IFSS outperform GES by a noticeable margin. This is quite a surprise as IFSS was not designed to outperform the underlying BN structure learning algorithm (here GES) but only to be scalable.

Table 2. Average performance of IFSS (with GES as underlying BN structure learning algorithm) and GES

	GES				IFSS			
	κ	TPR	FPR	WAcc	Time	κ	TPR	FPR	WAcc	Time
ASIA	0.959	1.000	0.050	0.975	0.10	0.959	1.000	0.050	0.975	0.06
ASIA8	0.867	1.000	0.024	0.988	27.87	0.834	1.000	0.031	0.984	1.29
ALARM	0.916	0.875	0.000	0.938	6.79	0.916	0.875	0.000	0.938	2.56
INSULIN	0.840	0.933	0.094	0.920	15.16	0.870	0.933	0.063	0.935	7.83
INSURANCE	0.663	0.700	0.063	0.819	5.81	0.858	0.860	0.019	0.921	2.60
HAILFINDER	0.589	0.571	0.037	0.767	48.71	0.517	0.471	0.016	0.727	6.19

In Table 3, the same indexes of accuracy are reported; the aim is to recover the MB given by GES (and not the true MB anymore). For instance, a True Positive is a variable given by GES and found by IFSS, *etc.*. IFSS is very close to GES in most cases. Some significant differences are observed on Hailfinder between IFSS and GES, output of IFSS is closer to the true MB than output of GES (see in Table Table 2). Moreover, the last column indicates the time saving when IFSS is used instead of GES.

5.2   Scalability

In this section, experiments demonstrate the ability of IFSS to solve a real world FSS problem involving thousands of features. We consider the THROM-BIN database which was provided by DuPont Pharmaceuticals for KDD Cup 2001. It is exemplary of a real drug design [13]. The training set contains 1909 instances characterized by $139,351$ binary features. The features describe the three-dimensional properties of the compounds. Each compound is labelled with

Table 3. Average performance of IFSS where the task is to recover the variables output by GES

	IFSS against GES				
	κ	TPR	FPR	WAcc	Time saving
ASIA	1.000 ± 0.000	1.000 ± 0.000	0.000 ± 0.000	1.000 ± 0.000	0.04 ± 0.04
ASIA8	0.900 ± 0.107	0.940 ± 0.102	0.014 ± 0.014	0.963 ± 0.055	26.58 ± 1.72
ALARM	1.000 ± 0.000	1.000 ± 0.000	0.000 ± 0.000	1.000 ± 0.000	4.23 ± 2.21
INSULIN	0.959 ± 0.049	0.968 ± 0.037	0.007 ± 0.021	0.980 ± 0.023	7.34 ± 1.31
INSURANCE	0.730 ± 0.058	0.863 ± 0.031	0.111 ± 0.037	0.876 ± 0.025	3.21 ± 2.47
HAILFINDER	0.490 ± 0.093	0.530 ± 0.200	0.062 ± 0.058	0.734 ± 0.077	42.52 ± 33.73

one out of two classes, either it binds to the target site or not. The task of KDD Cup 2001 was to learn a classifier from $1,909$ given compounds (learning data) in order to predict binding affinity and, thus, the potential of a compound as anti-clotting agent. The classifiers submitted to KDD Cup 2001 were evaluated on the remaining 634 compounds (testing data) as the weighted average (WAcc) of the accuracy on true binding compounds and the accuracy on true non-binding compounds. The THROMBIN database is challenging for three reasons. First, it has a huge number of features. Second, the learning data are extremely imbalanced: Only 42 out of the 1909 compounds bind. Third, the testing data are not sampled from the same probability distribution as the learning data, because the compounds in the testing data were synthesized based on the assay results recorded in the learning data. Scoring higher than 60% accuracy is impressive as noted in [6].

IFSS, with GES as the MB learner, was run 61 times in the time we have disposed for our experiments, with a prior over structures arbitrary fixed to $10^{-16 \times f}$, where f is the number of free parameters in the DAG. The outputs were used as input of Naive Bayesian Classifier, and a classification on the test data was perfomed. As shown in Figure 3, IFSS scores between 36% (really bad) to 71% with an average 55% and only 46 runs of IFSS score more than 50% weighted accuracy, i.e. the random classifier. These results are comparable to MBOR [7] and IAMB [14] that achieve respectively 53% (over 10 runs) and 54% (both over 114 runs). This is however worse than PCMB [6] that achieves 63% (over 114 runs). Of course, we have no idea what GES scores on such data since GES do not scale to such high-dimensional database. Note that each launch of IFSS lasted approximately 3 hours, which is the same order of magnitude as the other algorithms mentioned above.

Nonetheless, the best MB over 61 runs consists of five variables 3392, 10695, 23406, 79651 and 85738. This MB is depicted in Figure 2. It scores 71,1% which is impressive according to [13,6]. It worth mentioning that J. Cheng, the winner of the KDD cup 2001, only scores 71.1% accuracy and 68.4% weighted accuracy with four variables: 10695, 16794, 79651 and 91839. He used a Bayesian classifier to assess the accuracy of his feature set. It is shown in Figure 1. As may be seen, two variables are common with the winner's selection. IFFS outputs the THROMBIN MB in about 220 minutes on our laptop (2.6GHz Intel®

Core^{TM} 2 Duo with 1 GB of RAM). Of course, this time is highly dependent our MATLAB® implementation, and may significantly be reduced if written in C/C++ for instance.

The Figure 5 represents the ROC curves of the classifier given by IFSS with the best MB as input. The area under ROC curve is a well-known performance measurement. The ROC curve is the 2-D plot of sensitivity and 1-specificity acquired by applying a sequence of arbitrary cut-off threshold to the probabilities generated by the predictive model. A clear difference is observed between the ROC curve on the test set (in plain line) and the ROC curve on the training set (in dotted lineline, obtained by 10-fold cross validation). The reason is that the testing data was not sampled from the same probability distribution as the learning data, hence the difficulty of the task. The area under curve (AUC) is 0.6978 on the test set. This classifier scores 69% (and 71% when constructing a naive BN with the same variables) which seems highly competitive compared to PCMB [15] and IAMB [14] that achieves respectively 63% and 54% as shown in [6]. Table 4 reports the scores obtained with the best MB classifiers constructed from the sets of variables given by the respective algorithms.

Table 4. Results of classifiers with the output-model of the algorithm, the naive bayesian network model, the support vector machine classifier and the random forest classifier

	IFSS					Cheng				
	κ	TPR	FPR	Acc	WAcc	κ	TPR	FPR	Acc	WAcc
Output model	0.420	0.467	0.085	0.809	0.691	0.316	0.633	0.264	0.711	0.684
NaiveBN	0.437	0.547	0.120	0.801	0.713	0.297	0.600	0.258	0.708	0.671
SVM	0.464	0.500	0.076	0.823	0.712	0.312	0.313	0.056	0.795	0.629
RForest	0.439	0.513	0.099	0.809	0.707	0.312	0.313	0.056	0.795	0.629

5.3 Robustness

When using FSS on data sets with large number of features, but a relatively small number of samples, not only model performance but also robustness of the FSS process is important. For instance, in microarray analysis, domain experts clearly prefer a stable gene selection as in most cases these genes are subsequently analyzed further, requiring much time and effort [16]. With such high-dimensional databases, all FSS algorithms are subject to some variability. Surprisingly, the robustness of FSS techniques has received relatively little attention so far in the literature. As noted in [2], robustness can be regarded from different points of view: perturbation at the instance level (e.g. by removing or adding samples), at the feature level (e.g. by adding noise to features), or variation of the parameter of the FSS algorithm, or a combination of them. Here, we focus on the robustness of FSS selector as the variation of the output with respect to a random permutation of the variables. We consider again the 61 times runs of IFSS on THROMBIN data. A simple ensemble technique proposed in [2,16] works by aggregating the feature rankings provided by the FSS selector

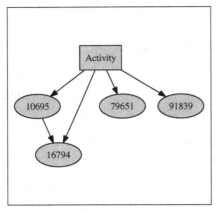

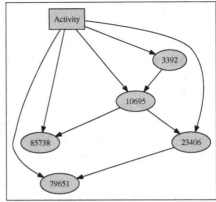

Fig. 1. BN of the KDD Cup winner   **Fig. 2.** Best MB output by IFSS

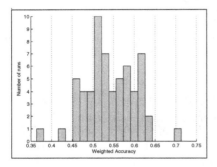

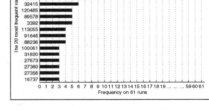

Fig. 3. Weighted accuracies of 61 runs of IFSS

Fig. 4. Frequencies of twenty most frequent variables over 61 runs of IFSS

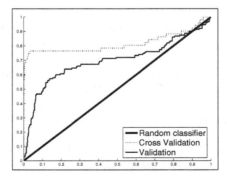

Fig. 5. ROC curves of the best MB output by IFSS on test set and training set using 10-fold cross-validation

into a final consensus ranking weighted by frequency. The variables returned by IFSS mostly differ by one or two variables. The top 20 ranked variables are shown in Figure 4 in decreasing order of frequency in the output of IFSS. As we can see, the variables 79651 and 10695 were always selected. These variables are also present among the four features of the winner of the KDD cup in 2001, and variable 79651 is always present in the top 10 MB output by KIAMB (see [6]). The third most frequent feature, namely 91839, is also one of the four features of the winner of the KDD cup.

Following [17], we take a similarity based approach where feature stability is measured by comparing the 61 outputs of IFSS. We use the Jaccard index as the similarity measure between two subsets S_1 and S_2. The more similar the outputs, the higher the stability measure. The overall stability can be defined as the average over all pairwise similarity comparisons between the $n = 61$ MBs:

$$I_{tot} = \frac{\sum_{i=1}^{n} \sum_{j=i+1}^{n} I(S_i, S_j)}{n(n-1)} \quad \text{with} \quad I(S_i, S_j) = \frac{|S_i \bigcap S_j|}{|S_i \bigcup S_j|}$$

An average of 0.336 (with a standard deviation of 0.116) was obtained.

6   Conclusion

We discussed a new scalable feature subset selection procedure. This procedure combines incrementally the outputs of non-scalable search-and-score Bayesian network structure learning methods that are run on much smaller sets of variables. The method was shown to be highly efficient in terms of both running time and accuracy. Future substantiation through more experiments with other BN learning algorithms are currently being undertaken and comparisons with other FSS techniques will be reported in due course.

References

1. Nilsson, R., Peña, J., Björkegren, J., Tegnér, J.: Evaluating feature selection for svms in high dimensions. In: European Conference on Machine Learning and Principles and Practice of Knowledge Discovery in Databases, ECML PKDD (2006)
2. Saeys, Y., Abeel, T., Van de Peer, Y.: Robust feature selection using ensemble feature selection techniques. In: Daelemans, W., Goethals, B., Morik, K. (eds.) ECML PKDD 2008, Part II. LNCS (LNAI), vol. 5212, pp. 313–325. Springer, Heidelberg (2008)
3. Tang, Y., Zhang, Y., Huang, Z.: Development of two-stage svm-rfe gene selection strategy for microarray expression data analysis. IEEE-ACM Transactions on Computational Biology and Bioinformatics 4, 365–381 (2007)
4. Ma, S., Huang, J.: Penalized feature selection and classification in bioinformatics. Briefings in Bioinformatics 5, 392–403 (2008)
5. Hua, J., Tembe, W., Dougherty, E.: Performance of feature-selection methods in the classification of high-dimension data. Pattern Recognition 42, 409–424 (2009)

6. Peña, J., Nilsson, R., Björkegren, J., Tegnér, J.: Towards scalable and data efficient learning of markov boundaries. International Journal of Approximate Reasoning 45(2), 211–232 (2007)
7. Rodrigues de Morais, S., Aussem, A.: A novel scalable and data efficient feature subset selection algorithm. In: Daelemans, W., Goethals, B., Morik, K. (eds.) ECML PKDD 2008, Part II. LNCS (LNAI), vol. 5212, pp. 298–312. Springer, Heidelberg (2008)
8. Tsamardinos, I., Brown, L.E., Aliferis, C.F.: The max-min hill-climbing bayesian network structure learning algorithm. Machine Learning 65(1), 31–78 (2006)
9. Yaramakala, S., Margaritis, D.: Speculative markov blanket discovery for optimal feature selection. In: IEEE International Conference on Data Mining, pp. 809–812 (2005)
10. Neapolitan, R.E.: Learning Bayesian Networks. Prentice-Hall, Englewood Cliffs (2004)
11. Pearl, J.: Probabilistic Reasoning in Intelligent Systems: Networks of Plausible Inference. Morgan Kaufmann, San Francisco (1988)
12. Chickering, D.M.: Optimal structure identification with greedy search. Journal of Machine Learning Research 3, 507–554 (2002)
13. Cheng, J., Hatzis, C., Hayashi, H., Krogel, M., Morishita, S., Page, D., Sese, J.: KDD Cup 2001 Report. In: ACM SIGKDD Explorations, pp. 1–18 (2001)
14. Tsamardinos, I., Aliferis, C.F., Statnikov, A.R.: Algorithms for large scale markov blanket discovery. In: FLAIRS Conference, pp. 376–381 (2003)
15. Peña, J.M., Björkegren, J., Tegnér, J.: Scalable, efficient and correct learning of markov boundaries under the faithfulness assumption. In: Godo, L. (ed.) ECSQARU 2005. LNCS (LNAI), vol. 3571, pp. 136–147. Springer, Heidelberg (2005)
16. Aussem, A., Rodrigues de Morais, S., Perraud, F., Rome, S.: Robust gene selection from microarray data with a novel Markov boundary learning method: Application to diabetes analysis. In: European Conference on Symbolic and Quantitative Approaches to Reasoning with Uncertainty ECSQARU 2009 (to appear, 2009)
17. Kalousis, A., Prados, J., Hilario, M.: Stability of feature selection algorithms: a study on high-dimensional spaces. Knowl. Inf. Syst. 12 (2007)

Feature Extraction and Selection from Vibration Measurements for Structural Health Monitoring

Janne Toivola and Jaakko Hollmén

Helsinki University of Technology, Department of Information and
Computer Science, P.O. Box 5400, FI-02015 TKK, Espoo, Finland
{jannetoivola,Jaakko.Hollmen}@tkk.fi

Abstract. Structural Health Monitoring (SHM) aims at monitoring
buildings or other structures and assessing their condition, alerting about
new defects in the structure when necessary. For instance, vibration measurements can be used for monitoring the condition of a bridge. We investigate the problem of extracting features from lightweight wireless
acceleration sensors. On-line algorithms for frequency domain monitoring are considered, and the resulting features are combined to form a
large bank of candidate features. We explore the feature space by selecting random sets of features and estimating probabilistic classifiers for
damage detection purposes. We assess the relevance of the features in
a large population of classifiers. The methods are assessed with real-life
data from a wooden bridge model, where structural problems are simulated with small added weights.

Keywords: structural health monitoring, damage detection, feature extraction, feature selection, wireless sensor network.

1 Introduction

Recent development of wireless sensor technology has opened new possibilities
for damage identification in large civil structures. For example, the condition of
buildings, bridges, or cranes can be monitored continuously and automatically,
reducing the need for periodic manual inspections. The monitoring process is
called Structural Health Monitoring (SHM) [4]. Practical limitations of costly
and error-prone wiring of sensors has been overcome by radio technology and
portable power supplies [4,8].

In this work, we consider an experimental setting where the vibration of a
wooden model bridge is monitored with wired accelerometers and damages are
simulated by imposing minute changes by attaching small additional weights on
the bridge. The model structure and measurement setting are described in [7],
where the acceleration measurements were first used. Since the current work is
done in an off-line setting, i.e. first gathering all the raw acceleration values to a
centralized database, we are able to test computationally demanding approaches
to solve the problem. However, we wish to come up with a solution that is feasible
in a practical setting in a resource-constrained wireless sensor network.

N. Adams et al. (Eds.): IDA 2009, LNCS 5772, pp. 213–224, 2009.
© Springer-Verlag Berlin Heidelberg 2009

To reduce the amount of irrelevant variability and the amount of data transmitted through the supposed wireless network, we propose feature extraction methods for monitoring the vibrations in the frequency domain. Since we don't have an analytical model of the structure or the variability caused by damages, we consider machine learning methods to estimate probabilistic models and use them for damage classification. The models are based on the feature extraction results computed locally on each sensor node and some simplistic assumptions allowing computational efficiency.

From machine learning point of view, this work considers supervised learning in a group classification setting [4]. This provides experimentally tested feature extraction and a preliminary step in reaching our long-term goal: applying unsupervised methods in a novelty detection setting for detecting the existence of damages. After detecting a damage, the sensor network can presumably be assigned to perform more thorough (and energy consuming) measurements, possibly leading to more accurate analytical results of the extent or position of the damage. Anyway, before a damage occurs, the sensor network should strive to minimize its power consumption and the amount of transmitted data.

The presented methods rely on the traditional way of removing redundancy in the data: considering the signals in a transformation domain and ignoring the parts that seem irrelevant from the task point of view. For example, monitoring vibration amplitudes only on certain selected frequencies allows us to ignore most of the spectrum and concentrate the computational resources and our analysis on the relevant part of the data. This is similar to the approach used in condition monitoring (CM) of rotating machinery [4], but the structures considered here have more complex properties and the relevance of the proposed features is unknown. Thus, methods for exploring relevant features are also presented.

The rest of this paper is organized as follows. Section 2 introduces the problem of monitoring structural health based on vibration measurements and Section 3 presents the frequency domain feature extraction methods considered in this work. Section 4 proposes an approach for exploring the feature space and assessing the attainable damage classification performance with a probabilistic model. Supposedly, features providing good classification performance will also be relevant for novelty detection purposes in the future. Experiments and their results are reported in Section 5 and the paper is summarized in Section 6.

2   Structural Health Monitoring Using Vibration Measurements

Structural health monitoring deals with assessing the condition of a given structure. There are no direct and practical ways of measuring the condition of large structures, but a structural change may well be reflected in the physical measurements. Civil structures like buildings, bridges, and cranes, experience vibrations due to their interaction with the environment (wind, cars, earthquakes, etc.). For instance, a damage in a bridge may well change the properties of the vibrations in the structure.

Such vibration properties may be captured with acceleration measurements. In the current paper, we concentrate on vibration monitoring of a bridge structure using acceleration sensors. The main contribution of this work is to find ways to extract vibration profiles for monitoring and a method for assessing their relevance in detecting changes in a structure.

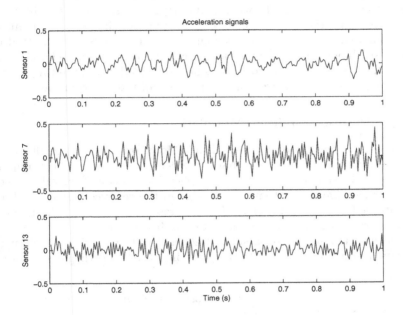

Fig. 1. Acceleration measurements in time domain: some of the several concurrent acceleration measurements at different locations of the structure

A short excerpt of vibration measurements from three acceleration sensors is illustrated in Figure 1. We denote the original time series i of acceleration measuments x from sensor s by $x_s^i = \{x_s[1], x_s[2], \ldots, x_s[N_i - 1], x_s[N_i]\}$.

Modeling each incoming sample of an acceleration measurement directly in the time domain is practically impossible in the supposed lightweight sensor network nodes considered in this work. Firstly, the amplitude of vibration changes randomly due to the environmental effects and the sources of vibration are not measured themselves. The quantities of interest (structural integrity) is expected to be manifested in the dependencies between the measurements of different sensors. Secondly, the feature extraction algorithm needs to run on-line, considering each sample of acceleration at a time, computing intermediate results, and finally discarding the original data. This is because the wireless nodes, considered for our use in the near future, don't facilitate large ring buffers familiar from digital signal processors: the available random access memory (RAM) remains very scarce in the current microcontrollers.

3   Online Monitoring of a Single Frequency

Widely known algorithms like Fast Fourier Transform (FFT) [9] can be used
for the frequency domain analysis of signals. They require a relatively large
buffer for storing the intermediate results, since the whole spectrum is considered
simultaneously. To achieve a frequency resolution below 1 Hz, one would need to
use more than 256-point FFT[1] when monitoring with sampling rate of 256 Hz.

We assume that there is no memory space for perfoming, say, 512-point FFT
on a sensor node, and that the phenomenon of interest is concentrated on a
relatively small portion of the vibration spectrum. In addition, we have observed
that the changes in vibration frequencies are very small, thus requiring relatively
accurate monitoring. Next, we consider a couple of possible solutions.

3.1   Quadrature Amplitude Modulation

Radio receivers using Quadrature Amplitude Modulation (QAM) [2,10] are based
on monitoring a narrow frequency band and detecting changes in the amplitude
and phase of the signal. Obviously, the application domain of digital radio communications is different in that the changes in the received signal are discrete
and controlled by the transmitter. In the present application, the monitored
quantities are continuous and are expected to drift slowly.

The idea of monitoring on a single frequency f begins with correlating the
acceleration measurements $x_s[n]$ with pure sine waves of orthogonal phases:

$$c_s(f) = \frac{1}{N} \sum_{n=1}^{N} x_s[n] \cdot \cos(2\pi(f/f_S)n + \phi_s), \qquad (1)$$

and

$$s_s(f) = \frac{1}{N} \sum_{n=1}^{N} x_s[n] \cdot \sin(2\pi(f/f_S)n + \phi_s), \qquad (2)$$

where f_S is the sampling frequency and the additional phase difference ϕ_s denotes the fact that wireless sensors have independent clocks. The amplitude of
vibration X_s can then be computed as

$$X_s(f) = \sqrt{c_s(f)^2 + s_s(f)^2}. \qquad (3)$$

To make it more suitable for online computing, the following exponentially
decaying window can be used (this can also be considered as the lowpass filter
required in QAM):

$$\tilde{c}_s(f, 0) = 0 \qquad (4)$$

$$\tilde{c}_s(f, n) = (1 - \epsilon) \cdot \tilde{c}_s(f, n - 1) + \epsilon \cdot x_s[n] \cdot \cos(2\pi(f/f_S)n), \qquad (5)$$

---

[1] Accuracy of FFT depends on the length of the considered time window, which also
determines the memory requirements.

where ϵ governs the effective window length of the method. There is a trade-off between accuracy (selectivity between adjacent frequencies) and the rate of convergence: small ϵ results in long windowing and slow response to changes, but also higher frequency resolution.

One of the benefits is that $X_s(f)$ is insensitive to phase differences (ϕ_s) and also small time differences between sensor nodes (caused by sub-ideal synchronization). As in QAM, also the phase information can be computed from the intermediate values c_s and s_s.

This method also resembles Discrete Cosine Transformation (DCT) and Discrete Sine Transformation (DST) [12], where

$$c_s[k] = \sqrt{\frac{2}{N}} \sum_{n=1}^{N} x_s[n] \cdot \cos\left(\frac{\pi k(2n+1)}{2N}\right), \tag{6}$$

and

$$s_s[k] = \sqrt{\frac{2}{N+1}} \sum_{n=1}^{N} x_s[n] \cdot \sin\left(\frac{\pi(k+1)(n+1)}{N+1}\right), \tag{7}$$

and the frequency bin k can be selected according to the monitoring frequency f as $k \approx 2N\frac{f}{f_s} > 0$.

3.2 The Goertzel Algorithm

The above algorithm suffers from the burden of synthesizing cosine and sine signals. A method called the Goertzel algorithm [5,9,13] is able to monitor a single narrow frequency band with even fewer requirements.

The iteration steps of the algorithm can be written as

$$v_k[0] = v_k[-1] = 0, \tag{8}$$
$$v_k[n] = x_s[n] + 2\cos(2\pi k/N) \cdot v_k[n-1] - v_k[n-2], \forall n \in [1, N] \tag{9}$$
$$|X[k]|^2 = v_k^2[N] + v_k^2[N-1] - 2\cos(2\pi k/N) \cdot v_k[N] \cdot v_k[N-1], \tag{10}$$

where $v_k[n]$, $v_k[n-1]$, and $v_k[n-2]$ are the only intermediate results needed for computing the signal power $|X[k]|^2$ at frequency bin k. This can be chosen according to the desired monitoring frequency f as $k \approx N\frac{f}{f_s}$.

The algorithm has several advantages. The cosine is computed only once, and the following computation is in terms of simple multiplications and additions. It is more efficient than FFT when only few frequency bins are needed: for K bins, Goertzel requires $O(KN)$ operations while FFT takes $O(N\log(N))$. For example, if $N = 512$, Goertzel is more (time) efficient if $K \approx 9$.

On the other hand, this algorithm does not provide phase information, which might turn out to be relevant from SHM point of view and the wireless sensor nodes might be synchronized well enough to utilize it. There exist also transformations (e.g. Hadamard and Haar [12]) which can be computed without multiplications, thus being more efficient to implement on wireless sensor nodes.

4 Feature Selection and Classification

Feature selection refers to the process of selecting a subset of features out of the original feature set. An introductory review article [6] divides feature selection methods into three classes: filter-based, wrapper-based, and embedded feature selection methods. In filter-based feature selection, a simplified method is used to select the features for a more complex model. In wrapper-based approach to feature selection, the model itself is used to evaluate the relevance of chosen features. The embedded methods of feature selection select the features as a part of the model estimation. In the current paper, we consider wrapper-based feature selection, since the feature selection is based on the same model as the detection itself.

On the other hand, our long term goal of implementing unsupervised novelty or drift detection algorithms for damage detection purposes prevents us from selecting a final set of relevant features in advance. In a general case, we cannot know what is relevant before the changes appear. Still, the current study provides more insights to the problem.

4.1 Candidate Features

As the final space of candidate features, we consider all pairs of sensors and the ratio of their acceleration amplitudes on a given frequency. While the acceleration amplitudes are computed by the above algorithms, supposedly on-line in a wireless sensor network, the information from several sensors can be combined in a centralized fashion.

In particular, we propose (frequency specific) *transmissibility* which is defined as the ratio of acceleration amplitudes measured by two sensors, s_1 and s_2:

$$T(s_1, s_2, f) = \frac{X_{s_1}(f)}{X_{s_2}(f)}, \tag{11}$$

where f is the monitoring frequency.

This is related to the structure as a medium for vibrations traveling through it: transmissibility describes how well an impulse travels from s_2 to s_1. This point-of-view applies while considering a single impulse from a single source – one might imagine traffic or other multiple sources of vibration making the situation more complex.

The features can also be considered as properties of *mode shapes*. Structures tend to have certain resonance frequencies corresponding to *standing waves* or modes. A simplified situation of this is shown in Figure 2, which shows the vibration amplitudes of a single mode at each point on a homogeneous bar. The ratio of amplitudes measured by two stationary sensors stays constant despite the changes in the overall amplitude of the vibration.

The situation becomes more complex, when the stiffness of the structure is not uniform (not a homogeneous bar) and more vibration modes are involved. This is why we need several sensors for monitoring smaller parts of the structure, several monitoring frequencies to cover each of the modes, and a feature selection approach to explore the large amount of combinations.

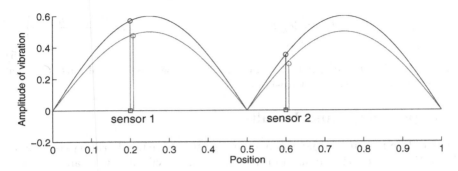

Fig. 2. A mode shape of one standing wave on a simple bar. Amplitude ratio between two sensors stay constant despite the change in amplitude.

4.2 Probabilistic Model for Damage Detection

To assess and demonstrate the relevance of the features, a probabilistic model was used for damage classification. The assumption is that the features found to be relevant in a supervised machine learning setting (classification) would also be relevant in an unsupervised setting (novelty detection). Another reason for benchmarking with a probabilistic model is the ability to cope with missing data in the future applications.

A naive Bayes classifier assumes conditional independence of the features[2] T_d given the class label C [3,1]. Thus, the joint probability distribution decomposes according to

$$P(T_1, \ldots, T_D, C) = P(C) \prod_{d=1}^{D} P(T_d \mid C). \tag{12}$$

Also, we assume that the probability density function of $P(T_d \mid C = c)$ is distributed according to a normal distribution with parameters $\mu_{c,d}$ and $\sigma_{c,d}^2$

$$P(T_d \mid C = c) \propto \frac{1}{\sigma_{c,d}} \exp\left(-\frac{(T_d - \mu_{c,d})^2}{2\sigma_{c,d}^2}\right) \tag{13}$$

Estimation of the parameters is done in the framework of maximum likelihood by estimating the mean and standard deviation for each feature and each class separately. The posterior probability of a class can be used as a decision variable, and can be readily calculated with the Bayes' theorem

$$P(C \mid T_1 \ldots T_D) = \frac{P(C)P(T_1 \ldots T_D \mid C)}{P(T_1 \ldots T_D)}. \tag{14}$$

In Maximum A Posteriori (MAP) classification setting, the denominator is the same for all classes. We can can discard the constant, and base the decision on the following log posterior:

---

[2] T_d refers to transmissibility $T(s_1, s_2, f)$ with some selected parameters $d \sim s_1, s_2, f$.

$$\log P(C = c \mid T_1 \ldots T_D) \propto \log P(C = c) + \sum_{d=1}^{D} \left(-\log \sigma_{c,d} - \frac{(T_d - \mu_{c,d})^2}{2\sigma_{c,d}^2} \right).$$
$$(15)$$

Finally, the classifier can detect a damage if the log posterior of class $C =$"no damage" is below a certain threshold.

5   Experiments and Results

We used data measured from a wooden bridge structure in laboratory conditions. The setting was previously used in [7]. The bridge is vibrated with an electrodynamic shaker to simulate random stimulation from the environment. In order to simulate damages, small weights were attached to the structure during a selected set of the measurements.

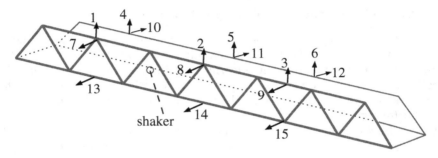

Fig. 3. The structure and the placement of the sensors and the shaker

The bridge was equipped with 15 acceleration sensors, which measured the vibration of the structure as shown in Figure 3. The sampling frequency was 256 Hz and sampling time for each measurement x^i was 32 seconds. The data set consists of 2509 of these measurements.

First, we extracted the amplitude features with the QAM and Goertzel algorithms mentioned above. A uniformly spaced set of frequencies was selected: $\{2, 4, 6, \ldots, 120\}$ Hz, in total, 60 monitoring frequencies below the half of the sampling frequency.

Then, we trained the classifiers using labeled training data set (x^i, c^i), which consisted of every other measurement ($i \in \{1, 3, 5, \ldots, 2509\}$). Finally, we used the rest of the data as a validation set in studying the damage detection performance by using Receiver Operating Characteristic (ROC) curves [11], which visualize the trade-off between sensitivity and specificity for all decision thresholds.

There are 105 possible pairs of sensors (s_1, s_2), and multiplied with the 60 monitoring frequencies the final feature space consists of 6300 features. We continued with selecting D of these by random to see how much information is contained in the random combination of D transmissibility features. The random selection and classifier training was repeated to examine the statistics of the performance.

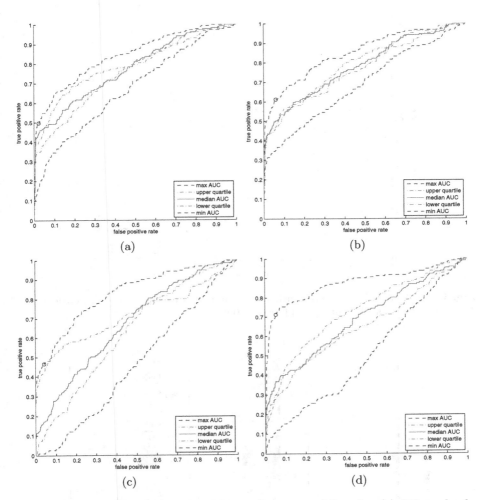

Fig. 4. ROC curves of damage detection performance while using (a) 512 randomly selected features by QAM, (b) 512 random features by the Goertzel algorithm, (c) 32 random features by QAM, and (d) 32 random features by the Goertzel algorithm

Figure 4 shows ROC curves from four populations of 1000 feature sets (and classifiers): QAM feature extraction vs. the Goertzel method, and a small set of features vs. a large set of features. The Goertzel algorithm seemed to provide marginally better results throughout the whole experiment. The damage detection performance, and overall "usefulness" of the feature set and classifier method, can be measured in terms of the area under the ROC curve (AUROC) and the seemingly best classifiers relied on $D = 16$ to $D = 32$ features, as shown in Figure 5.

As an example, we can choose one of the thousands of classifiers (and its feature set): the one which provided the best AUROC in Figure 5 and the best

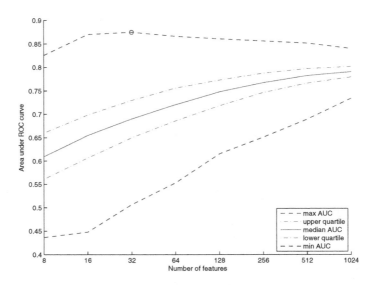

Fig. 5. Area under ROC curves with various numbers of randomly selected features. The maximum is at 32 features and area of 0.87.

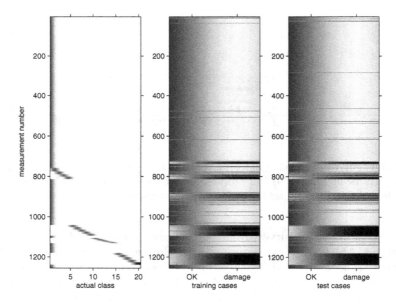

Fig. 6. The correct classification of measurements versus training set and test set damage detection. A good set of 32 features by the Goertzel algorithm and the naive Bayes classifier was used. Class 1 is 'OK' – others are damages.

ROC curve in Figure 4(d). It reaches 71% true positive ratio with mere 5.3% false positive ratio with a certain detection threshold. The detection performance is demonstrated in Figure 6. False negatives seem to happen mostly with the smaller weights (damages) and certain damage locations. For some reason, also false positives are concentrated on consecutive sets of measurements.

Although the features used in the classifiers were selected randomly, some of the classifiers seem to produce reasonable detection performance. We can investigate the selected features in the best performing classifers among the population of random classifiers. Focusing on the most prevalent features in the feature sets, some features occur persistently in many of the best classifers. For instance, features 4286 and 4287 occur 24 and 13 times, respectively. Both of the features measure transmissibility between sensors 7 and 10 on neigboring frequency bins.

More quantitative and detailed examination of the feature sets may lead to a parsimonious list of features that can be used for efficient damage detection in the structural health monitoring setting. Also different kinds of classifiers should be considered in addition to the naive Bayes classifier.

6   Summary and Conclusions

We have investigated a feature extraction problem in the context of structural health monitoring application. Acceleration sensors measure the vibration of a structure, for instance, a bridge. A large number of frequency features can be extracted with frequency monitoring algorithms by combining pairs of measurements to form energy-invariant feature representations. We tried features randomly and estimated probabilistic detection models from labeled data. We summarized the attainable detection performance in terms of ROC curves and the area under the ROC curves.

The proposed methods were able to detect damages even when most of the original data was discarded by the feature extraction and selection process. This provided a baseline result for developing more advanced methods.

Acknowledgments

The current work has been done in the project "Intelligent Structural Health Monitoring System" (ISMO), funded by Helsinki University of Technology (TKK) in Finland and its Multi-disciplinary Institute in Digitalisation and Energy (MIDE). Many of the ideas in this work have arisen as the result of discussions within the ISMO project group, lead by Jyrki Kullaa.

References

1. Bishop, C.M.: Pattern Recognition and Machine Learning. Information Science and Statistics. Springer, New York (2006)
2. Couch, I., Leon, W.: Digital and analog communication systems. Prentice-Hall, Upper Saddle River (2001)

3. Duda, R.O., Hart, P.E., Stork, D.G.: Pattern Classification, 2nd edn. John Wiley & Sons, Chichester (2001)
4. Farrar, C.R., Worden, K.: An introduction to structural health monitoring. Philosophical Transactions of the Royal Society A 365, 303–315 (2007); Published online December 12, 2006
5. Goertzel, G.: An algorithm for evaluation of finite trigonometric series. American Mathematical Monthly 65, 34–35 (1958)
6. Guyon, I., Elisseeff, A.: An introduction to variable and feature selection. Journal of Machine Learning Research 3(1), 1157–1182 (2003)
7. Kullaa, J.: Elimination of environmental influences from damage-sensitive features in a structural health monitoring system. In: Balageas, D.L. (ed.) Proceedings of the First European Workshop on Structural Health Monitoring 2002, pp. 742–749. DEStech Publications Inc., Onera (2002)
8. Lynch, J.P., Sundararajan, A., Law, K.H., Kiremidjian, A.S., Carryer, E.: Embedding damage detection algorithms in a wireless sensing unit for operational power efficiency. Smart Materials and Structures 13(4), 800–810 (2004)
9. Mitra, S.K.: Digital Signal Processing: A Computer-Based Approach, 2nd edn. McGraw-Hill, New York (2002)
10. Proakis, J.G., Salehi, M.: Digital communications. McGraw-Hill, New York (2008)
11. Swets, J.A.: Measuring the accuracy of diagnostic systems. Science 240(4857), 1285–1293 (1988)
12. Theodoridis, S., Koutroumbas, K.: Pattern Recognition. Elsevier, San Diego (2003)
13. Wang, W., Gao, Z., Huang, L., Yao, Y.: Spectrum sensing based on Goertzel algorithm. In: 4th International Conference on Wireless Communications, Networking and Mobile Computing, 2008. WiCOM 2008, October 2008, pp. 1–4 (2008)

Zero-Inflated Boosted Ensembles for Rare Event Counts*

Alexander Borisov[1], George Runger[2], Eugene Tuv[1],
and Nuttha Lurponglukana-Strand[2]

[1] Intel, Chandler, AZ
[2] Industrial and Systems Engineering, Arizona State University, Tempe, AZ

Abstract. Two linked ensembles are used for a supervised learning problem with rare-event counts. With many target instances of zero, more traditional loss functions (such as squared error and class error) are often not relevant and a statistical model leads to a likelihood with two related parameters from a zero-inflated Poisson (ZIP) distribution. In a new approach, a linked pair of gradient boosted tree ensembles are developed to handle the multiple parameters in a manner that can be generalized to other problems. The result is a unique learner that extends machine learning methods to data with nontraditional structures. We empirically compare to two real data sets and two artificial data sets versus a single-tree approach (ZIP-tree) and a statistical generalized linear model.

Introduction

The analysis of count data is of primary interest in many applications such as health, traffic, engineering, and so forth. When the target attribute for a model is a rare event count the appropriate loss function distinguishes the problem from traditional supervised categorical or numerical targets. For example, data that models accident counts from the design attributes of intersections and traffic signals is expected to contain zeros for many intersections. Similarly, data that models the number of products viewed on a Web site from the design attributes of the site are also expected to contain many zeros.

In such cases squared error and class error loss are not usually relevant metrics for model quality. However, a statistical likelihood function can be a useful starting point for an effective machine learning solution. A Poisson regression model is commonly used to explain the relationship between counts (non-negative integers) and input attributes [1]. However, it is often the case that the target data contains more zeros than can be accounted for in Poisson regression.

Lambert [2] proposed a mixture of a Poisson distribution and the distribution with a point mass at zero, called zero-inflated Poisson (ZIP) regression, to handle zero-inflated count data for defects in a manufacturing process. Since then many

---

* This material is based upon work supported by the National Science Foundation under Grant No. 0743160.

N. Adams et al. (Eds.): IDA 2009, LNCS 5772, pp. 225–236, 2009.
© Springer-Verlag Berlin Heidelberg 2009

extensions or modified ZIP models were elaborated. For example, [3] proposed a Markov zero-inflated Poisson regression (MZIP), [4] introduced multivariate ZIP models, [5] proposed a tree-based approach for Poisson regression, [6] used semi-parametric ZIP in animal abundance studies, [7] proposed a weighted ZIP, and [8] used a zero-inflated generalized Poisson (ZIGP) regression model for over-dispersed count data (increased variance). ZIP regression is not only applied in manufacturing, but it is also widely used in many other areas such as public health, epidemiology, sociology, psychology, engineering, agriculture, etc. ([4], [9], [10]).

The statistical literature has focused on linear models and numerical input attributes, but more flexible machine learning models would be valuable. A ZIP-tree model was introduced by [5]. They modified the splitting criteria of the classification and regression tree (CART) algorithm [11] to use the zero-inflated Poisson (ZIP) likelihood error function instead of a residual sum of squares. Each terminal node of a ZIP tree is assigned its own ZIP distribution parameters (zero inflation probability p and Poisson distribution parameter λ). To improve upon a single tree model, ensemble methods are used, especially boosted tree ensembles (AdaBoost [12]) or gradient boosted trees [13]).

A unique approach here uses two linked tree ensembles to model the ZIP distribution parameters p, λ. Although ZIP regression represents our application, the linked ensembles are a new contribution to the machine learning literature. The algorithm minimizes ZIP log-likelihood loss function by a gradient descent method similar to the one proposed for multi-class logistic regression (MCLRT) [13]. Our algorithm uses the log-link function for λ and the logit-link function for p as proposed for standard ZIP regression [2]. The first ensemble predicts $\log(p/1 - p)$, the second one $-\log(\lambda)$ (where log stands for natural logarithm here and in the following). Each subsequent tree in both ensembles is fitted to the corresponding gradient of the loss function by means of the standard CART algorithm with the variance reduction split criteria.

The linked ensembles for ZIP regression is compared to a single ZIP tree and generalized linear models (GLM's) from statistics on actual data. The ZIP ensemble, ZIP tree and ZIP regression (GLM) are also compared on several artificial data sets with ZIP distribution parameters generated as known functions of the inputs. We studied the effect of different underlying target models, noise levels, and parameter settings on the performance of single tree and ensemble model. It is shown that the linked ensembles result in lower log-likelihood (error) value on an independent test set and better approximation for ZIP distribution parameters.

Rare event counts are an important form of data structure and machine learning tools for such models are lacking. The method here illustrates the flexibility and the performance increase that can be obtained from linked ensembles matched to relevant training objectives. Section 1 provides background on Poisson regression. Section 2 summarizes gradient boosting and Section 3 presents our linked ensembles. Section 4 provides simulated and actual experiments and Section 5 provides conclusions.

1 Previous Work: ZIP Regression and ZIP Tree

The Poisson distribution parameters depend on the values of input variables as:

$$y_i \sim \begin{cases} 0, \text{ with probability } p_i, \\ \text{Poisson}(\lambda_i), \text{ with probability } 1 - p_i, \quad i = 1 \ldots n, \end{cases}$$

where n is the number of instances and y_i is the target for i-th instance. This model implies that

$$P(y_i = k) = P(p_i, \lambda_i, k) = \begin{cases} p_i + (1 - p_i)e^{-\lambda_i}, k = 0, \\ (1 - p_i)e^{-\lambda_i}\lambda_i^k/k!, k = 1, 2, \ldots. \end{cases}$$

For a ZIP regression model the parameters λ, p are obtained from the linear combinations of inputs via *log-* and *logit*-link functions:

$$\log(\lambda_i) = \beta x_i, \text{ and } \log it(p_i) = \log\left(\frac{p_i}{1 - p_i}\right) = \gamma x_i.$$

Here x_i is the input feature vector for observed row i (and for simplicity of notation we always assume that a constant variable $x = 1$ is added as the first input variable to take the intercept term into account), β, γ are vectors of coefficients to be fit. The ZIP model is usually fit using maximum likelihood estimation. The log-likelihood function for this model is:

$$L(\beta, \gamma, y) = \sum_{i=1}^{n} L(\beta, \gamma, y_i) = \sum_{i=1}^{n} \log P(\beta, \gamma, y_i)$$

$$= \sum_{y_i=0} \log\left(\exp(\gamma x_i) + \exp(-e^{\beta x_i})\right) + \sum_{y_i>0} (y_i \beta x_i - \exp(\beta x_i))$$

$$- \sum_{i=1}^{n} \log\left(1 + \exp(\gamma x_i)\right) - \sum_{y_i>0} \log(y_i!),$$

Log-likelihood can be maximized using the Newton-Raphson method, but usually the expectation-maximization (EM) algorithm [14] is used because of the complexity of calculating the Hessian in the Newton-Raphson method (the log of the double exponent makes it complex), and the better robustness of EM. As mentioned by [2], the EM always converges (local minimum) but Newton-Raphson failed in some experiments. Different models for the p parameter of ZIP distribution were compared by [2]. In addition to the independent linear model for p and λ, p was modeled as a function of λ. That can be a reasonable assumption in many cases and simplifies fitting the model. The same ZIP regression model was applied by [15] to decayed, missing and filled teeth (DMFT) data. They used a piecewise constant model for the p parameter. Both [2] and [15] considered using mixture models (with the most popular a mixture of Poisson and negative binomial distributions), but claimed that such models are more difficult to fit and usually provide worse predictions than ZIP models.

The ZIP likelihood was used as a splitting criterion for a decision tree by [5]. They modified the CART algorithm to use the negative ZIP likelihood as an impurity measure in a node. The negative ZIP likelihood of the data in node T can be expressed as

$$L_{ZIP}(T) = L_{ZIP}(p, \lambda, y) = -n_0 \cdot \log \left(p + (1-p)e^{-\lambda} \right)$$

$$-(n - n_0) \cdot (\log(1-p) - \lambda) - \sum_{x_i \in T} y_i \cdot \log \lambda + \sum_{x_i \in T, y_i > 0} \log(y_i!),$$

where p, λ are estimates of Poisson distribution parameters in node T.

The new splitting criterion is based on the difference of the ZIP likelihood in the left and the right child nodes from the ZIP likelihood in the parent node. The expression for split weight can be written as

$$\phi(s, T) = L_{ZIP}(T) - L_{ZIP}(T_L) - L_{ZIP}(T_R)$$

where T is the parent node, T_L, T_R are the left and right children of T, and s is the split in node T. The same best split search strategy can be applied as in the CART algorithm with this revised splitting criterion.

To compute the ZIP likelihood function, the parameters p and λ in a tree node are determined as maximum likelihood estimates. For ZIP likelihood, there is no closed form solution for these MLEs (one needs to solve a complex optimization problem). Therefore, [15] fit the λ parameter using zero-truncated Poisson distribution (thus taking only samples with $y_i > 0$ in node into account):

$$\frac{\lambda}{1 - e^{-\lambda}} = \bar{y} = \text{mean}(y_i | y_i > 0, x_i \in T).$$

After an estimate for the λ parameter is obtained, p can be estimated from the known proportion of zero-class samples in the node:

$$p = \frac{n_T^0 / n_T - e^{-\lambda}}{1 - e^{-\lambda}},$$

where n_T^0 is the number of zero count instances and n_T is total number of instances in node T.

2   Boosting Framework

The gradient boosting framework is now described more formally. Given training data $\{y_i, x_i\}_{i=1...n}, x_i = \{x_{i1}, \ldots, x_{iK}\} \in X, y_i \in Y$, where n is the number of data instances, K is the number of input attributes. Our goal is to find a function $F^*(x) : X \to Y$ that minimizes the expected value of the specified loss function $L(y, F(x))$ over the joint distribution of x, y values:

$$F^*(x) = \arg \min_{F(x)} E_{x,y} \, L(y, F(x)).$$

Here the expectation term cannot usually be computed directly as the joint distribution of x, y is not known. So in practice it is replaced with expected risk, i.e.:

$$F^*(x) = \arg\min_{F(x)} \sum_{i=1}^{n} L(y_i, F(x_i)).$$

Boosting uses an additive model to approximate $F^*(x)$: $F^*(x) = \sum_{m=0}^{M} h(x, a_m)$, where the function $h(x, a_m)$ is some simple function ("base learner") of parameter vector a. Base learner parameters $a_m, m = 1 \ldots M$ are fit in a forward stepwise manner. It starts from some initial approximation $F_0(x)$, then proceeds as follows:

$$a_m = \arg\min_{a} \sum_{i=1}^{n} L\left(y_i, F_{m-1}(x_i) + h(x_i, a)\right), \qquad (1)$$

$$F_m(x) = F_{m-1}(x) + h(x, a_m).$$

Gradient boosting [13] solves the optimization problem (1) using the stepwise steepest descent method. The function $h(x, a)$ is fit by least squares:

$$a_m = \arg\min_{a} \sum_{i=1}^{n} (\tilde{y}_{im} - h(x_i, a))^2$$

to the current "pseudo-residuals" or "pseudo-responses" :

$$\tilde{y}_{im} = -\left[\frac{\partial L(y_i, F(x_i))}{\partial F(x_i)}\right]_{F(x)=F_{m-1}(x)}. \qquad (2)$$

A gradient boosted tree (GBT) is the specialization of this approach to the case where the base learner is a regression tree:

$$h(x, \{R_{lm}\}_{l=1\ldots L}) = \sum_{l=1}^{L} \gamma_{lm} \cdot I(x \in R_{lm}),$$

where L is the number of terminal regions (nodes) R_{lm}, and γ_{lm} is the response (mean) in node R_{lm}. The tree is fit using the CART algorithm [11] for the regression case with variance reduction criteria. This gives rise to the following algorithm:

Algorithm 1: Gradient boosted tree

1. $F_0(x) = \arg\min_{\gamma} \sum_{i=1}^{n} L(y_i, \gamma)$

2. For $m = 1$ to M do:
3. $\tilde{y}_{im} = -\left[\frac{\partial L(y_i, F(x_i))}{\partial F(x_i)}\right]_{F(x)=F_{m-1}(x)}, i = 1 \ldots n$
4. $\{R_{lm}\}_{l=1\ldots L} = L-$ terminal node tree
5. $\gamma_{lm} = \arg\min_{\gamma} \sum_{x_i \in R_{lm}} L(y_i, F_{m-1}(x_i) + \gamma)$
6. $F_m(x) = F_{m-1}(x) + \nu \cdot \gamma_{lm} I(x \in R_{lm})$
7. End for

Here ν is a "shrinkage rate" or "regularization parameter" that controls the learning rate of the algorithm. Smaller values for shrinkage (0.01-0.1) reduce over-fitting, thus building models with better generalization ability. Usually only a randomly selected percentage of samples (about 60%) are used to learn a tree on step 4 (bootstrapping). This speeds up the model building and also reduces over-fitting.

A particularly interesting case of Algorithm 1 is two-class logistic regression (that also has a multi-class generalization that we omit). It is derived from the GBT framework with a CART tree as a base learner, and the negative binomial log-likelihood as the loss function. Assume that the response is binary, $y \in \{-1, 1\}$, and the loss function is negative binomial log-likelihood : $L(y, F) = \log(1 + \exp(-2yF))$, where F is a two-class logistic transform:

$$F(x) = \frac{1}{2} \log \left[\frac{\Pr(y = 1|x)}{\Pr(y = -1|x)} \right]. \tag{3}$$

Each tree approximates the log-odd of class 1 probability, and the pseudo-response derived from equation (2) or step 3 of Algorithm 1 is $\tilde{y}_{im} = 2y_i/(1 + \exp(2y_i F_{m-1}(x_i))$.

The optimization problem in step 5 cannot be solved in closed form, so a single Newton-Raphson step approximation is used, which leads to the following expression

$$\gamma_{lm} = \frac{-\sum_{x_i \in R_{lm}} \frac{\partial L(y_i, F_{m-1}(x_i)+\gamma)}{\partial \gamma}}{\sum_{x_i \in R_{lm}} \frac{\partial^2 L(y_i, F_{m-1}(x_i)+\gamma)}{\partial \gamma^2}} = \frac{-\sum_{x_i \in R_{lm}} \frac{\partial L(y_i, f)}{\partial f} \Big|_{f=F_{m-1}(x_i)}}{\sum_{x_i \in R_{lm}} \frac{\partial^2 L(y_i, f)}{\partial f^2} \Big|_{f=F_{m-1}(x_i)}} \tag{4}$$
$$= \sum_{x_i \in R_{jm}} \tilde{y}_{im} / \sum_{x_i \in R_{jm}} |\tilde{y}_{im}| \cdot (2 - |\tilde{y}_{im}|).$$

After the ensemble model $F_M(x)$ for log-odds is built, probabilities of classes can be derived from (3):

$$p_+(x) = \Pr(y = 1|x) = 1/(1 + \exp(-2F_M(x))),$$
$$p_-(x) = \Pr(y = -1|x) = 1/(1 + \exp(2F_M(x))).$$

These can also be used for classification. The predicted response is the one having higher probability.

Both the prediction accuracy and the execution speed of gradient boosting can be substantially improved with dynamic feature selection which promotes more relevant features [16]. This also reduces the over-fitting effect. Furthermore, feature selection can be applied initially to reduce the dimensionality of the problem. The artificial contrast ensemble (ACE) method [17] integrates GBT in a comprehensive feature selection algorithm.

To increase the robustness of GBT algorithm, influence trimming [13] can be applied when selecting samples for building a subsequent tree. Samples with low influence for estimated base learner parameters can be omitted in tree construction. Influence can be estimated as follows. We estimate the response in a terminal node on step 5 of Algorithm 1 via the equation

$$\sum_{x_i \in R_{lm}} \partial L(y_i, F_{m-1}(x_i) + \gamma)/\partial \gamma = 0.$$

The influence of the i-th instance on the solution can be gauged by the second derivative of the loss function, i.e.,

$$w_i = w(x_i) = \left.\frac{\partial^2 L(y_i, F_{m-1}(x_i)+\gamma)}{\partial\gamma^2}\right|_{\gamma=0} = \left.\frac{\partial^2 L(y_i, f)}{\partial f^2}\right|_{F_{m-1}(x_i)} = |\tilde{y}_{im}| \cdot (2 - |\tilde{y}_{im}|).$$

Influence trimming not only speeds up the tree construction, but also improves robustness of the Newton-Raphson method step in equation (4) by preventing small denominator values for a tree node. The denominator is proportional to the sum of sample influences in the node. Influence trimming deletes all observations with $w_i < w_{l(\alpha)}$, where $l(\alpha)$ is the solution to $\sum_{i=1}^{l(\alpha)} w_{(i)} = \alpha \cdot \sum_{i=1}^{N} w_i$. Here the weights w_i's sorted in ascending order, and α is usually chosen in the range $[0.05, 0.2]$.

3   ZIP-Boosted Ensemble

A unique characteristic of ZIP regression is the inter-related parameters (p, λ). Two ensembles are simultaneously used to approximate the transformed Poisson regression parameters. A negative ZIP-likelihood is used as the loss function, and the CART model as a base learner. From the transformations log-link for p and logit-link for λ used by [2] we obtain:

$$\mu = \log\left(p/(1-p)\right), p = e^\mu/(1+e^\mu), \ \nu = \log(\lambda), \lambda = e^\nu.$$

The first ensemble fits a model for $\mu(x)$, and the second one for $\nu(x)$. The initial value for ν is estimated from a zero truncated Poisson distribution of the target:

$$\frac{\lambda_0}{1-e^{-\lambda_0}} = \bar{y} = \text{mean}(y_i|y_i > 0), \nu_0 = \log(\lambda_0),$$

and μ_0 as

$$p_0 = \frac{n_0/n - e^{-\lambda_0}}{1-e^{-\lambda_0}}, \mu_0 = \text{logit}(p_0),$$

where n_0 is the number of zero-class instances $(y_i = 0)$.

The loss function to be minimized takes the form

$$L(y, p, \lambda) = \sum_{i=1}^{n} L(y_i, p_i, \lambda_i) = -\sum_{y_i=0} \log\left(p_i + (1-p_i)e^{-\lambda_i}\right)$$
$$- \sum_{y_i>0} (\log(1-p_i) - \lambda_i) - \sum_{y_i>0} y_i \log \lambda_i + \sum_{y_i>0} \log(y_i!),$$

where we denote $p_i = p(x_i), \lambda_i = \lambda(x_i)$ to simplify notation. The last term is not dependent on the model and can be dropped. In the other terms,

$$L(y, p, \lambda) = L(y, \mu, \nu) = \sum_{i=1}^{n} L(y_i, \mu_i, \nu_i)$$
$$= -\sum_{y_i=0} \left(\log(e^{\mu_i} + \exp(-e^{\nu_i}) - \log(1+e^{\mu_i})\right) + \sum_{y_i>0} \left(\log(1+e^{\mu_i}) + e^{\mu_i}\right)$$
$$- \sum_{y_i>0} y_i\nu_i,$$

where $\mu_i = \mu(x_i), \nu_i = \nu(x_i)$.

Pseudo-responses are calculated as follows. The pseudo-responses for the p-ensemble are

$$\tilde{\mu}_{im} = - \left[\frac{\partial L(y_i, \mu_i, \nu_i)}{\partial \mu_i} \right]_{\mu_i = \mu_{m-1}(x), \nu_i = \nu_{m-1}(x)}$$

$$= \left[\frac{\partial L(y_i, p_i, \lambda_{m-1}(x_i))}{\partial p_i} \right]_{p_i = p_{m-1}(x)} \cdot \left[\frac{\partial p(\mu_i)}{\partial \mu_i} \right]_{\mu_i = \mu_{m-1}(x)}$$

$$= \begin{cases} \frac{(e^{-\lambda_i} - 1)}{p_i + (1 - p_i) e^{-\lambda_i}} \cdot p_i(1 - p_i), & y_i = 0, \\ 1/(1 - p_i) \cdot p_i(1 - p_i) = p_i & y_i > 0, \end{cases}$$

where $p_i = p_{m-1}(x_i), \lambda_i = \lambda_{m-1}(x_i)$. Here the pseudo-response is expressed in terms of p_i, λ_i to simplify notation.

The pseudo-responses for the λ-ensemble are derived in the same way. Note that $\partial p(\mu)/\partial \mu = e^\mu/(1 + e^\mu)^2 = p(1 - p), \partial \lambda(\nu)/\partial \nu = e^\nu = \lambda$ and

$$\tilde{\nu}_{im} = \begin{cases} \frac{\lambda_i e^{-\lambda_i}}{p_i/(1 - p_i) + e^{-\lambda_i}}, & y_i = 0, \\ \lambda_i - y_i, & y_i > 0. \end{cases}$$

The predicted target for a node is an optimization problem (in step 5 of algorithm 1) that is solved via a single step of Newton-Raphson. Unfortunately in our case the Hessian (second derivative) can sometimes be negative. Such occasions are rare and possibly indicate over-fitting or "self-contradictory" data (i.e., cases when data instances with similar x values have very different (p, λ) values). A negative Hessian means that the target function is not concave and thus cannot be approximated by a second-order polynomial. In such a case we use one step of steepest descent instead of the Newton-Raphson step. The second derivatives for the p-tree (which are summands in the denominator in (4)) are:

$$\frac{\partial^2 L(y_i, \mu_{m-1}(x_i) + \gamma, \nu_{m-1}(x_i))}{\partial \gamma^2} \bigg|_{\gamma = 0} = \frac{\partial^2 L(y_i, \mu_i, \nu_{m-1}(x_i))}{\partial \mu_i^2} \bigg|_{\mu_i = \mu_{m-1}(x_i)} = \tilde{\tilde{\mu}}_{im}$$

$$= \begin{cases} -\frac{(1 - p_i)^2 e^{-\lambda_i} - p_i^2}{(p_i + (1 - p_i) e^{-\lambda_i})^2} \cdot p_i(1 - p_i) \cdot (1 - e^{-\lambda_i}), & y_i = 0, \\ p_i(1 - p_i), & y_i > 0. \end{cases}$$

Similarly, for the λ-tree:

$$\frac{\partial^2 L(y_i, \mu_{m-1}(x_i), \nu_{m-1}(x_i) + \gamma)}{\partial \gamma^2} \bigg|_{\gamma = 0} = \frac{\partial^2 L(y_i, \mu_{m-1}(x_i), \nu_i)}{\partial \nu_i^2} \bigg|_{\nu_i = \nu_{m-1}(x_i)} = \tilde{\tilde{\nu}}_{im}$$

$$= \begin{cases} \lambda_i(1 - p_i) \cdot \frac{1 - p_i + p_i e^{\lambda_i} \cdot (1 - \lambda_i)}{(p_i + (1 - p_i) e^{\lambda_i})^2}, & y_i = 0, \\ \lambda_i, & y_i > 0. \end{cases}$$

The formula (4) for the "optimal" response in a p-tree terminal node is

$$\gamma_{lm}^1 = \begin{cases} \frac{\sum_{x_i \in R_{jm}} \tilde{\mu}_{im}}{\sum_{x_i \in R_{jm}} \tilde{\tilde{\mu}}_{im}}, & \text{if } \sum_{x_i \in R_{jm}} \tilde{\tilde{\mu}}_{im} > \varepsilon = 10^{-6}, \\ \sum_{x_i \in R_{jm}} \frac{\tilde{\mu}_{im}}{n(R_{jm})}, & \text{otherwise,} \end{cases}$$

where $n(R_{jm})$ is the count of training instances in node R_{jm}. Similarly, for the λ-tree

$$\gamma_{lm}^2 = \begin{cases} \frac{\sum_{x_i \in R_{jm}} \tilde{\nu}_{im}}{\sum_{x_i \in R_{jm}} \tilde{\tilde{\nu}}_{im}}, \text{if } \sum_{x_i \in R_{jm}} \tilde{\tilde{\nu}}_{im} > \varepsilon, \\ \sum_{x_i \in R_{jm}} \frac{\tilde{\nu}_{im}}{n(R_{jm})}, \text{otherwise.} \end{cases}$$

The algorithm is of substantial interest for rare-event count data and has been used successfully in the semiconductor industry. Several tricks were used to improve the numerical stability of the algorithm. To prevent μ_i, ν_i from causing numerical overflows (or underflows) we simply threshold by a reasonable constant. That is, if x denotes either μ or ν require $\exp(x) < fmax/2$ (or $x < \log(fmax/2)$) and for underflow $x > -\log(fmax/2)$ where $fmax$ is the maximum floating point value.

We also adopted the influence trimming strategy to prevent a small absolute Hessian in a tree node. We found that one cannot remove instances where the second derivative of the loss function is negative because it can severely harm the performance of the algorithm. However, one can trim samples with a small absolute second derivative in a p-tree. Consequently, we do not use influence trimming for the λ-tree (as small absolute second derivatives of the loss function are not likely to happen there), but apply influence trimming with weights $w_i = p_i(1 - p_i)$ for the p-tree in the way described earlier for two-class logistic regression.

4   Evaluation

First we validate our ZIP boosting algorithm and compare its performance with our implementation of the ZIP tree on two artificial data sets. Both data sets are generated from known models for the ZIP distribution parameters (p, λ) with a small amount of random noise added as

$$p = p(x_1, x_2) \cdot (1 + \varepsilon \cdot u_1), u_1 \in U(-1, 1),$$
$$\lambda = \lambda(x_1, x_2) \cdot (1 + \varepsilon \cdot u_2), u_2 \in U(-1, 1).$$

Then target value y_i is generated from a ZIP distribution with parameters $(p_i = p(x_{1i}, x_{2i}), \lambda_i = \lambda(x_{1i}, x_{2i}))$. In all three experiments three values for the noise level $\varepsilon = 0, 0.2, 0.5$ are used. The size of all data sets is 10000 samples.

The first data set uses a linear model for (p, λ) as

$$p = 0.2 + 0.6 \cdot (0.3x_1 + 0.7x_2), \ \lambda = 1.5 + 7 \cdot (0.6x_1 + 0.4x_2).$$

The second data set uses a more complex, highly nonlinear model

$$\text{logit}(p) = 2\sin(20x_1) + 3x_2 \cdot (x_2 - 0.5), \ \log(\lambda) = \sin(30x_1) + 3x_2.$$

In all experiments the model complexity (which is the pruning step for the tree and the number of iterations for GBT) is selected based on the best CV error.

Table 1. Comparisons of ZIP tree, ZIP GBT, and GLM on two artificial data sets

Data	Noise (ε)	Base error	Model	Train error	CV error	δp	$\delta\lambda$	$\delta\lambda_{rel}$
LINEAR	0	1.801	TREE	1.663	1.690	0.043	0.355	0.074
			GBT	1.653	1.675	0.027	0.182	0.038
			GLM	1.673	1.673	0.01	0.196	0.043
	0.2	1.859	TREE	1.707	1.736	0.043	0.416	0.092
			GBT	1.702	1.721	0.032	0.179	0.040
			GLM	1.705	1.705	0.01	0.225	0.048
	0.5	1.873	TREE	1.744	1.775	0.040	0.441	0.093
			GBT	1.733	1.754	0.032	0.234	0.049
			GLM	1.75	1.75	0.018	0.183	0.041
NON-LINEAR	0	2.920	TREE	1.535	1.675	0.146	3.105	0.403
			GBT	1.360	1.413	0.058	1.844	0.255
			GLM	2.08	2.08	0.249	4.783	1.012
	0.2	3.037	TREE	1.594	1.735	0.156	3.027	0.423
			GBT	1.425	1.492	0.064	1.810	0.247
			GLM	2.148	2.149	0.25	4.757	1.022
	0.5	3.310	TREE	1.774	1.925	0.154	3.112	0.394
			GBT	1.577	1.663	0.073	1.812	0.253
			GLM	2.34	2.332	0.253	4.754	1.014

For artificial data sets, the following parameters are used. The ZIP TREE used tree_depth = 6, min_split = 50, min_bucket = 20 and ZIP GBT used nit = 1000, tree_depth = 3, min_split = 400, min_bucket = 200, shrinkage = 0.01, infl_trimming = 0.1. Here tree_depth is a maximum tree depth (so that a node is not split if it is at the specified depth), min_split is a minimum size of the node that will be split (so that if it has less observations it is NOT split), min_bucket is a minimum size of the terminal node (so that the split is not accepted if it creates a terminal node with smaller size), nit is a maximum number of iterations for an ensemble, shrinkage is the ν parameter (regularization) on step 6 of Algorithm 1, infl_trimming is the α threshold for influence trimming.

The base error column in Table 1 shows the negative ZIP log-likelihood for the best constant model. The training and CV-error (5-fold) are also calculated for the negative ZIP log-likelihood. Furthermore, δp is the average absolute difference in the estimated p parameter ($\delta p = \sum_{i=1}^{n} |p(x_{1i}, x_{2i}) - \hat{p}(x_{1i}, x_{2i})|/n$ where $\hat{p}(x_1, x_2)$ is the prediction from the model), $\delta\lambda$ is the average absolute difference in the estimated λ parameter ($\delta\lambda = \sum_{i=1}^{n} |\lambda(x_{1i}, x_{2i}) - \hat{\lambda}(x_{1i}, x_{2i})|/n$), and $\delta\lambda_{rel}$ is the average relative difference in the estimated λ parameter ($\delta\lambda_{rel} = \sum_{i=1}^{n} |1 - \hat{\lambda}(x_{1i}, x_{2i})/\lambda(x_{1i}, x_{2i})|/n$). These three latter values show how well ZIP distribution parameters are approximated by the model.

For the examples in Table 1 GBT is always superior to a single tree in terms of training error, CV error and parameter estimation error. The data subsets in the CV results for the GBT and single tree are identical and the standard deviations are small. All differences are statistically significant at a 5% level.

Also, one can see that over-fitting (the difference between CV and train errors) is much smaller for GBT, especially for larger noise levels and more complex models. For the linear model GLM is expected to perform well. The data subsets in the CV results for the GBT and GLM are not identical but because of the large sample sizes all differences are again statistically significant at a 5% level. But even for the linear model GBT provides nearly equivalent performance from a practical perspective, and it substantially excels for the nonlinear model.

We also compared the performance of ZIP GBT, ZIP tree, and GLM on actual data sets well known in the statistical literature in Table 2. Lambert [2] studied the behavior of the linear algorithm on soldering data from AT&T Bell Labs. We were not able to find this data, but there is another similar soldering data (SOLDER) that is shipped with the free R software package Rpart that we used. The second is DMFT (decayed, missing and filled teeth) data set used by [15]. Parameters of both algorithms were adjusted manually to minimize cross-validation error. ZIP TREE used tree_depth = 6, min_split = 15, min_bucket = 10 and ZIP GBT used nit = 1000, tree depth = 3, min_split = 30, min_bucket = 20, shrinkage = 0.02(0.005 for DMFT), infl_trimming = 0.1. On the SOLDER data set, ZIP GBT is much better than a single tree in terms of cross-validated log-likelihood, and better than GLM. The data subsets in the CV results for the GBT and a single tree are identical, but randomized for GLM. From the CV standard deviations (in the last column) all differences are statistically significant at a level much smaller than 5%. For DMFT the data is not very informative and consequently the algorithms all perform similarly. The data subsets in the CV folds were selected as before and the CV error for ZIP GBT is statistically different at 5%.

Table 2. Comparisons of ZIP tree, ZIP GBT, and GLM on actual data sets

Data	Base error	Model	Train error	CV error	CV stdev
SOLDER	4.464	TREE	2.493	2.714	0.043
SOLDER		GBT	1.510	1.818	0.012
SOLDER		GLM	1.819	1.934	0.019
DMFT	1.789	TREE	1.548	1.563	0.010
DMFT		GBT	1.519	1.550	0.003
DMFT		GLM	1.556	1.568	0.005

5 Conclusion

This work introduces a new modeling technique based on two linked ensembles for a supervised learning problem with rare-event counts. The methodology extends machine learning to supervised problems where the usual loss functions are not relevant. The approach provides an outline so that, as needed for the statistical objective, multiple linked ensembles can be integrated. The tree based ensembles provide the learner with substantial benefits over the more traditional, statistical GLM models. The ensemble easily handle mixed data, high dimensions, and interactions. For simulated examples with nonlinear models the algorithm's performance (both in terms of the log-likelihood value and the prediction of the ZIP distribution parameters as a function of inputs) is superior to the performance of a ZIP tree and GLM models.

References

1. Cameron, A., Trivedi, P.: Regression analysis of count data. Cambridge University Press, Cambridge (1998)
2. Lambert, D.: Zero-inflated poisson regression with an application to defects in manufacturing. Technometrics 34(1), 1–14 (1992)
3. Wang, P.: Markov zero-inflated poisson regression models for a time series of counts with excess zeros. Journal of Applied Statistics 28(5), 623–632 (2001)
4. Li, C., Lu, J., Park, J., Kim, K., Brinkley, P., Peterson, J.: Multivariate zero-inflated poisson models and their applications. Technometrics 41(1), 29–38 (1999)
5. Lee, S., Jin, S.: Decision tree approaches for zero-inflated count data. Journal of applied statistics 33(8), 853–865 (2006)
6. Chiogna, M., Gaetan, C.: Semiparametric zero-inflated poisson models with application to animal abundance studies. Environmetrics 18, 303–314 (2007)
7. Hsu, C.: A weighted zero-inflated poisson model for estimation of recurrence of adenomas. Statistical Methods in Medical Research 16, 155–166 (2007)
8. Famoye, F., Singh, K.: Zero-inflated generalized poisson regression model with an application to domestic violence data. Journal of Data Science 4, 117–130 (2006)
9. Cheung, Y.: Zero-inflated models for regression analysis of count data: a study of growth and development. Statistics in Medicine 21, 1461–1469 (2002)
10. Yau, K., Lee, A.: Zero-inflated poisson regression with random effects to evaluate an occupational injury prevention programme. Statistics in Medicine 20, 2907–2920 (2001)
11. Breiman, L., Friedman, J., Olshen, R.A., Stone, C.J.: Classification and Regression Trees. Chapman and Hall/CRC, New York (1984)
12. Freund, Y., Schapire, R.: A decision-theoretic generalization of on-line learning and an application to boosting. Journal of Computer and System Sciences 55, 119–139 (1997)
13. Friedman, J.H.: Greedy function approximation: A gradient boosting machine. The Annals of Statistics 29(5), 1189–1232 (2001)
14. Hastie, T., Tibshirani, R., Friedman, J.: Elements of statistical learning. Springer, New York (2001)
15. Bohning, D., Dietz, E., Schlattman, P., Mendonca, L., Kirchner, U.: The zero-inflated poisson model and the decayed, missing and filled teeth index in dental epidemiology. Journal of the Royal Statistcal Society: Series A 162(2), 195–209 (1999)
16. Borisov, A., Eruhimov, V., Tuv, E.: Tree-based ensembles with dynamic soft feature selection. In: Guyon, I., Gunn, S., Nikravesh, M., Zadeh, L. (eds.) Feature Extraction Foundations and Applications. Studies in Fuzziness and Soft Computing. Springer, Heidelberg (2006)
17. Tuv, E., Borisov, A., Runger, G., Torkkola, K.: Best subset feature selection with ensembles, artificial variables, and redundancy elimination. Journal of Machine Learning Research (2008) (to appear)

Mining the Temporal Dimension
of the Information Propagation

Michele Berlingerio[1], Michele Coscia[2], and Fosca Giannotti[3]

[1] IMT-Lucca, Lucca, Italy
[2] Dipartimento di Informatica, Pisa, Italy
{name.surname}@isti.cnr.it
[3] ISTI-CNR, Pisa, Italy

Abstract. In the last decade, Social Network Analysis has been a field in which the effort devoted from several researchers in the Data Mining area has increased very fast. Among the possible related topics, the study of the information propagation in a network attracted the interest of many researchers, also from the industrial world. However, only a few answers to the questions "How does the information propagates over a network, why and how fast?" have been discovered so far. On the other hand, these answers are of large interest, since they help in the tasks of finding experts in a network, assessing viral marketing strategies, identifying fast or slow paths of the information inside a collaborative network. In this paper we study the problem of finding frequent patterns in a network with the help of two different techniques: TAS (Temporally Annotated Sequences) mining, aimed at extracting sequential patterns where each transition between two events is annotated with a typical transition time that emerges from input data, and Graph Mining, which is helpful for locally analyzing the nodes of the networks with their properties. Finally we show preliminary results done in the direction of mining the information propagation over a network, performed on two well known email datasets, that show the power of the combination of these two approaches.

1   Introduction

In the last decade, the interest in Social Network Analysis topics from researchers in the Data Mining area has increased very fast. Much effort has been devoted, for example, in the Community Discovery, Leader Detection and Network Evolution problems [7,18,4,17]. Another topic that has attracted much interest recently is how the information propagates over a network [10,1,14,13]. This problem has been studied from several points of view: statistics, modeling, mining are few of the approaches that have been applied so far in this direction. However, only a few answers to the questions "How does the information propagates over a network, why and how fast?" have been discovered so far. On the other hand, these answers are of large interest, since they help in the tasks of finding experts in a network, assessing viral marketing strategies, identifying fast or slow paths

N. Adams et al. (Eds.): IDA 2009, LNCS 5772, pp. 237–248, 2009.
© Springer-Verlag Berlin Heidelberg 2009

of the information inside a collaborative network, and so on. In this paper we study the problem of finding frequent patterns in a network focusing in two aspects:

- The temporal dimension intrinsically contained in the flow of information: why certain topics are spread faster than others? What is the distribution of the temporal intervals among the "hops" that the information passes through?
- The causes of the information propagation: why certain discussions are passed over while others stop in two hops? What are the characteristics of the nodes that pass the information?

As one can notice, the two dimensions of our focus are orthogonal to each other: certain nodes with certain characteristics may let a particular kind of information spread faster or slower than other nodes, or compared to information with other characteristics. The combination of the two aspects finds several possible application in real life. Among all of them, we believe that Viral Marketing can be powerfully enhanced by such kind of analysis. Companies willing to advertise a new product in their network of users may discover that giving a certain kind of information or special offers to a particular set of selected nodes may result in a cheaper or more effective advertisement campaign.

In this paper, we study the above problem on the well known Enron email dataset [12], and the 20 Newsgroups dataset [11,16], and with the help of two different techniques: TAS (Temporally Annotated Sequences) mining, which is a paradigm aimed at extracting sequential patterns where each transition between two events is annotated with a typical transition time that emerges from input data, and Graph Mining, which is helpful for locally analyzing the nodes of the networks with their properties.

The contribution of this paper can be summarized as follows: we show how to extract useful information from a network in order to mine the information propagation, in the format of a graph where nodes are users and edges are words used as email subjects, and a set of timestamped sequences of emails grouped by threads; we show how to apply the two techniques above to a real-life dataset; we present the preliminary results obtained by applying the two algorithms on graphs extracted from the datasets, and on the sequences of exchanged email, showing a general methodology that can be applied in any sort of network where an exchange of information is present.

The rest of the paper is organized as follows: section 2 presents some work related to our problem; section 3 defines what is the problem under investigation and which kind of data we want to analyze; section 4 shows the preliminary results obtained during our analysis of the datasets; section 5 briefly summarizes the results of our work and some possible future work.

2   Related Work

During the last years, several approaches have been proposed addressing the problem of analyzing how the information propagates in a network.

In [10], the authors summarize three papers focusing on finding communities and analyzing the small world phenomenon by means of statistical approaches. In [1], the authors describe general categories of information epidemics and create a tool to infer and visualize the paths specific infections take through the network by means of statistical tools and Support Vector Machines. In [6], the model of timestamped graph and digraph are introduced in order to study the influence in a network. In [14] the problem of Viral Marketing is analyzed with several different statistical approaches.

Among several other possible works, we believe that [13,15] are the closest to our work: they focus on the temporal behavior in the network, and in the characteristic of the users in the network.

However, to the best of our knowledge, this is the first time that TAS mining and Graph Mining are used in conjunction in order to tackle the problem of finding frequent patterns of information propagation together with their causes in a network.

3 Problem Definition

We are given a dataset $\mathcal{D}$ of activities in a network, from which we can extract both a network of users $\mathcal{U}$ as a graph $\mathcal{G}$ and a flow of any kind of information (emails, documents, comments, instant messages, etc.) as a set of timestamped sequences $\mathcal{S}$. Examples of such datasets can be a set of emails exchanged among people, the logs of an instant messaging service, the logs of a social networking system, the content of a social bookmarking site, and so on. In this dataset we are interested in finding frequent patterns of information propagation, and we want to let the causes of such patterns emerge from the data. This can be done by applying a framework for extracting temporally annotated sequences, as shown in section 4, which allow to find such causes modeled as itemsets. We then want to compare these rich patterns with the local patterns found in the graph $\mathcal{G}$, to see how the characteristics of the nodes interact both with the information spread and with the interactions of the nodes with their local communities in the network.

We assume $\mathcal{D}$ to contain at least the information about:

- a set of users with their characteristics (such as: gender, country, age, typical discussed topics, degree, betweenness and closeness centrality computed over the network, and so on)
- a timestamped set of sequences of actions performed by the above users that involve the propagation of a certain kind of information (such as: exchange of emails, posts in a forum, instant messages, comments in a blog, and so on)

From the first, we can build several kinds of graphs that can be analyzed with classical graph mining techniques. In order to mine and analyze the local communities of the nodes with the focus on the spread of information, we want to build such graphs on the basis of the information exchanged among the nodes.

As an example, the nodes of the graph can be the users of a mailserver, while there is an edge between two nodes if the nodes exchanged an email. The edge can be then labeled with the typical words used in the communications, that may be also grouped semantically or by statistical properties. Depending on the characteristics of the users and the way we consider them connected among each other, we are able to perform Social Network Analysis of the original network from several different points of view. For example, we may want to use as vertex labels the gender, the country and the age if we are analyzing a so called web social network, while we may want to use structural properties such as the degree, the closeness centrality, the betweenness or the clustering coefficient, if we are analyzing a network of a company. Each different combination of properties would result in a different kind of analysis.

From the second, we can derive a set of timestamped sequences to use as an input of the TAS mining paradigm (see Section 4), in order to be able to extract sequences of itemsets (i.e. characteristics of the users) that are found frequent in the data, together with frequent temporal annotations for them.

The entire analysis will be an interactive and iterative loop of the following steps:

1. Building a graph $\mathcal{G}$ of users in $\mathcal{U}$, connected by edges representing typical words or topics discussed by or among them
2. Assigning labels to the users in $\mathcal{U}$ according to their semantical (such as age, gender, newsgroup of major activity, preferred topic, etc.) and statistical (computed in $\mathcal{G}$, such as betweenness, closeness centrality, etc.) characteristics, collecting them in a set $\mathcal{L}$
3. Assigning labels to the edges in $\mathcal{G}$ according to their semantical (such as semantical cluster, etc.) or statistical (such as frequency of the stemmed word or topic in the subjects, etc.) characteristics, collecting them in a set $\mathcal{W}$
4. Extracting the flows of information in $\mathcal{D}$, grouped by any property to use as transaction identifier (thread, email subject, conversation ID, ..), and building a set of temporally annotated sequences $\mathcal{S}$, containing both the information on the users involved in each flow (represented as itemsets of labels in $\mathcal{L}$), and the temporal information about the flow (usually found as timestamps in seconds since the Epoch)
5. Extracting frequent Temporally Annotated Sequences $\mathcal{T}$ from $\mathcal{S}$, representing the frequent flows of information, and containing both the temporal dimension of the patterns, and the characteristics of the users involved
6. Extracting frequent subgraphs from $\mathcal{G}$ with the help of classical Graph Mining, that represent the local communities of nodes together with their characteristics and typical words or topics used
7. Analyzing the results produced in 4 in order to find frequent items (users' caracteristics) associated with typical fast or slow transition times, then analyze the patterns produced in 6 in order to find patterns containing nodes with the same characteristics as labels: these patterns will tell if the users with these characteristics are the best ones in spreading fast the type of information described by the graph patterns

Steps 1, 2, and 3, are clearly crucial and may vary the analysis that will be performed. By setting different labelings for the edges in $\mathcal{G}$ and including or excluding different characteristics as vertex labels in 6, the analyst may drive the search for frequent patterns in different directions. Please note that, due to the techniques available nowadays to the best of our knowledge, while the TAS mining framework allows for the use of itemsets, which represent a set of characteristics, there appears not to exist a graph miner able to handle more than one label per edge. Hence, step 3 basically implies that we have to produce different input graphs for every kind of analysis we want to perform.

4   Case Study

4.1   Dataset

We used for our experiments two e-mail datasets. The first one is the Enron email dataset[12]. This dataset contains 619,446 email messages complete with senders, recipients, cc, bcc, and text sent and received from 158 Enron's employees. This dataset is characterized by an exceptional wealth of information, and it allows to track flows of communication, together with their associated subjects and the complete data regarding the exchange of information. We took from the entire dataset the "from", "to", "cc", "bcc", "subject" and "date" fields in each email in the "sent" folder of every employee. We took only the emails that were sent to other Enron employees, removing the outgoing emails. We also performed basic cleaning by removing emails with empty subjects, noise, and so on. After the cleaning stage, the number of remaining emails was about 12,000. We refer to it as the "Enron" dataset.

The second dataset consists of Usenet articles collected from 20 different newsgroups about general discussions on politics and religion, technical discussions on computers and hardware, general discussions on hobbies and sports, general discussion on sciences, and a newsgroup for items on sale, and was first used in [11,16]. Over a period of time, 1000 articles were taken from each of the newsgroups, which makes an overall number of 20,000 documents in this collection. Except for a small fraction of the articles, each document belongs to exactly one newsgroup. We took from each sent email the "from", "to" and "date" field. After a cleaning stage, the number of remaining emails was about 18,000. We refer to it as the "Newsgroup" dataset.

4.2   Tools

For our analysis, we used the MiSTA software [9,8], which extracts frequent Temporally Annotated Sequences from a dataset of timestamped sequences, and that has been successfully applied in several contexts [3,2]; we also used a single graph miner in order to find frequent subgraphs of a large graph, implementing a Minimum Image Support function as described in [5].

All the experiments were conducted on a machine equipped with 4 processors at 3.4GHz, 8GB of RAM, running the Ubuntu 8.04 Server Edition, and took from seconds to minutes for the TAS mining, and from minutes to hours for the graph mining.

4.3 Steps of Analysis

We then followed the steps described in section 3 in order to perform our analysis. In the following steps, the subscripts E and N indicate whether the sets refer to the Enron or Newsgroup datasets, respectively.

As step 1, we built the graph $\mathcal{G}_{\mathcal{E}}$ for the Enron dataset by taking the users as nodes and connecting two nodes with edges representing the subjects of emails exchanged between them. For the Newsgroup dataset, we built $\mathcal{G}_{\mathcal{N}}$ by taking the users as nodes and connecting two nodes with edges representing the subjects for which both users posted a message to the newsgroups.

As step 2, we labeled the users $\mathcal{U}_{\mathcal{E}}$ and $\mathcal{U}_{\mathcal{N}}$ following five different possible labeling, according to their structural characteristics in the graphs $\mathcal{G}_{\mathcal{E}}$ and $\mathcal{G}_{\mathcal{N}}$: the degree (the number of ties to other nodes in the network, referred as "DEG"), the closeness centrality (i.e., the inverse of the distance in number of edges of the node from all other nodes in the network, referred as "CL"), the betweenness centrality (i.e., the number of geodesic paths that pass through the node, referred as "BET") and two different clustering annotations (the first, referred as "CC1", is the triadic closure ratio, while the second, referred as "CC2", is a modified version that privileges the 2-neighborhood clustering). Table 1 shows the labeling according to the real values of these variables. For users in $\mathcal{U}_{\mathcal{N}}$ we also performed a labeling according to the newsgroup in which the user was most active, assigning thus 20 possible labels for each node.

As step 3, the edges in $\mathcal{G}_{\mathcal{E}}$ and $\mathcal{G}_{\mathcal{N}}$ have been assigned a label according to various criteria. Both for the Enron and the Newsgroup datasets, the most frequent words in the subjects were manually clustered by their semantic in 5 different clusters per dataset. Each edge was then labeled with the most frequent cluster

Table 1. The labels assigned to the users in the datasets

	Enron				
Label	Degree	Closeness	Betweenness	CC1	CC2
1	$[0-5]$	$[0-0.21[$	$[0-0.0015[$	0	0
2	$[6-15]$	$[0.21-0.2329[$	$[0.0015-0.0046[$	$]0-0.2[$	$]0-35e^{-6}[$
3	$[16-33]$	$[0.2329-0.2513[$	$[0.0046-0.013[$	$[0.2-0.34[$	$[35e^{-6}-14e^{-5}[$
4	$[34-75]$	$[0.2513-0.267[$	$[0.013-0.034[$	$[0.34-0.67[$	$[14e^{-5}-61e^{-5}[$
5	$[76-+\infty[$	$[0.267-1]$	$[0.034-1]$	$[0.67-1]$	$[61e^{-5}-1[$

	Newsgroup				
Label	Degree	Closeness	Betweenness	CC1	CC2
1	$[0-15]$	0	0	0	0
2	$[16-39]$	$]0-0.12[$	$]0-0.0002[$	$]0-0.42[$	$]0-0.00015[$
3	$[40-84]$	$[0.12-0.145[$	$[0.0002-0.001[$	$[0.42-0.61[$	$[0.00015-0.00085[$
4	$[85-154]$	$[0.0002-0.001[$	$[0.001-0.002[$	$[0.61-1[$	$[0.00085-0.005[$
5	$[155-+\infty[$	$[0.1632-1]$	$[0.002-1]$	1	$[0.005-1[$

Table 2. The dataset statistics

Graph	n	e	$\bar{k}$	#Components	GiantComponent	$\bar{C}$	ℓ	Diameter
Enron S	3731	9543	5.11	30	98.01%	0.17	4.52199	15
Newsgroup S	1457	12560	17.24	151	64.51%	0.78	4.02730	11
Newsgroup F	3923	31632	16.12	249	82.41%	0.73	4.42142	17

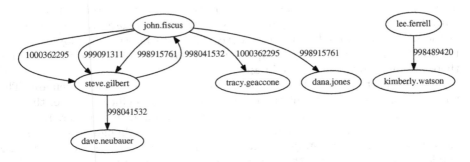

Fig. 1. Example of mail flow for the subject "2002 capital plan"

among its words (ignoring the words not belonging to any cluster). The edges corresponding to subjects for which none of the contained words was frequent or was not belonging to any of the clusters were removed from the graphs. We refer to the graphs created in this way as "Enron S" and "Newsgroup S". For the Newsgroup dataset we performed also a different labeling: all the words were divided in three frequency classes and the edges were then labeled accordingly. We refer to this graph as "Newsgroup F". Finally, in each graph, multiple edges between two nodes have been collapsed into a single edge labeled with its more frequent label. Table 2 shows some statistical properties of the graphs generated, in which: n is the number of nodes, e the number of edges, $\bar{k}$ the average degree, *#Components* the number of components, *GiantComponent* the size of the largest component of the graph (percentage of the total number of nodes), $\bar{C}$ the average clustering coefficient of the graph (between 0 and 1), ℓ the average length of the shortest paths in the graph and *Diameter* the length of the longest shortest path in the graph.

For the step 4, in order to build our $\mathcal{S}_{\mathcal{E}}$ and $\mathcal{S}_{\mathcal{N}}$ for the TAS mining paradigm, we grouped all the emails by subject, keeping the timestamp given by the mail-server to every email. Figure 1 is a graphical representation of the flow of emails in Enron with initial subject "2002 capital plan". In order to give this in input to the software, we processed each of these flow by splitting it in all the possible sequences of emails passed from an user to the others, following the natural temporal ordering. This last step was not necessary for Newsgroup, as the emails were sent only to one recipient, namely the newsgroup. The complete set of these timestamped sequences constituted then our $\mathcal{S}_{\mathcal{E}}$ and $\mathcal{S}_{\mathcal{N}}$.

Steps 5 and 6 produced the results in the following paragraph.

4.4   Results

Although the focus in this paper is only to show the power of the combination of the two techniques we used in our analysis, we can make some interpretation of some of the resulting patterns, that clearly show the differences between the two datasets.

Graph Mining

The Enron graph represents interactions in the working environment of a company, from which we can infer particular considerations regarding possible stages of the workflow followed by the employees. Contacts between employees are direct (not thus as in the newsgroup case), and are very often one-to-many (i.e. there are many recipients in cc in an email).

The first pattern extracted, Figure 2a, represents an exchange of emails. The labels on the nodes represent the level of Clustering Coefficient 2, i.e. the tendency of an employee to create a working group around him or her. It can be noticed that employees with high CC2 have a frequent exchange of emails among each other with several subjects. At a certain moment, one of these high CC2 employees has a contact with a lower CC2 employee (a node outside the central part of the graph being maybe a specific member of a work group) with a different subject (label 4 vs other labels in the pattern). This pattern may represent the mechanism by which members acting as "bridges" between groups detect, with a mutual exchange of knowledge, who can solve a problem.

Figure 2b, where the label "<5" means any label lower than 5, is a generalization of Figure 2a. We found several patterns that follows this generalization, thus we consider the phenomenon described above quite interesting in this case study.

Another interesting pattern in the Enron graph is represented in Figure 2c, where labels are assigned to nodes according to the first definition of clustering coefficient (CC1). One can see that nodes with a low clustering coefficient (discovered to be synonymous of high degree, in our case study) tend to behave in contrast to the value of such a coefficient, as the nodes are found in a frequent clique. This happens because these nodes have a very high degree (i.e. they represent managers and directors), and they are all connected by edges labeled with 1, which represents subjects regarding high-level decisions in the company. In other words, cliques among managers are frequent only if they are speaking about high-level topics.

We conclude the discussion on frequent graph pattern noting that, among the results obtained in the Newsgroup dataset by using as labels for the edges the

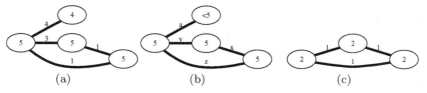

Fig. 2. Subgraphs found in Enron dataset

frequencies of the words composing the subjects, and the value of the CC2 as labels for the nodes, users with a specific CC2 tend to speak about a specific class of subjects with other users having the same CC2, while they speak about other subjects when talking to people with a different CC2. Examples of this behavior are patterns in Figure 3a and 3b.

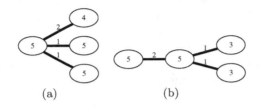

(a)   (b)

Fig. 3. Subgraphs found in Newsgroup dataset

Fig. 4. An example of TAS found

TAS

We now present some considerations derived from the analysis of the most frequent temporal sequences extracted from the two datasets. Figure 4 is a graphical representation of a possible extracted pattern, saying that a user with DEG=4 and CC2=5 replied to an email 10 to 14 time units (5 minutes each) afterwards, to a recipient with CC2=5, which replied to the same email 278 to 284 time units afterwards, to a user with CL=5.

Consider graphs in Figure 5. The 5a and 5b graphs were generated by analyzing the average response time of the most frequent sequences (i.e. the most representative) according to different characteristics (Degree, Closeness, Betweenness and CC2) of the sender. First, we can notice the difference of reaction times, found to be higher in average in the Enron dataset. This can be explained by considering the different nature of the exchange of knowledge in a working environment: there are not (frequent) immediate answers, since, after an email, usually there are several steps of gathering of information, meetings and brainstorming, thus enlarging the time needed for providing a response. On the opposite side, users within a social community usually only need to read all the messages before answering, leading to usual short response time.

Regarding the Enron dataset, Figure 5a reveals an important piece of information regarding the response time of the employees with an high degree of betweenness centrality: having such an high centrality for an employee means that many shortest paths in internal communications pass through that employee. TAS mining revealed that this tends to result in much higher response times, due to the additional working burden that employees of this type have to face. The knowledge that can be extracted from this analysis is to avoid, if

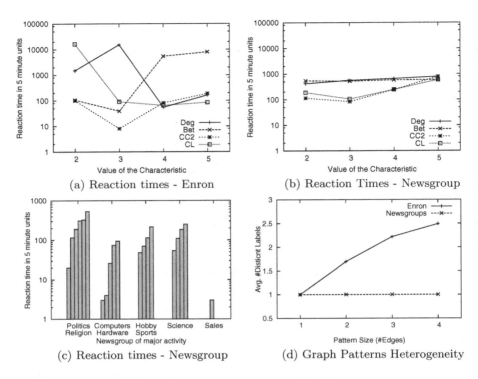

(a) Reaction times - Enron   (b) Reaction Times - Newsgroup

(c) Reaction times - Newsgroup   (d) Graph Patterns Heterogeneity

Fig. 5. Quantitative Analysis of the Results

possible, the structural hubs when there is the need to speed up a communication. This correlation between reaction times and betweenness was not found frequent in the Newsgroup dataset.

Regarding the Newsgroup dataset, Figure 5b shows another difference of behavior from the Enron dataset: the higher regularity of growth of the reaction time for users with higher degree. The degree grows as the user follows many different discussions and, especially, when these discussions involve an increasing number of users. The results showed that the typical response times go up because these discussions are probably the most controversial and interesting ones, and in order to follow them, much time and attention have to be spent. These considerations are not necessarily true in a business context: an employee with a high degree (many different contacts) often answers very quickly.

Another consideration can be done w.r.t the "newsgroup" labeling. Consider Figure 5(c). In the x axis we have the newsgroup of major activity of the users (i.e., a possible value of the "newsgroup" label for the nodes), clustered by main topics (x labels), while in the y the reaction times as found in the frequent TAS. Each of the 20 bars represents one particular newsgroup. As we can notice, there are differences in the reaction times according to the main topic of the newsgroups. While politics and religion seem to be general "relaxed" discussion topics, technical discussions in computers and hardware find more reactive answers. The most reactive is the newsgroup where people put items for sale: the first offer is generally set after 5-10 minutes, as we can see from the figure.

Based on the above consideration, we can give a "draft" of what could be done in order to perform step 7 of the general approach described in section 3: once found that the "Sale" label could be a characteristic related to the speed of the users, it is possible to go back on the results of the graph mining and see if that label was found frequent, possibly in the center of a large subgraph pattern where other nodes have different labels. If the frequency of this pattern is found high, one can argue that passing the information to "Sale" nodes would result in a faster and effective spread of information. In this case study, the meaning of the "Sale" label does not really suggest anything special, but the focus here is to give an idea of the potentialities of the general approach followed.

Finally, consider Figure 5d that shows a direct comparison between the two datasets. It shows the degree of heterogeneity of the communication (i.e. the number of different semantic labels associated with the edges of the frequent patterns) compared with the volume of communication (number of edges in the pattern). From the graph it can be inferred that the business environment shows a greater heterogeneity in the communications: while in the Enron dataset employee tend to speak about different topics with their neighbors, in the Newsgroup dataset close users speak about the same topics. This seems quite easy to explain: employees usually manage more than one different situation, while users in newsgroups tend to be clustered by newsgroup, and hence by topics.

5   Conclusions and Future Work

We have shown a general methodology to mine the information propagation in a network where users exchange information. We have described how to extract useful information from such a network in order to be able to use a combination of two powerful techniques, namely TAS mining and graph mining, in order to find frequent patterns of propagation of information that involve also the possible causes of this propagation. We have shown how this combination can help in finding frequent temporal behaviors in the network together with the characteristics of the users, and what are the roles of these users in the network. We have presented preliminary results of a case study on real-life datasets and we have provided a possible interpretation of some of them.

These first results are encouraging and open the way for a powerful methodology that can help applications such as Viral Marketing.

In the future we plan to extend this analysis to other datasets, where we will be able to use more characteristics of the words and different characteristics of the users, such as the country, the gender, the age and so on.

References

1. Li Zhang, L.A., Adamic, R.M., Lukose, E.A.: Implicit structure and the dynamics of blogspace. Communications of the ACM: CACMa publ. of the Association for Computing Machinery 47(12), 35–39

2. Berlingerio, M., Bonchi, F., Giannotti, F., Turini, F.: Mining clinical data with a temporal dimension: a case study. In: Proc. of The 1st Intern.Conf. on Bioinf. and Biomed. (2007)
3. Berlingerio, M., Bonchi, F., Giannotti, F., Turini, F.: Time-annotated sequences for medical data mining. In: Proc. of The Intern. Workshop of Data Min. in Medicine (2007)
4. Borgwardt, K.M., Kriegel, H.-P., Wackersreuther, P.: Pattern mining in frequent dynamic subgraphs. In: IEEE International Conference on Data Mining, pp. 818–822 (2006)
5. Bringmann, B., Nijssen, S.: What is frequent in a single graph? In: Washio, T., Suzuki, E., Ting, K.M., Inokuchi, A. (eds.) PAKDD 2008. LNCS (LNAI), vol. 5012, pp. 858–863. Springer, Heidelberg (2008)
6. Cheng, E., Grossman, J.W., Lipman, M.J.: Time-stamped graphs and their associated influence digraphs. Discrete Appl. Math. 128(2-3), 317–335 (2003)
7. Desikan, P., Srivastava, J.: Mining temporally changing web usage graphs. In: Mobasher, B., Nasraoui, O., Liu, B., Masand, B. (eds.) WebKDD 2004. LNCS (LNAI), vol. 3932, pp. 1–17. Springer, Heidelberg (2006)
8. Giannotti, F., Nanni, M., Pedreschi, D.: Efficient mining of temporally annotated sequences. In: Proc. of the 6th SIAM Intern. Conf. on Data Min. (2006)
9. Giannotti, F., Nanni, M., Pedreschi, D., Pinelli, F.: Mining sequences with temporal annotations. In: Proc. of the 2006 ACM Symp. on Applied Comp. (SAC), pp. 593–597 (2006)
10. Huberman, B.A., Adamic, L.A.: Information dynamics in the networked world (October 2003)
11. Joachims, T.: A probabilistic analysis of the rocchio algorithm with tfidf for text categorization. In: Fisher, D.H. (ed.) Proceedings of ICML 1997, 14th International Conference on Machine Learning, Nashville, US, pp. 143–151. Morgan Kaufmann Publishers, San Francisco (1997)
12. Klimt, B., Yang, Y.: The enron corpus: A new dataset for email classification research. In: Boulicaut, J.-F., Esposito, F., Giannotti, F., Pedreschi, D. (eds.) ECML 2004. LNCS (LNAI), vol. 3201, pp. 217–226. Springer, Heidelberg (2004)
13. Kossinets, G., Kleinberg, J., Watts, D.: The structure of information pathways in a social communication network (June 2008)
14. Leskovec, J., Adamic, L.A., Huberman, B.A.: The dynamics of viral marketing (September 2005)
15. Liben-Nowell, D., Kleinberg, J.: Tracing information flow on a global scale using Internet chain-letter data. Proceedings of the National Academy of Sciences 105(12), 4633–4638 (2008)
16. Mitchell, T.: Machine Learning. McGraw-Hill Education (ISE Editions) (October 1997)
17. Sun, J., Faloutsos, C., Papadimitriou, S., Yu, P.S.: Graphscope: parameter-free mining of large time-evolving graphs. In: KDD 2007: Proceedings of the 13th ACM SIGKDD international conference on Knowledge discovery and data mining, pp. 687–696. ACM, New York (2007)
18. Tantipathananandh, C., Berger-Wolf, T., Kempe, D.: A framework for community identification in dynamic social networks. In: KDD 2007: Proceedings of the 13th ACM SIGKDD international conference on Knowledge discovery and data mining, pp. 717–726. ACM Press, New York (2007)

Adaptive Learning from Evolving Data Streams

Albert Bifet and Ricard Gavaldà

Universitat Politècnica de Catalunya, Barcelona, Spain
{abifet,gavalda}@lsi.upc.edu

Abstract. We propose and illustrate a method for developing algorithms that can adaptively learn from data streams that drift over time. As an example, we take Hoeffding Tree, an incremental decision tree inducer for data streams, and use as a basis it to build two new methods that can deal with distribution and concept drift: a sliding window-based algorithm, Hoeffding Window Tree, and an adaptive method, Hoeffding Adaptive Tree. Our methods are based on using change detectors and estimator modules at the right places; we choose implementations with theoretical guarantees in order to extend such guarantees to the resulting adaptive learning algorithm. A main advantage of our methods is that they require no guess about how fast or how often the stream will drift; other methods typically have several user-defined parameters to this effect.

In our experiments, the new methods never do worse, and in some cases do much better, than CVFDT, a well-known method for tree induction on data streams with drift.

1 Introduction

Data streams pose several challenges on data mining algorithm design. Limited use of resources (time and memory) is one. The necessity of dealing with data whose nature or distribution changes over time is another fundamental one. Dealing with time-changing data requires in turn strategies for detecting and quantifying change, forgetting stale examples, and for model revision. Fairly generic strategies exist for detecting change and deciding when examples are no longer relevant. Model revision strategies, on the other hand, are in most cases method-specific.

Most strategies for dealing with time change contain hardwired constants, or else require input parameters, concerning the expected speed or frequency of the change; some examples are *a priori* definitions of sliding window lengths, values of decay or forgetting parameters, explicit bounds on maximum drift, etc. These choices represent preconceptions on how fast or how often the data are going to evolve and, of course, they may be completely wrong. Even more, no fixed choice may be right, since the stream may experience any combination of abrupt changes, gradual ones, and long stationary periods. More in general, an approach based on fixed parameters will be caught in the following tradeoff: the user would like to use large parameters to have more accurate statistics (hence, more precision) during periods of stability, but at the same time use small parameters to be able to quickly react to changes, when they occur.

Many ad-hoc methods have been used to deal with drift, often tied to particular algorithms. In this paper, we propose a more general approach based on using two primitive

© Springer-Verlag Berlin Heidelberg 2009

design elements: change detectors and estimators. The idea is to encapsulate all the statistical calculations having to do with detecting change and keeping updated statistics from a stream an abstract data type that can then be used to replace, in a black-box way, the counters and accumulators that typically all machine learning and data mining algorithms use to make their decisions, including when change has occurred.

We believe that, compared to any previous approaches, our approach better isolates different concerns when designing new data mining algorithms, therefore reducing design time, increasing modularity, and facilitating analysis. Furthermore, since we crisply identify the nuclear problem in dealing with drift, and use a well-optimized algorithmic solution to tackle it, the resulting algorithms more accurate, adaptive, and time- and memory-efficient than other ad-hoc approaches. We have given evidence for this superiority in [3, 2, 4] and we demonstrate this idea again here.

We apply this idea to give two decision tree learning algorithms that can cope with concept and distribution drift on data streams: Hoeffding Window Trees in Section 4 and Hoeffding Adaptive Trees in Section 5. Decision trees are among the most common and well-studied classifier models. Classical methods such as C4.5 are not apt for data streams, as they assume all training data are available simultaneously in main memory, allowing for an unbounded number of passes, and certainly do not deal with data that changes over time. In the data stream context, a reference work on learning decision trees is the Hoeffding Tree or Very Fast Decision Tree method (VFDT) for fast, incremental learning [7]. The methods we propose are based on VFDT, enriched with the change detection and estimation building blocks mentioned above.

We try several such building blocks, although the best suited for our purposes is the ADWIN algorithm [3], described in Section 4.1. This algorithm is parameter-free in that it automatically and continuously detects the rate of change in the data streams rather than using apriori guesses, thus allowing the client algorithm to react adaptively to the data stream it is processing. Additionally, ADWIN has rigorous guarantees of performance (a theorem). We show that these guarantees can be transferred to decision tree learners as follows: if a change is followed by a long enough stable period, the classification error of the learner will tend, and the same rate, to the error rate of VFDT.

We test on Section 6 our methods with synthetic datasets, using the SEA concepts, introduced in [11], and two sets from the UCI repository, Adult and Poker-Hand. We compare our methods among themselves but also with CVFDT, another concept-adapting variant of VFDT proposed by Domingos, Spencer, and Hulten [10]. A one-line conclusion of our experiments would be that, because of its self-adapting property, we can present datasets where our algorithm performs much better than CVFDT and we never do much worse. Some comparison of time and memory usage of our methods and CVFDT is included.

A longer version of this paper [5] with additional theoretical results and experiments, is available from the first author webpage.

2   A Methodology for Adaptive Stream Mining

The starting point of our work is the following observation: In the data stream mining literature, most algorithms incorporate one or more of the following ingredients: windows to remember recent examples; methods for detecting distribution change in the input; and methods for keeping updated estimations for some statistics of the input. Our claim is that by basing mining algorithms on well-designed, well-encapsulated modules for these tasks, one can often get more generic and more efficient solutions than by using ad-hoc techniques as required. Similarly, we will argue that our methods for inducing decision trees are simpler to describe, adapt better to the data, perform better or much better, and use less memory than the ad-hoc designed CVFDT algorithm, even though they are all derived from the same VFDT mining algorithm.

A similar approach was taken in [2] and in [4]. In [2] a general framework for change detection and prediction was presented. It is shown that change detection may be monitored using appropriate statistics like statistical charts, CUSUM, and EWMA. CUSUM charts may be very competitive, but they need that users choose the correct values for the parameters. In [4] using our approach, simple adaptive closed-tree mining adaptive algorithms are given. Using a general methodology to identify closed patterns based in Galois Lattice Theory, three closed tree mining algorithms were developed: an incremental one INCTREENAT, a sliding-window based one, WINTREENAT, and finally one that mines closed trees adaptively from data streams, ADATREENAT.

3   Incremental Decision Trees: Hoeffding Trees

Classical decision tree learners such as ID3, C4.5, and CART assume that all training examples can be stored simultaneously in main memory, and are thus severely limited in the number of examples they can learn from. In particular, they are not applicable to data streams, where potentially there is no bound on number of examples and these arrive sequentially.

Domingos and Hulten [7] developed Hoeffding trees, an incremental, anytime decision tree induction algorithm that is capable of learning from massive data streams, assuming that the distribution generating examples does not change over time.

Hoeffding trees exploit the fact that a small sample can often be enough to choose an optimal splitting attribute. This idea is supported mathematically by the Hoeffding bound, which quantifies the number of observations (in our case, examples) needed to estimate some statistics within a prescribed precision (in our case, the goodness of an attribute). VFDT (Very Fast Decision Trees) is the implementation of Hoeffding trees, with a few heuristics added, described in [7]; we basically identify both in this paper.

4   Decision Trees on Sliding Windows

We propose a general method for building incrementally a decision tree based on a keeping sliding window of the last instances on the stream. To specify one such method, we specify how to:

- place one or more change detectors at every node that will raise a hand whenever something worth attention happens at the node
- create, manage, switch and delete alternate trees
- maintain estimators of only relevant statistics at the nodes of the current sliding window

We call *Hoeffding Window Tree* any decision tree that uses Hoeffding bounds, maintains a sliding window of instances, and that can be included in this general framework. Figure 1 shows the pseudo-code of HOEFFDING WINDOW TREE.

HOEFFDING WINDOW TREE($Stream, \delta$)

1 ▷ Let HT be a tree with a single leaf(root)
2 ▷ Init sufficient node statistics at root
3 **for** each example (x, y) in Stream
4   **do** HWTREEGROW($(x, y), HT, \delta$)

HWTREEGROW($(x, y), HT, \delta$)

1   ▷ Sort (x, y) to leaf l using HT
2   ▷ Update sufficient node statistics
3   at leaf l and nodes traversed in the sort
4   **if** this node has an alternate tree T_{alt}
5   HWTREEGROW($(x, y), T_{alt}, \delta$)
6   ▷ Compute information gain G for each attribute
7   **if** G(Best Attr.)$-G$(2nd best)$> \epsilon = \sqrt{R^2 \ln(1/\delta)/(2n)}$[a]
8   **then**
9   ▷ Split leaf on best attribute
10 **for** each branch
11   **do** ▷ Start new leaf
12   and initialize sufficient node statistics
13 **if** one accuracy change detector has detected change
14   **then**
15 ▷ Create an alternate tree with the new best attribute at its root, if there is none
16 **if** existing alternate tree is more accurate
17   **then**
18 ▷ replace current node with alternate tree

---

[a] Here δ' should be the *Bonferroni correction* of δ to account for the fact that many tests are performed and we want all of them to be simultaneously correct with probability $1 - \delta$. It is enough e.g. to divide δ by the number of tests performed so far. The need for this correction is also acknowledged in [7], although in experiments the more convenient option of using a lower δ was taken. We have followed the same option in our experiments for fair comparison.

Fig. 1. Hoeffding Window Tree algorithm

4.1   HWT-ADWIN: Hoeffding Window Tree Using ADWIN

Recently, we proposed an algorithm termed ADWIN [3] (for Adaptive Windowing) that is an estimator with memory and change detector. We use it to design HWT-ADWIN, a new Hoeffding Window Tree that uses ADWIN as a change detector. The main advantage of using a change detector as ADWIN is that as it has theoretical guarantees we can extend this guarantees to the learning algorithms. ADWIN keeps a variable-length window of recently seen items, with the property that the window has the maximal length statistically consistent with the hypothesis "there has been no change in the average value inside the window".

ADWIN is parameter- and assumption-free in the sense that it automatically detects and adapts to the current rate of change. Its only parameter is a confidence bound δ, indicating how confident we want to be in the algorithm's output, inherent to all algorithms dealing with random processes.

Also important for our purposes, ADWIN does not maintain the window explicitly, but compresses it using a variant of the exponential histogram technique in [6]. This means that it keeps a window of length W using only $O(\log W)$ memory and $O(\log W)$ processing time per item, rather than the $O(W)$ one expects from a naïve implementation.

4.2 CVFDT

As an extension of VFDT to deal with concept change Hulten, Spencer, and Domingos presented Concept-adapting Very Fast Decision Trees CVFDT [10] algorithm. We review it here briefly and compare it to our method.

CVFDT works by keeping its model consistent with respect to a sliding window of data from the data stream, and creating and replacing alternate decision subtrees when it detects that the distribution of data is changing at a node. When new data arrives, CVFDT updates the sufficient statistics needed to compute most heuristic measures, including information gain at its nodes by incrementing the counts n_{ijk} corresponding to the new examples and decrementing the counts n_{ijk} corresponding to the oldest example in the window, which is effectively forgotten. CVFDT is a Hoeffding Window Tree as it is included in the general method previously presented.

Two external differences among CVFDT and our method is that CVFDT has no theoretical guarantees (as far as we know), and that it uses a number of parameters, with default values that can be changed by the user - but which are fixed for a given execution. Besides the example window length, it needs:

1. T_0: after each T_0 examples, CVFDT traverses all the decision tree, and checks at each node if the splitting attribute is still the best. If there is a better splitting attribute, it starts growing an alternate tree rooted at this node, and it splits on the currently best attribute according to the statistics in the node.
2. T_1: after an alternate tree is created, the following T_1 examples are used to build the alternate tree.
3. T_2: after the arrival of T_1 examples, the following T_2 examples are used to test the accuracy of the alternate tree. If the alternate tree is more accurate than the current one, CVDFT replaces it with this alternate tree (we say that the alternate tree is promoted).

The default values are $T_0 = 10,000, T_1 = 9,000$, and $T_2 = 1,000$. One can interpret these figures as the preconception that often about the last $50,000$ examples are likely to be relevant, and that change is not likely to occur faster than every $10,000$ examples. These preconceptions may or may not be right for a given data source.

The main internal differences of HWT-ADWIN respect CVFDT are:

– The alternates trees are created as soon as change is detected, without having to wait that a fixed number of examples arrives after the change. Furthermore, the more abrupt the change is, the faster a new alternate tree will be created.

- HWT-ADWIN replaces the old trees by the new alternates trees as soon as there is evidence that they are more accurate, rather than having to wait for another fixed number of examples.

These two effects can be summarized saying that HWT-ADWIN adapts to the scale of time change in the data, rather than having to rely on the *a priori* guesses by the user.

5   Hoeffding Adaptive Trees

In this section we present Hoeffding Adaptive Tree as a new method that evolving from Hoeffding Window Tree, adaptively learn from data streams that change over time without needing a fixed size of sliding window. The optimal size of the sliding window is a very difficult parameter to guess for users, since it depends on the rate of change of the distribution of the dataset.

In order to avoid to choose a size parameter, we propose a new method for managing statistics at the nodes. The general idea is simple: we place instances of estimators of frequency statistics at every node, that is, replacing each n_{ijk} counters in the Hoeffding Window Tree with an instance A_{ijk} of an estimator.

More precisely, we present three variants of a *Hoeffding Adaptive Tree* or HAT, depending on the estimator used:

- HAT-INC: it uses a linear incremental estimator $\hat{x}_k = (1 - 1/N)\hat{x}_{k-1} + 1/N \cdot x_k$.
- HAT-EWMA: it uses an Exponential Weight Moving Average (EWMA) ($\alpha = .01$) $\hat{x}_k = (1 - \alpha)\hat{x}_{k-1} + \alpha \cdot x_k$.
- HAT-ADWIN : it uses an ADWIN estimator. As the ADWIN instances are also change detectors, they will give an alarm when a change in the attribute-class statistics at that node is detected, which indicates also a possible concept change.

The main advantages of this new method over a Hoeffding Window Tree are:

- All relevant statistics from the examples are kept in the nodes. There is no need of an optimal size of sliding window for all nodes. Each node can decide which of the last instances are currently relevant for it. There is no need for an additional window to store current examples. For medium window sizes, this factor substantially reduces our memory consumption with respect to a Hoeffding Window Tree.
- A Hoeffding Window Tree, as CVFDT for example, stores only a bounded part of the window in main memory. The rest (most of it, for large window sizes) is stored in disk. For example, CVFDT has one parameter that indicates the amount of main memory used to store the window (default is 10,000). Hoeffding Adaptive Trees keeps all its data in main memory.

5.1   Example of Performance Guarantee

In this subsection we show a performance guarantee on the error rate of HAT-ADWIN on a simple situation using discrete attributes. Roughly speaking, it states that after a distribution and concept change in the data stream, followed by a stable period, HAT-ADWIN will start, in reasonable time, growing a tree identical to the one that VFDT would grow if starting afresh from the new stable distribution. Statements for more complex scenarios are possible, including some with slow, gradual, changes, but require more space than available here.

Theorem 1. *Let D_0 and D_1 be two distributions on labelled examples. Let S be a data stream that contains examples following D_0 for a time T, then suddenly changes to using D_1. Let t be the time that until VFDT running on a (stable) stream with distribution D_1 takes to perform a split at the node. Assume also that VFDT on D_0 and D_1 builds trees that differ on the attribute tested at the root. Then with probability at least $1 - \delta$:*

- *By time $t' = T + c \cdot V^2 \cdot t \log(tV)$, HAT-ADWIN will create at the root an alternate tree labelled with the same attribute as VFDT(D_1). Here $c \leq 20$ is an absolute constant, and V the number of values of the attributes.[1]*
- *this alternate tree will evolve from then on identically as does that of VFDT(D_1), and will eventually be promoted to be the current tree if and only if its error on D_1 is smaller than that of the tree built by time T.*

If the two trees do not differ at the roots, the corresponding statement can be made for a pair of deeper nodes.

Lemma 1. *In the situation above, at every time $t + T > T$, with probability $1 - \delta$ we have at every node and for every counter (instance of ADWIN) $A_{i,j,k}$*

$$|A_{i,j,k} - P_{i,j,k}| \leq \sqrt{\frac{\ln(1/\delta')\,T}{t(t+T)}}$$

where $P_{i,j,k}$ is the probability that an example arriving at the node has value j in its ith attribute and class k.

Observe that for fixed δ' and T this bound tends to 0 as t grows.

To prove the theorem, use this lemma to prove high-confidence bounds on the estimation of $G(a)$ for all attributes at the root, and show that the attribute *best* chosen by VFDT on D_1 will also have maximal $G(best)$ at some point, so it will be placed at the root of an alternate tree. Since this new alternate tree will be grown exclusively with fresh examples from D_1, it will evolve as a tree grown by VFDT on D_1.

6   Experimental Evaluation

We tested Hoeffding Adaptive Trees using synthetic and real datasets. In the experiments with synthetic datasets, we use the SEA Concepts [11]. In the longer version of this paper [5], a changing concept dataset based on a rotating hyperplane is also used. In the experiments with real datasets we use two UCI datasets [1] Adult and Poker-Hand from the UCI repository of machine learning databases. In all experiments, we use the

---

[1] This value of t' is a very large overestimate, as indicated by our experiments. We are working on an improved analysis, and hope to be able to reduce t' to $T + c \cdot t$, for $c < 4$.

Table 1. SEA on-line errors using discrete attributes with 10% noise

	CHANGE SPEED		
	1,000	10,000	100,000
HAT-INC	16.99%	16.08%	14.82%
HAT-EWMA	16.98%	15.83%	**14.64** %
HAT-ADWIN	16.86%	**15.39%**	14.73 %
HAT-INC NB	16.88%	15.93%	14.86%
HAT-EWMA NB	**16.85%**	15.91%	14.73 %
HAT-ADWIN NB	16.90%	15.76%	14.75 %
CVFDT $|W| = 1,000$	19.47%	*15.71%*	15.81%
CVFDT $|W| = 10,000$	17.03%	17.12%	*14.80%*
CVFDT $|W| = 100,000$	*16.97%*	17.15%	17.09%

values $\delta = 10^{-4}$, $T_0 = 20,000$, $T_1 = 9,000$, and $T_2 = 1,000$, following the original CVFDT experiments [10].

In all tables, the result for the best classifier for a given experiment is marked in **boldface,** and the best choice for CVFDT window length is shown in *italics.*

We included an improvement over CVFDT (which could be made on the original CVFDT as well). If the two best attributes at a node happen to have exactly the same gain, the tie may be never resolved and split does not occur. In our experiments this was often the case, so we added an additional split rule: when $G(best)$ exceeds by three times the current value of $\epsilon(\delta, \ldots)$, a split is forced anyway.

We have tested the three versions of Hoeffding Adaptive Tree, HAT-INC, HAT-EWMA($\alpha = .01$), HAT-ADWIN, each with and without the addition of Naïve Bayes (NB) classifiers at the leaves. As a general comment on the results, the use of NB classifiers does not always improve the results, although it does make a good difference in some cases; this was observed in [8], where a more detailed analysis can be found.

First, we experiment using the SEA concepts, a dataset with abrupt concept drift, first introduced in [11]. This artificial dataset is generated using three attributes, where only the two first attributes are relevant. All three attributes have values between 0 and 10. We generate 400,000 random samples. We divide all the points in blocks with different concepts. In each block, we classify using $f_1 + f_2 \leq \theta$, where f_1 and f_2 represent the first two attributes and θ is a threshold value. We use threshold values 9, 8, 7 and 9.5 for the data blocks. We inserted about 10% class noise into each block of data.

We test our methods using discrete and continuous attributes. The on-line errors results for discrete attributes are shown in Table 1. On-line errors are the errors measured each time an example arrives with the current decision tree, before updating the statistics. Each column reflects a different speed of concept change. We observe that CVFDT best performance is not always with the same example window size, and that there is no optimal window size. The different versions of Hoeffding Adaptive Trees have a very similar performance, essentially identical to that of CVFDT with optimal window size for that speed of change. More graphically, Figure 2 shows its learning curve using continuous attributes for a speed of change of $100,000$. Note that at the points where the concept drift appears HWT-ADWIN, decreases its error faster than CVFDT, due to the fact that it detects change faster.

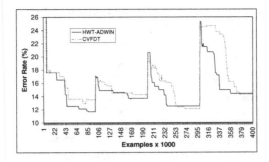

Fig. 2. Learning curve of SEA Concepts using continuous attributes

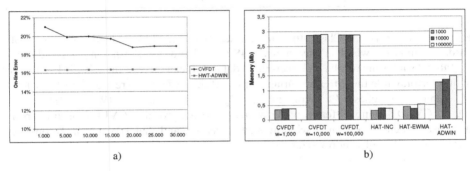

a)   b)

Fig. 3. a) On-line error on UCI Adult dataset, ordered by the *education* attribute and b) Memory used on SEA Concepts experiments

We test Hoeffding Adaptive Trees on two real datasets in two different ways: with and without concept drift. We tried some of the largest UCI datasets [1], and report results on Adult and Poker-Hand. For the Covertype and Census-Income datasets, the results we obtained with our method were essentially the same as for CVFDT (ours did better by fractions of 1% only) – we do not claim that our method is always better than CVFDT, but this confirms our belief that it is never much worse.

An important problem with most of the real-world benchmark data sets is that there is little concept drift in them [12] or the amount of drift is unknown, so in many research works, concept drift is introduced artificially. We simulate concept drift by ordering the datasets by one of its attributes, the *education* attribute for Adult, and the first (un-named) attribute for Poker-Hand. Note again that while using CVFDT one faces the question of which parameter values to use, our method just needs to be told "go" and will find the right values online.

The Adult dataset aims to predict whether a person makes over 50k a year, and it was created based on census data. Adult consists of 48,842 instances, 14 attributes (6 continuous and 8 nominal) and missing attribute values. The Poker-Hand dataset consists of 1,025,010 instances and 11 attributes. Each record of the Poker-Hand dataset is an example of a hand consisting of five playing cards drawn from a standard deck of 52. Each card is described using two attributes (suit and rank), for a total of 10 predictive

Table 2. On-line classification errors for CVFDT and Hoeffding Adaptive Trees on Poker-Hand data set

	NO DRIFT	ARTIFICIAL DRIFT
HAT-INC	38.32%	39.21%
HAT-EWMA	39.48%	40.26%
HAT-ADWIN	38.71%	41.85%
HAT-INC NB	41.77%	42.83%
HAT-EWMA NB	24.49%	**27.28%**
HAT-ADWIN NB	**16.91%**	33.53%
CVFDT $|W| = 1,000$	49.90%	49.94%
CVFDT $|W| = 10,000$	*49.88%*	*49.88 %*
CVFDT $|W| = 100,000$	49.89%	52.13 %

attributes. There is one Class attribute that describes the "Poker Hand". The order of cards is important, which is why there are 480 possible Royal Flush hands instead of 4.

Table 2 shows the results on Poker-Hand dataset. It can be seen that CVFDT remains at 50% error, while the different variants of Hoeffding Adaptive Trees are mostly below 40% and one reaches 17% error only. In Figure 3 we compare HWT-ADWIN error rate to CVFDT using different window sizes. We observe that CVFDT on-line error decreases when the example window size increases, and that HWT-ADWIN on-line error is lower for all window sizes.

7   Time and Memory

In this section, we discuss briefly the time and memory performance of Hoeffding Adaptive Trees. All programs were implemented in C modifying and expanding the version of CVFDT available from the VFML [9] software web page. We have slightly modified the CVFDT implementation to follow strictly the CVFDT algorithm explained in the original paper by Hulten, Spencer and Domingos [10]. The experiments were performed on a 2.0 GHz Intel Core Duo PC machine with 2 Gigabyte main memory, running Ubuntu 8.04.

Consider the experiments on SEA Concepts, with different speed of changes: $1,000$, $10,000$ and $100,000$. Figure 3 shows the memory used on these experiments. HAT-INC and HAT-EWMA, are the methods that use less memory. The reason for this fact is that they don't keep examples in memory as CVFDT, and that they don't store ADWIN data for all attributes, attribute values and classes, as HAT-ADWIN. We have used the default $10,000$ for the amount of window examples kept in memory, so the memory used by CVFDT is essentially the same for $W = 10,000$ and $W = 100,000$, and about 10 times larger than the memory used by HAT-INC memory.

Figure 4 shows the number of nodes used in the experiments of SEA Concepts. We see that the number of nodes is similar for all methods, confirming that the good results on memory of HAT-INC is not due to smaller size of trees.

Finally, with respect to time we see that CVFDT is still the fastest method, but HAT-INC and HAT-EWMA have a very similar performance to CVFDT, a remarkable fact

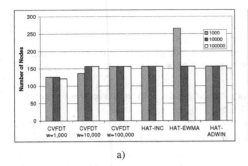 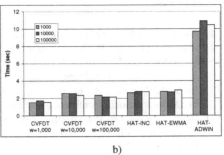

a)   b)

Fig. 4. a) Number of Nodes used on SEA Concepts experiments and b) Time on SEA Concepts experiments

given that they are monitoring all the change that may occur in any node of the main tree and all the alternate trees. HAT-ADWIN increases time by a factor of 4, so it is still usable if time or data speed is not the main concern.

8 Conclusions and Future Work

We have presented a general adaptive methodology for mining data streams with concept drift, and and two decision tree algorithms. We have proposed three variants of Hoeffding Adaptive Tree algorithm, a decision tree miner for data streams that adapts to concept drift without using a fixed sized window. Contrary to CVFDT, they have theoretical guarantees of performance, relative to those of VFDT.

In our experiments, Hoeffding Adaptive Trees are always as accurate as CVFDT and, in some cases, they have substantially lower error. Their running time is similar in HAT-EWMA and HAT-INC and only slightly higher in HAT-ADWIN, and their memory consumption is remarkably smaller, often by an order of magnitude.

We can conclude that HAT-ADWIN is the most powerful method, but HAT-EWMA is a faster method that gives approximate results similar to HAT-ADWIN. An obvious future work is experimenting with the exponential smoothing factor α of EWMA methods used in HAT-EWMA.

We would like to extend our methodology to ensemble methods such as boosting, bagging, and Hoeffding Option Trees.

Acknowledgments

Partially supported by the EU PASCAL2 Network of Excellence (FP7-ICT-216886), and by projects SESAAME-BAR (TIN2008-06582-C03-01), MOISES-BAR (TIN2005-08832-C03-03). Albert Bifet was supported by a FI grant through the SGR program of Generalitat de Catalunya.

References

[1] Asuncion, D.N.A.: UCI machine learning repository (2007)

[2] Bifet, A., Gavaldá, R.: Kalman filters and adaptive windows for learning in data streams. In: Todorovski, L., Lavrač, N., Jantke, K.P. (eds.) DS 2006. LNCS (LNAI), vol. 4265, pp. 29–40. Springer, Heidelberg (2006)

[3] Bifet, A., Gavaldà, R.: Learning from time-changing data with adaptive windowing. In: SIAM International Conference on Data Mining (2007)

[4] Bifet, A., Gavaldà, R.: Mining adaptively frequent closed unlabeled rooted trees in data streams. In: 14th ACM SIGKDD International Conference on Knowledge Discovery and Data Mining (2008)

[5] Bifet, A., Gavaldà, R.: Adaptive parameter-free learning from evolving data streams. Technical report, LSI-09-9-R, Universitat Politècnica de Catalunya, Barcelona, Spain (2009)

[6] Datar, M., Gionis, A., Indyk, P., Motwani, R.: Maintaining stream statistics over sliding windows. SIAM Journal on Computing 14(1), 27–45 (2002)

[7] Domingos, P., Hulten, G.: Mining high-speed data streams. In: Knowledge Discovery and Data Mining, pp. 71–80 (2000)

[8] Holmes, G., Kirkby, R., Pfahringer, B.: Stress-testing hoeffding trees. In: Jorge, A.M., Torgo, L., Brazdil, P.B., Camacho, R., Gama, J. (eds.) PKDD 2005. LNCS (LNAI), vol. 3721, pp. 495–502. Springer, Heidelberg (2005)

[9] Hulten, G., Domingos, P.: VFML – a toolkit for mining high-speed time-changing data streams (2003), http://www.cs.washington.edu/dm/vfml/

[10] Hulten, G., Spencer, L., Domingos, P.: Mining time-changing data streams. In: KDD 2001, San Francisco, CA, pp. 97–106. ACM Press, New York (2001)

[11] Street, W.N., Kim, Y.: A streaming ensemble algorithm (sea) for large-scale classification. In: KDD 2001, pp. 377–382. ACM Press, New York (2001)

[12] Tsymbal, A.: The problem of concept drift: Definitions and related work. Technical Report TCD-CS-2004-15, Department of Computer Science, University of Dublin, Trinity College (2004)

An Application of Intelligent Data Analysis Techniques to a Large Software Engineering Dataset

James Cain[1], Steve Counsell[2], Stephen Swift[2], and Allan Tucker[2]

[1] Quantel Limited, Turnpike Road, Newbury, Berkshire, RG14 2NX, UK
James.Cain@Quantel.com
[2] School of Information Systems, Computing and Mathematics, Brunel University,
Uxbridge, UB8 3PH, UK
{Steve.Counsell,Stephen.Swift,Allan.Tucker}@Brunel.ac.uk

Abstract. Within the development of large software systems, there is significant value in being able to predict changes. If we can predict the likely changes that a system will undergo, then we can estimate likely developer effort and allocate resources appropriately. Within object oriented software development, these changes are often identified as refactorings. Very few studies have explored the prediction of refactorings on a wide-scale. Within this paper we aim to do just this, through applying intelligent data analysis techniques to a uniquely large and comprehensive software engineering time series dataset. Our analysis show extremely promising results, allowing us to predict the occurrence of future large changes.

Keywords: Software Engineering, Time Series Analysis, Hidden Markov Models.

1 Introduction

An emerging discipline in the software engineering (SE) community and the subject of many studies in the recent past is that of refactoring. A refactoring is defined as a change made to code which changes the internal behaviour of that code while preserving its external behaviour. In other words, refactoring is the process of changing how something is done but not what that something does. In this sense it differs from re-engineering of code, where both the 'how' and the 'what' may change.

The benefits of being able to predict refactorings are many. First, if we view refactoring as addressing poorly written code, then predicting the poorly written parts of code allows us to allocate time and effort to those areas. This has significant implications for the likely fault-profiles in different versions, since poorly written code is often the most fault-prone. Second, predicting refactorings would allow us interleave or combine refactoring effort with regular maintenance activity. An open research issue at present is the relationship between refactorings and 'other' changes made as part of a developer's activity. Third, predicting types of refactorings will also provide a means of prioritising which refactorings should be accomplished in that time. For example should we primarily focus on renaming or restructuring refactorings? Finally, one issue that arises during a refactoring is the need to re-test

N. Adams et al. (Eds.): IDA 2009, LNCS 5772, pp. 261–272, 2009.
© Springer-Verlag Berlin Heidelberg 2009

after each refactoring. If we can predict refactorings then that will also inform an understanding of the test requirements associated with those refactorings and decisions to be made as a result by project managers on developer resource allocation.

This paper aims to make use of a large dataset from a software company that contains information about different versions of a software system over time, in order to predict potential future refactorings using Hidden Markov Models.

Refactoring is usually associated with Java and other mainstream object-oriented (OO) programming languages. The initial research into refactoring is credited to Opdyke in his PhD Thesis [12]. Opdyke used C++ as a basis of his suggestions for improving the design of code. The seminal text which has also been instrumental in motivating refactoring-based studies is that of Fowler [7] which uses Java as the refactoring vehicle. On the basis that all Java classes have methods and fields that operate on those methods, one of the most common refactorings that we might want to undertake (and is listed in Fowler's text) is to 'Rename [a] Field'. The purpose of undertaking such a refactoring would be to communicate the intention and purpose of that field to a developer more clearly. There is clear value in changing the name of a field from let's say just 'x' to 'x-coordinate' as a concrete example. The list of 72 specific refactorings in Fowler's text describes both simple and complex refactorings. Often, the motivation for completing a refactoring is to reduce coupling dependencies between classes; coupling is widely considered in the software engineering community to be a contributor to faults and so refactoring has a strong tie with the potential for eliminating faults. Coupling [16] can be defined as the inter-relatedness of (and dependencies between) classes in a system. Cohesion on the other hand is defined as the intra-relatedness of the elements internal to a class and takes no account of the external dependencies. Developers should always strive for low coupling and high cohesion as sound software engineering practice. Coupling and Cohesion can be extended to collections of related classes (e.g. those that form part of the same subsystem or modules), where we have Coupling measured between different modules and Cohesion measured between the classes within a module [16].

There are number of motivating factors for why we would want to use refactoring. First, by simplifying code and making it more understandable, we are potentially postponing the decay of code that is a feature of most systems as they age. The problem of decaying code is that it takes increasing amounts of time and effort to repair the faults associated with poorly written code. Refactoring offers a mechanism for addressing and even reversing that decay. Fowler suggests that refactoring should be undertaken 'mercilessly' over the life of a system. Second, no system remains the same in terms of the user requirements made of it. The dynamics of a system in a changing environment mean that new functionality has to be added and old functionality offered by the system needs to be changed. If consistent refactoring is done to a system, then accommodating change is relatively straightforward since the developer (acting in their role as maintainer of a system) will find making changes to the system easier. Maintenance of a software system typically accounts for at least 75% of a developer's time – there are therefore compelling reasons why refactoring should be attempted.

Of course, if refactoring was easy or there was unlimited time for refactoring, then developers would refactor. However, in many systems, short term loss for long-term gain (which refactoring offers) is simply not practiced. A final motivation for refactoring is that developers become more familiar with the design and code they work with if they are consistently scrutinising that code for refactoring opportunities. Again, time is the limiting factor.

Whilst there have been many studies into the nature and behaviour of refactorings [9, 10] which have made significant progress in the area, a wide range of issues and problems still exist and are left for researchers to explore. To our knowledge, although some work on the prediction of refactorings using standard statistical techniques and metrics has been undertaken [5, 17], no study has previously looked at the prediction of refactorings using Intelligent Data Analysis, let alone prediction of refactorings based on a version-by-version analysis of a large proprietary system. Many of the aforementioned studies used open-source software as a basis. The research described in this paper therefore contributes on at least two levels.

This paper is organised as follows, section 2 looks at the methods we used in this paper along with the data collection and pre-processing, section 3 describes the experiments we carried out, section 4 discusses the results of the experiments and finally section 5 draws some conclusions and maps out the future direction of this research project.

2 Methods

This section describes the methods employed in this paper, along with data collection and data pre-processing.

2.1 Data Creation

The program code that is the focus of this paper was provided by the international company Quantel Limited. Quantel is a world-leading developer of innovative, high performance content creation and delivery systems across television and film post production along with broadcast content creation. It is over 30 years old and has offices worldwide. It has been the recipient of many awards including Oscars, Emmys and the prestigious MacRobert Award, presented by The Royal Academy of Engineering, the UK's premier award for innovation in engineering. It supplies products to many of the world's leading media companies, such as ESPN, BBC, Sky, and Fox and its products have been used to make some of the world's greatest films such as Slumdog Millionaire.

The data source that is the focus for this paper is the processed source code of a library that is part of an award winning[1] product line architecture that has received over a decade of development, and multiple person centuries of effort. The product line architecture has delivered over 15 distinct products in that time, the library under analysis is the persistence engine used by all these products. The entire code base currently runs to over 12 million lines of C++; the subset that we analyse in this paper is over 0.5 million lines of C++. Microsoft Visual C++ produces two types of data files that contain type information: The Debug Symbol Information Program Database (PDB files) and the Browser Database (BSC files). The Bowser Database from Visual C++ [2] is offered by every version of Microsoft Visual C++. It is able to

---

[1] A list of recent awards won by Quantel can be found at: http://www.quantel.com/list.php?a=News

generate a Browser Database that can be used to navigate all the source code in a project. Every version of Visual C++ (including Visual C++ .NET) has had a corresponding API for programmatic access to the Bowser Database. However it only contains symbol information, not full type information.

The Debug Symbol Information files contain all the type information in a system so that debuggers can interpret global, stack and heap locations and map them back to the types that they represent. This file format is undocumented by Microsoft [13], however in March 2002 an API was released by Microsoft that allowed access to (some of) the debug type information without undue reverse engineering [15]. More details on extracting type information using the DIA SDK can be found in [3]. The PDB files for each version of the code have been archived, and was analysed using bespoke software that interfaced with the PDB files using the DIA SDK.

Each PDB was checked into a revision control system. The data was collected over the period 17/10/2000 to 03/02/2005, given a total of 503 PDBs in total. To ensure anonymity, all of the class names (types) in all the PDBs were sorted into an alphabetically ordered master class table, this was used as a global index, converting each class name to a globally unique ID. The system contained a total of 6120 classes. Not all of the 6120 classes exist at each time point, in fact at any one time there are between 29 and 1626 active classes. Classes generally appear at a time point, and then "disappear" at a later time point. We note that if a class is renamed then it will appear in the dataset as a new class with a new identifier, and hence some of the "appearances" and "disappearances" could be explained by this. Currently we have no way to detect this phenomenon, but we are looking into resolving this as part of future research.

The dataset consists of five time series of directed graphs with integer edge weights; the absence of an edge weight implies a weight of zero. Each of the five graphs represents a relationship between classes as described in Table 1. Each node in the graph represents a class as described above; the edges represent the strength of the relationship. For example, if class 1 and class 2 have an edge weight of 3 in the Attributes graph then class 1 uses class 2 as an attribute 3 times; an edge weight of zero (no edge) would mean that there is no relationship.

Table 1. Class Relationships used to define the Graphs

Class Relationship	Description
Attributes (A_t)	The types used as data members in a class
Bases (B_t)	The types used as immediate base classes
Inners (I_t)	Any type declared inside the scope of a class
Parameters (P_t)	Types used as parameters to member functions of a class
Returns (R_t)	Types used as return values from member functions of a class

2.2 Pre-processing the Graphs

The graphs themselves each consist of a 6120 by 6120 relationship matrix, each being highly sparse (for the reasons discussed above). Initial analysis showed that none of the graphs over the 5 types of relationships were fully connected, i.e. each graph consisted of a number of disconnected sub graphs. This may seem unusual, since each class should be indirectly related to all other classes, otherwise they would not be part of the same application. This is true if each type of graph is combined for each time slice, but not when each type of relationship is considered on its own. The fact that the graphs can be decomposed into sub graphs allows us to easily conduct a rudimentary modularisation [8] of each relationship graph, thus giving us an approximate module structure. We note that this is a fairly "trivial" clustering or modularisation, but use these terms to avoid confusion. We do not conduct a more computational decomposition of the larger sub-graphs, since the sizes of the sub-graphs range from 1 to 1446, fairly continuously (except for the very smallest of sizes) as can be seen in Fig. 1. There would be no point decomposing the smaller sub-graphs, whilst there may be value decomposing larger sub-graphs, however where this line is drawn would be arbitrary, hence we have decided to work with the sub-graphs as they are. This also allows us to avoid running a very larger number of clustering or modularisation experiments and also allows us to avoid all of the issues involved with choosing an appropriate technique.

In order to forecast the size of a change, we need to pre-process the graphs into a metric that can be used to indicate the size of a change; this would ideally give us a series of univariate time series, which would be much easier to model than a time series of graphs. The aim is to forecast when major refactorings occur. One way to do this would be to measure how much change has been made between time slices, i.e. the size of the difference between two consecutive graphs; how we do this is discussed in the next sections.

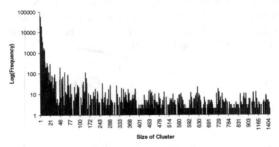

Fig. 1. Plot of cluster size against Log frequency for the decomposition of each graph into sub-graphs across all graphs

2.2.1 Notation

Each graph is represented as an adjacency matrix, and ordered according to time, as indicated in the abbreviations in table 1. For example, A_1 is the first graph for Attributes, and B_{503} is the last graph for Bases. We shall refer to the modularisation or clustering of a graph (using the simple procedure above) as $C(G)$, where G is a graph, for example $C(A_5)$ is the modularisation of the fifth Attributes graph. G_t will be used to denote a graph at time t where we have not specified the type of the graph (the notation covers all possibilities). We will define $G(a,b)$ or $A_t(a,b)$ (for a specific type of graph for example) as the edge weight between class a and class b within the graph specified (G or A_t in this example). A modularisation (or clustering) will be represented as a list of lists, where $C_m(G)$ is the mth cluster of $C(G)$ and $C_{m,k}(G)$ is the kth variable (class) of the mth cluster of G. The notation $|C(G)|$ will denote the number of clusters within $C(G)$ and similarly $|C_m(G)|$ will denote the size of cluster/module $C_m(G)$.

2.2.2 Homogeneity

Homogeneity [4] is a cluster validation index that is widely used in the data analysis community. The index reflects the compactness of a cluster, by giving an average measure of how far each cluster member is from the cluster centre. For the purposes of this paper, the centre cannot be easily measured, due to the nature of the graphs; hence we will use a more suitable measure. We will measure Homogeneity in two ways, the first will be on a cluster by cluster basis (equation (2)), and will be the average of all of the edge weights between the members in a cluster, we will then define the Homogeneity of a clustering arrangement as the average Homogeneity of all of the clusters (equation (1)). It can easily be seen that this measure is strongly related to software engineering measure of cohesion. Homogeneity $(H(C(G)))$ is more formally defined in equations (1) and (2). We use Homogeneity to define the stability or "worth" of each graph within the dataset, high values of Homogeneity reflects that all of the members of each sub-graph of each graph are highly related to each other. Low values of Homogeneity will show that the members have little relationship to each other. As previously mentioned, it is assumed that software engineers refactor to increase cohesion and decrease coupling. We will model the Homogeneity of each graph rather than the graph itself, the aim being to measure the internal stability of each of the graphs. If the Homogeneity steadily deteriorates, then the cohesion of software is deteriorating, we hypothesise that such a pattern may precede a major change or refactoring in the program code, i.e. the internal code structure gets steadily worse until a major change "has" to be made.

$$H(C(G)) = \frac{1}{|C(G))|} \sum_{j=1}^{|C(G))|} H(C_j(G)) \; . \tag{1}$$

$$H(C_j(G)) = \frac{1}{|C_j(G) \, \| \, C_j(G) - 1|} \sum_{a=1}^{|C_j(G) \| C_j(G)|} \sum_{\substack{b=1 \\ a \neq b}} G(C_{j,a}(G), C_{j,b}(G)) \; . \tag{2}$$

2.2.3 Jaccard Similarity Coefficient

The Jaccard Similarity Coefficient (JSC) is a simple metric that is used to define the overlap between two sets of objects, and ranges between 0, no overlap or members in common and 1, meaning identical sets. We intend to use JSC to measure how different two sets of graphs are (in terms of refactoring), given the number of classes active at one time point (for a graph), and the number of classes active at the next time point, the smaller the JSC, the more classes that have been added or deleted, which is an indirect measure of refactoring, i.e. the larger the difference in classes between two consecutive graphs, the larger the number of changes that have been made. JSC is formally defined in equation (3), where the function $V(G)$ simply returns a set of all the classes (IDs) that are in the modularisation (clustering) of a graph.

$$V(G_t) = \bigcup_{j=1}^{|C(G_t)|} C_j(G_t)$$

$$(3)$$

$$J(G_t, G_{t+1}) = \frac{|V(G_t) \cap V(G_{t+1})|}{|V(G_t) \cup V(G_{t+1})|}.$$

2.3 Hidden Markov Models

The Hidden Markov Model (HMM) [14] will be used to model the time-series (the Homogeneity and JSC indices). The HMM is a temporal probabilistic model in which a discrete random class variable captures the state of a process. The HMM model parameters consist of two matrices, the first representing the transition probabilities of the (potentially hidden) class variable over time which we can interpret for the scope of this paper as the class representing a major refactoring event. Formally, the transition probability is $Pr(\Pi_t|\Pi_{t+1})$ where Π_t represents the state at time point t. In other words, it represents the probability of the current state given the state at the previous time point. The other matrix contains the probability distribution of the observed variables (here Attributes, Parameters, Return type, etc...) conditioned upon the class node, or formally $Pr(X_t|\Pi_t)$ where X_t represents the observed variables at time point t. HMMs can be used to perform a variety of operations including smoothing, monitoring and forecasting and for this study we focus on the one step-ahead forecasting of the class variable given the values of the observed variables at the previous time point: essentially we calculate $Pr(\Pi_{t+1}|X_t)$ by using the junction tree inference algorithm [11]. Fig. 2 illustrates the dependencies of the model graphically.

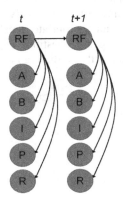

Fig. 2. The graphical model for the HMM used to model the time series data, A, B, I, P and R represent the Attributes, Bases, Inners, Parameters and Returns type variables, whilst RF represents the refactoring node.

3 Experiments

Within this paper, we assume that the variables extracted from the SE data are a temporal sequence that carries information which should be able to predict when a major refactoring event occurs. We test this by applying a machine learning approach to the data. More specifically, we use Hidden Markov Models [14] which are typically used to model temporal or sequential data with some hidden underlying process (here representing the major refactoring events). We adopt a bootstrapping approach [6] and ensure a balanced model by randomly sampling (with replacement) two consecutive cases from the data so that there are an equal number where a major refactoring event has taken place and where one has not. We then split these into a training set and test set of equal sizes for testing the accuracy of the HMM at predicting the hidden state at the following time point, given only the variables at the previous time. This entire process is repeated 100 times in order to calculate confidences in the prediction. The HMM requires the hidden state variable to be discrete, so we map the JSC variable to a discrete variable by taking the minimum of the individual JSC for each class relationship characteristic and applying a threshold at a value of 0.9. This represented a change of about 10% of the number classes between consecutive graphs.

4 Results and Discussion

One measure of the predictive accuracy of a classifier is through evaluating the area under the ROC curve (AUC). Note that a classifier that makes uniformly random predictions will generate an ROC along the diagonal. For all experiments the AUC was evaluated as 0.845, the curve can be seen in Fig. 3. This curve plots the sensitivity against the specificity of the predicted state (here major refactoring) for varying thresholds. It clearly indicates that we have obtained a very good degree of forecast accuracy with a fairly equal balance of false positives and negatives. Something we aim to follow up is how to decide an appropriate threshold for the refactoring forecast problem: in other words by asking the question "is predicting a major refactoring incorrectly more serious than incorrectly predicting no major refactoring?".

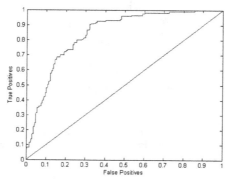

Fig. 3. The ROC curve of forecast accuracy for all HMM experiments

Another measure of accuracy is simply looking at the number of correct forecasts divided by the number of total forecasts. This is accuracy shown in the Box Whisker plot in Fig. 4. Here the forecast accuracy is averaged for each experiment (mean overall accuracy = 0.761). Fig. 4 shows clearly that the forecast accuracy is to a high degree of reliability with the quartile range spanning accuracies from 0.730 to 0.850. However it is worth noting that there were a small number of low accuracy predictions, as indicated in the figure.

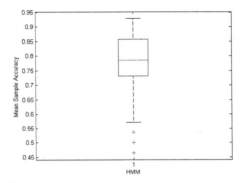

Fig. 4. Mean HMM forecast accuracy

The next analysis of the results involved querying the HMM network in order to see where the main influences occurred. Within Fig. 5, we have shown the Box Whiskers plot of some inference queries over the HMM models learnt from the data. Specifically it shows a comparison of the distributions for the mean of each class relationship characteristic generated during the bootstrap experiments where 1 represents a major refactoring event and 2 represents the absence of a major refactoring event. The y-axis is the expected posterior Homogeneity measure (cohesion) after applying inference.

The box plots for the different OO features show varying results. Those for Attributes show that only small changes in these values precede a refactoring. We would view changes to attributes as a common change that is regularly done as part of the maintenance process and not necessarily the precursor to major refactoring; this might explain the result in this case. Moreover, changes in attributes are, in most cases, hidden from every other outside class through feature encapsulation. The values for Bases show a similar profile. Previous research [1] has shown that changes to the inheritance hierarchy are rarely undertaken by developers because of the inherent complexity of large structures and high dependencies found within an inheritance hierarchy. Consequently, developers tend to avoid such changes. Three revealing results are those for Returns, Parameters and Inners. Addition of parameters and return types to the method signatures of a class is a common maintenance task, but unlike attributes affect the external coupling of the methods of a class to a greater extent. Often, addition of return types and parameters cause ripple effect elsewhere in the system. In other words, such changes are more likely to precede a major refactoring effort and this might explain the result for these two features.

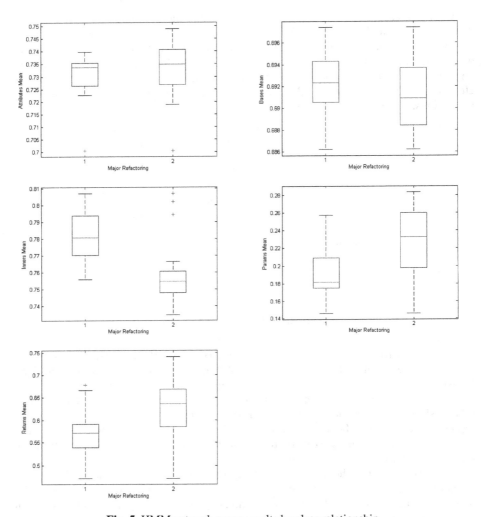

Fig. 5. HMM network query results by class relationship

The result for inner classes (Inners) is arguably the most interesting. A main aim of inner classes is to facilitate manipulation of "Action Listeners" to graphical interfaces and provide a mechanism for event handling. In theory, they should be an aid to maintenance, since the inner class is hidden outside its enclosing class and there is a close coupling relationship between inner classes and their enclosing classes.

However, many software engineers have cast doubt on the value of inner classes because of the deviance from standard class structure definition and the problem of understanding them as a result. (This is a particularly acute problem for less experienced developers unused to manipulating them.) The results from the study presented suggest that large changes in inner classes precede refactoring effort. In other words, inner classes may cause a problem that can only be resolved by significant refactoring effort - the nature of inner classes means that they will impact considerably on system behaviour. This result therefore casts more doubt on the value of using inner classes; if they do cause relatively high disruption to the structure of a system, then serious thought and consideration should be given prior to their use. The study thus shows a feature of the system studied that would be difficult to show empirically or using standard statistical techniques and has implications for the burden that software maintenance presents.

5 Conclusions

Within this paper we have shown how a time series of graphs can be modelled and forecast successfully using Hidden Markov Models. Additionally we have shown that the forecasts are of a high degree of accuracy. This paper has also shed light on some of the reasons behind the causes of refactorings within large software systems, and has opened up many avenues of potential research and collaboration. There is much scope for further research opportunities, at the moment we have only forecast in the short term, we will need to adapt our approach to increase the forecast horizon, if the approach is to be truly useful within an industrial context. Additionally we have had to discretise a number of the variables within our time series, we aim to investigate the use of continuous HMM models to see if we can further increase the accuracy.

One threat posed by the study is that we have considered refactoring in the context of all changes made to a system. Purists would argue that such changes are simply re-engineering (i.e. part of all maintenance activities) and not refactoring in its true semantic preserving sense. However, in defence of this threat, many of the changes we have captured in the system are either standard refactorings or part the mechanics of those refactorings and we thus see the study of such system maintenance activities as valid in that context as in any.

Acknowledgements

We would like to thank our collaborators at Quantel Limited for allowing us access to this uniquely large and interesting dataset, and providing us with their valuable expertise.

References

1. Advani, D., Hassoun, Y., Counsell, S.: Extracting Refactoring Trends from Open-source Software and a Possible Solution to the 'Related Refactoring' Conundrum. In: Proceedings of ACM Symposium on Applied Computing, Dijon, France (April 2006)
2. BSCKIT Browser Toolkits for Microsoft Visual C++, Microsoft support knowledge base article number Q153393, http://support.microsoft.com/
3. Cain, J.: Debugging with the DIA SDK, Visual System Journal (April 2004), http://www.vsj.co.uk/dotnet/display.asp?id=320
4. Chen, G., Banerjee, N., Jaradat, S.A., Tanaka, T.S., Ko, M.S.H., Zhang, M.Q.: Evaluation and Comparison of Clustering Algorithms in Analyzing ES Cell Gene Expression Data. Statistica Sinica 12, 241–262 (2002)

5. Demeyer, S., Ducasse, S., Nierstrasz, O.: Finding refactorings via change metrics. In: ACM Conference on Object Oriented Programming Systems Languages and Applications (OOPSLA), Minneapolis, USA, pp. 166–177 (2000)
6. Efron, B., Tibshirani, R.J.: An Introduction to the Bootstrap. Chapman & Hall, London (1993)
7. Fowler, M.: Refactoring: Improving the Design of Existing Code. Addison-Wesley, Reading (1999)
8. Mancoridis, S., Mitchell, B.S., Rorres, C., Chen, Y., Gansner, E.R.: Using automatic clustering to produce high-level system organizations of source code. In: Proceedings of the 6th International Workshop on Program Comprehension (IWPC 1998), Ischia, Italy, pp. 45–52. IEEE Computer Society Press, Los Alamitos (1998)
9. Mens, T., Tourwe, T.: A Survey of Software Refactoring. IEEE Transactions on Software Engineering 30(2), 126–139 (2004)
10. Mens, T., van Deursen, A.: Refactoring: Emerging Trends and Open Problems (2003), http://www.swen.uwaterloo.ca/~reface03/Papers/TomMens.pdf
11. Murphy, K.: Dynamic Bayesian Networks: Representation, Inference and Learning. PhD Thesis, UC Berkeley, Computer Science Division (July 2002)
12. Opdyke, W.: Refactoring Object-Oriented Frameworks. PhD thesis, University of Illinois at Urbana-Champaign (1992)
13. Pietrek, M.: Under the Hood. MSDN Magazine 17(3) (2002)
14. Rabiner, L.R.: A tutorial on HMM and selected applications in speech recognition. Proceedings of the IEEE 77(2), 257–286 (1989)
15. Schreiber, S.: Undocumented Windows 2000 Secrets, A Programmer's Cookbook. Addison-Wesley, Reading (2001)
16. Stevens, W., Myers, G., Constantine, L.: Structured design. IBM Systems Journal 13(2), 115–139 (1974)
17. Zhao, L., Hayes, J.: Predicting Classes in Need of Refactoring: An Application of Static Metrics. In: Proceedings of 2nd International PROMISE Workshop, Philadelphia, US (2006)

Which Distance for the Identification and the Differentiation of Cell-Cycle Expressed Genes?

Alpha Diallo, Ahlame Douzal-Chouakria, and Francoise Giroud

Université Joseph Fourier Grenoble 1
Laboratory TIMC-IMAG, CNRS UMR 5525,
Faculté de Médecine, 38706 LA TRONCHE Cedex, France
{Alpha.Diallo,Ahlame.Douzal,Francoise.Giroud}@imag.fr

Abstract. This paper addresses the clustering and classification of active genes during the process of cell division. Cell division ensures the proliferation of cells, but becomes drastically aberrant in cancer cells. The studied genes are described by their expression profiles (i.e. time series) during the cell division cycle. This work focuses on evaluating the efficiency of four major metrics for clustering and classifying gene expression profiles. The study is based on a random-periods model for the expression of cell-cycle genes. The model accounts for the observed attenuation in cycle amplitude or duration, variations in the initial amplitude, and drift in the expression profiles.

Keywords: Time series, distance, clustering, classification, gene expression profiles.

1 Introduction

Though most cells in our bodies contain the same genes, not all genes are active in every cell: genes are turned on (i.e. expressed) when needed. Expressed genes define the molecular pattern of a specific cell's function, and are organized into molecular-level regulation networks. To understand how cells achieve such specialization, it is necessary to identify which genes are involved in different types of cells. Moreover, it is helpful to know which genes are turned on or off in diseased versus healthy human tissues, and which genes are expressed differently in the two tissues, thus possibly causing the disease. DNA microarray technology allows us to monitor the expression levels of thousands of genes simultaneously during important biological processes to determine which ones are expressed in a specific cell type [7]. Clustering and classification techniques have proven helpful in understanding gene function, gene regulation, and cellular processes (e.g., [10], [17], [20], [21]). We distinguish at least two main approaches to clustering and classifying profiles or time series. First, the parametric approach consists of projecting time series into a given functional basis space, which corresponds to a polynomial ARIMA or a discrete Fourier transform approximation of the time series. Time series clustering and classification is then performed on the fitted coefficients (e.g., [2], [3], [8], [11], [18], [22]). The second approach

N. Adams et al. (Eds.): IDA 2009, LNCS 5772, pp. 273–284, 2009.
© Springer-Verlag Berlin Heidelberg 2009

is non-parametric, and consists of clustering or classifying time series based on their initial temporal descriptions. Thus, the challenge in this approach is how to include information on dependency between measurements (e.g., [1], [6], [9], [13], [19], [23]). Within the context of the non-parametric approach, we propose to evaluate the efficiency of four major metrics for the clustering and classification of gene expression. This study is based on a random-periods model for the expression of cell-cycle genes. The model accounts for observed biological variations, such as attenuation in cycle amplitude, drift in the expression profiles, and variations in the initial amplitude or the cycle duration. The remainder of the paper is organized as follows. The next section clarifies what constitutes gene expression data and introduces the biological problem of interest. Section 3 defines the four major metrics to be evaluated, and discusses their specifications. Section 4 indicates how the metrics will be compared through the clustering and the classification of genes. Section 5 presents the overall methods of evaluation based on a random-periods model, and discusses the results obtained.

2   Identification of Genes Expressed in the Cell Cycle

The biological problem of interest is the analysis of the progression of gene expression during the cell division process. Cell division is the main process in cell proliferation, and it consists of four main phases (G_1, S, G_2, and M) and three inter-phases (the G_1/S, G_2/M, and M/G_1 transition phases). The division process begins at the G_1 phase, during which the cell prepares for duplication (DNA pre-synthesis). Then comes the S phase, during which DNA is replicated (i.e., each chromosome is duplicated); this is followed by the G_2 phase, during which the cell prepares for cell division (DNA post-synthesis). Finally comes the mitosis phase, which is also called the M phase, during which the cell is divided into two daughter cells. During these four phases, genes are turned on and off at specific times, so one important aim in understanding cell proliferation is to identify those genes that are highly expressed in, and characteristic of, each phase of the cell cycle. This can help, for instance, to understand how hormonal treatment can induce cell proliferation by activating specific genes. To better our understanding of gene expression during the cell division process, DNA molecules representing many genes are placed in discrete spots organized in a line or column matrix, which is called a DNA microarray. Microarray technology allows us to determine which gene is represented by each spot, and to measure its expression level at specific points in the cell division cycle. Finally, each gene of interest is analyzed for its expression profile observed during one or more cell division cycles.

3   Proximity between Gene Expression Profiles

Let $g_1 = (u_1, ..., u_p)$ and $g_2 = (v_1, ..., v_p)$ be the expressions of two genes observed at time $(t_1, ..., t_p)$. The clustering and classification of gene expression data commonly involve Euclidean distance or the Pearson correlation coefficient. The following section defines four major metrics for gene expression analysis and their specifications in accounting for proximity in values or behavior.

3.1 Euclidean Distance

The Euclidean distance δ_E between g_1 and g_2 is defined as:

$$\delta_E(g_1, g_2) = \left(\sum_{i=1}^{p} (u_i - v_i)^2 \right)^{\frac{1}{2}}.$$

Based on the above definition, the closeness between two genes depends on the closeness of their values, regardless of their expression behavior. In other words, the Euclidean distance ignores the temporal dependence of the data.

3.2 Pearson Correlation Coefficient

Many works use the Pearson correlation coefficient as a behavior proximity measure. Without loss of generality, consider that g_1 and g_2 have values in $[0, N]$. The genes g_1 and g_2 exhibit similar behavior if over any observed period $[t_i, t_{i+1}]$, they increase or decrease simultaneously at the same rate. In contrast, g_1 and g_2 have opposite behavior if over any observed period $[t_i, t_{i+1}]$ where g_1 increases, g_2 decreases, and vice-versa, at the same rates (in absolute value). To illustrate the correlation coefficient specification, let us consider the following formula, based on the differences between the expression values:

$$\text{COR}(g_1, g_2) = \frac{\sum_{i,i'} (u_i - u_{i'})(v_i - v_{i'})}{\sqrt{\sum_{i,i'} (u_i - u_{i'})^2} \sqrt{\sum_{i,i'} (v_i - v_{i'})^2}}.$$

We see that the correlation coefficient is based on the differences between all pairs of values (i.e. observed at all the pairs of time (i, i')), which implicitly assumes the independence of the observed data. Consequently, the correlation coefficient can overestimate behavior proximity. For instance, in the case of a high tendency effect, as shown in Section 4, two genes with opposite behavior may have a relatively high, positive correlation coefficient.

3.3 Temporal Correlation Coefficient: A Behavior Proximity Measure

To overcome the limitations of the Pearson correlation coefficient, the temporal correlation coefficient introduced in [4] is considered, as it reduces the Pearson correlation coefficient to the first order differences:

$$\text{CORT}(g_1, g_2) = \frac{\sum_i (u_{(i+1)} - u_i)(v_{(i+1)} - v_i)}{\sqrt{\sum_i (u_{(i+1)} - u_i)^2} \sqrt{\sum_i (v_{(i+1)} - v_i)^2}},$$

with $\text{CORT}(g_1, g_2) \in [-1, 1]$. The value $\text{CORT}(g_1, g_2) = 1$ indicates that g_1 and g_2 exhibit similar behavior. The value $\text{CORT}(g_1, g_2) = -1$ indicates that g_1 and g_2 exhibit opposite behavior. Finally, $\text{CORT}(g_1, g_2) = 0$ expresses that the growth rates g_1 and g_2 are stochastically linearly independent, thereby identifying genes with different behavior that are neither similar nor opposite.

3.4   Behavior and Values Proximity Measure

For a proximity measure to cover both behavior and value proximities, the dissimilarity index D_k proposed in [5] is considered. It includes both the Euclidean distance, for proximity with respect to values, and the temporal correlation, for proximity with respect to behavior:

$$D_k(g_1, g_2) = f(\text{CORT}(g_1, g_2))\, \delta_E(g_1, g_2), \quad \text{with } f(x) = \frac{2}{1 + exp(k\, x)}, \quad k \geq 0.$$

This index is based on a tuning function $f(x)$ that modulates the proximity with respect to values according to the proximity with respect to behavior. An exponential function $f(x)$ is preferred to a linear form to ensure a nearly equal modulating effect for extreme values (i.e., CORT=-1, +1 and 0) and their nearest neighbors. In the case of genes with different behavior (i.e., with CORT near 0), $f(x)$ is near 1 whenever the value of k, and D_k is approximately equal to δ_E. However, if CORT $\neq 0$ (that is, non-different behavior), the parameter k modulates the contributions of both types of proximity, with respect to values and with respect to behavior, to the dissimilarity index D_k. As k increases, the contribution of proximity with respect to behavior, $1 - 2/(1 + exp(k\, |\text{CORT}|))$, increases, whereas the contribution of proximity with respect to values, $2/(1 + exp(k\, |\text{CORT}|))$, decreases. For instance, for $k = 0$ and $|\text{CORT}| = 1$ (similar or opposite behavior), the behavior proximity contributes 0% to D_k whereas the value proximity contributes 100% to D_k (the value of D_k is totally determined by δ_E). For $k = 2$ and $|\text{CORT}| = 1$, the behavior proximity contributes 76.2% to D_k whereas the value proximity contributes 23.8% to D_k (23.8% of the value of D_k is determined by δ_E, and the remaining 76.2% by CORT). Note that the widely-used dynamic time warping (see for instance [14], [15]) is not addressed in this work, as it is not appropriate for generating cell-cycle gene expression profiles. Indeed, the identification of genes expressed during the cell-cycle is mainly based on the time at which the genes are highly expressed. To best cluster or classify gene expression profiles, time should not be warped when evaluating proximities.

4   Metrics Comparison

A simulation study is performed to evaluate the efficiency of the metrics defined in Section 3. For the clustering process, the PAM (Partitioning Around Medoids) approach is used to partition the simulated genes into n clusters, n being the number of cell-cycle phases or inter-phases of interest. The PAM algorithm is preferred to the classical K-means for many reasons. It is more robust with respect to outliers, which are numerous in gene expression data. It also allows a more detailed analysis of the partition by providing clustering characteristics; in particular, it indicates whether each gene is well classified (i.e. highly expressed in a cell-cycle phase) or whether it lies on the boundary of the cluster (i.e. it is involved in a transition phase). For more details about the PAM algorithm, see Kaufman and Rousseeuw [12]. The efficiency of each metric in clustering gene expression profiles is evaluated through the goodness of the obtained partitions. Three criteria are measured: the average silhouette width (asw), the within-between ratio (wbr), and the corrected Rand index (RI). For the classification process, the 10-NN approach is used to classify gene expression profiles. The efficiency of each metric is evaluated through the estimated misclassification error rate.

5   Simulation Study

5.1   Random-Periods Model for Periodically Expressed Genes

We use gene expression profiles generated using the random-periods model proposed by Liu et al. [16] to study periodically expressed genes. This model allows us to simulate attenuation in the amplitude of periodic gene expression with regard to stochastic variations during the various phases of the cell-cycle, while also permitting us to estimate the phase of the cycle in which the gene is most frequently transcribed. The sinusoid function for characterizing the expected periodic expression of a cell-cycle gene g is

$$f(t, \theta_g) = a_g + b_g t + \frac{K_g}{\sqrt{2\pi}} \int_{-\infty}^{+\infty} cos(\frac{2\pi t}{Texp(\sigma z)} + \Phi_g)exp(-\frac{z^2}{2})dz,$$

where θ_g is explicitly $(K_g, T, \sigma, \Phi_g, a_g, b_g)$, specific to each gene g. Integration in the model computes the expected cosine across the lognormal distribution of periods, and thereby accounts for the aggregation of expression levels across a large number of cells. The parameter Φ_g corresponds to the cell-cycle phase during which the gene undergoes its peak level of transcription, with $\Phi_g = 0$ corresponding to the point when cells are first released to resume cycling. The parameter K_g is the initial amplitude of the periodic expression pattern. The parameters a_g and b_g account for any drift (intercepts and slopes, respectively) in a gene's background expression level, and T and σ are the parameters of the lognormal distribution of cell-cycle duration. The parameter σ governs the rate of attenuation in amplitude. If σ is zero, the duration of the cell-cycle does not vary, as cells remain synchronous through time, and the expression profile shows no attenuation in amplitude. Larger values of σ correspond to faster attenuation of the peak amplitude. Figure 1 illustrates the progression of gene expression during five cell-cycle phases.

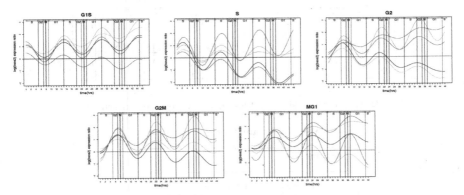

Fig. 1. Genes expression progression during five cell-cycle phases

5.2   Simulation Protocol

Based on the above random model and on the parameters specification given in [16], four experiments are simulated to study how each metric accounts for gene variations. The first experiment generates genes with varying initial amplitudes K_g varying in $[0.34, 1.33]$. The second experiment simulates genes with amplitude attenuation, with governed by σ, varying in $[0.054, 0.115]$. The third experiment varies the drift, with slopes $b_g \in [-0.05, 0.05]$ and intercepts $a_g \in [0, 0.8]$. The last experiment simulates genes with simultaneous variations of initial amplitude, amplitude attenuation during the cell-cycle, and drift. Figure 2 shows the variations generated across the four experiments for genes expressed in the G_1/S phases. The model parameter specifications of the four experiments are summarized in Table 1. For all simulations, T is fixed to 15, and Φ_g takes the values 0, 5.190, 3.823, 3.278, or 2.459 to simulate the expression profiles of the five classes G_1/S, S, G_2, G_2/M, or M/G_1, respectively. For each experiment $j \in \{1, ..., 4\}$, 10 samples S_{ij} $i \in \{1, ..., 10\}$ are simulated. Each sample S_{ij} is composed of 500 gene expression profiles (of length 47) with 100 genes for each of the five phases or inter-phases G_1/S, S, G_2, G_2/M, and M/G_1. The comparison of metrics is performed within each experiment through the clustering and the classification of 5000 simulated genes (i.e. 10 samples of 500 genes each).

5.3   Metrics Efficiency for Clustering Gene Expression Profiles

For each experiment and for each metric δ_E, COR, and CORT, a PAM algorithm is performed to partition each sample S_{ij} into 5 clusters (i.e. 5 cell-cycle phases and inter-phases). For instance, for the experiment j and for the metric δ_E, the

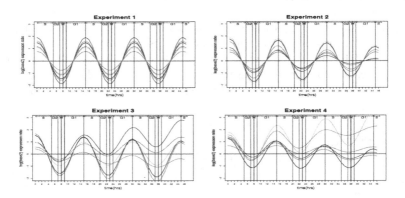

Fig. 2. G_1/S expression profiles through the four experiments

Table 1. Parameters specification

Experiment number	K_g	σ	b_g	a_g
1	[0.34, 1.33]	0	0	0
2	[0.34, 1.33]	[0, 0.115]	0	0
3	[0.34, 1.33]	0	[-0.05, 0.05]	[0, 0.8]
4	[0.34, 1.33]	[0, 0.115]	[-0.05, 0.05]	[0, 0.8]

Table 2. k* mean and variance

Adaptive	Exp1	Exp2	Exp3	Exp4
Clustering	(6,0)	(6,0)	(6,0)	(5.85,0.06)
Classification	(3,3.53)	(3,3.53)	(4.55,1.18)	(4.84,0.98)

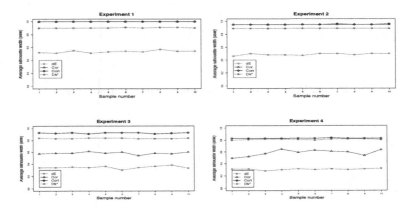

Fig. 3. Asw progression across the clustered samples

PAM algorithm is applied to the 10 samples $S_{1j}, ..., S_{10j}$ to extract the 10 partitions $P_{\delta_E}^{1j}, ..., P_{\delta_E}^{10j}$. For each partition, $P_{\delta_E}^{ij}$, three goodness criteria are measured: the average silhouette width (asw), the within/between ratio (wbr), and the corrected Rand index (RI). The corrected Rand index allows us to measure the proximity between $P_{\delta_E}^{ij}$ and the true partition (i.e. that defined by S_{ij}). Finally, the efficiency of the metric δ_E within the experiment j is summarized by the average values of the criteria asw, wbr, and RI of the 10 partitions $P_{\delta_E}^{1j}, ..., P_{\delta_E}^{10j}$. For the dissimilarity index D_k, the adaptive clustering proposed in [5] is applied. The adaptive clustering of S_{ij} consists of performing the PAM algorithm for several values of k from 0 to 6 (per a lag of 0.01) to find the value k* that yields the optimal partition $P_{D_{k*}}^{ij}$, using as goodness criteria the average silhouette width and the within/between ratio. Note that $k*$ provides the best contribution of the proximity with respect to values and with respect to behavior to the dissimilarity index, thus the learned D_{k*} is identified as best clustering S_{ij}. Table 2 gives, for each experiment, the mean and the variance $(\overline{k*}, var(k*))$ of k*. As in the case of the metrics δ_E, COR, and CORT, the efficiency of the metric D_k within the experiment j is summarized by the average values of the criteria asw, RI and wbr of the 10 partitions $P_{D_{k*}}^{1j}, ..., P_{D_{k*}}^{10j}$. Figures 3, 4, and 5 depict, for each experiment and for each metric, the progression of the criteria asw, wbr, and RI across the 10 clustered samples $S_{1j}, ..., S_{10j}$. Figure 6 shows for each metric the progression, across the four experiments, of the average values of the criteria asw (top), wbr (middle) and RI (bottom)

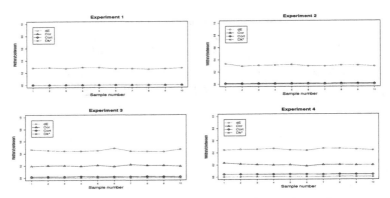

Fig. 4. Wbr progression across the clustered samples

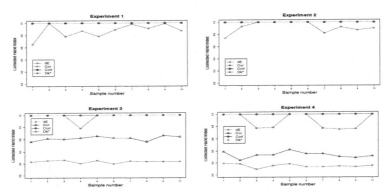

Fig. 5. RI progression across the clustered samples

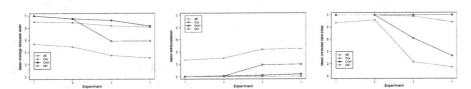

Fig. 6. Metric efficiency to cluster gene expression profiles

5.4   Metrics Efficiency for Classifying Gene Expression Profiles

For each experiment and for each metric δ_E, COR, and CORT, a 10-NN algorithm is performed to classify each sample S_{ij}. For instance, for the experiment j and for the metric δ_E, the 10-NN algorithm is applied to the 10 samples $S_{1j}, ..., S_{10j}$ to generate the 10 classifications $C_{\delta_E}^{1j}, ..., C_{\delta_E}^{10j}$. For each classification $C_{\delta_E}^{ij}$ the misclassification error rate is measured. The efficiency of the metric δ_E in classifying gene expression profiles within the experiment j, is summarized by the average misclassification error rates of the 10 classifications $C_{\delta_E}^{1j}, ..., C_{\delta_E}^{10j}$. For the

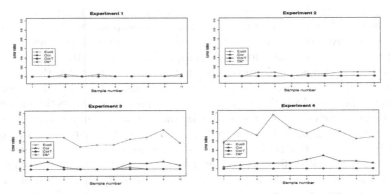

Fig. 7. Misclassification error rates progression through the classified samples

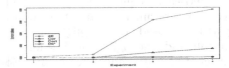

Fig. 8. Metric efficiency to classify gene expression profiles

dissimilarity index D_k, an adaptive classification is performed. It consists of performing the 10-NN algorithm on S_{ij} for several values of k from 0 to 6 (in increments of 0.01) to find the $k*$ that minimizes the misclassification error rate of $C_{D_k}^{ij}$. The efficiency of the metric D_k in classifying gene expression profiles within the experiment j, is summarized by the average misclassification error rates of the 10 classifications $C_{D_{k*}}^{1j}, ..., C_{D_{k*}}^{10j}$. Figure 7 depicts, for each of the four experiments, the progression of the misclassification error rates across the 10 classified samples. Figure 8 shows, for each metric, the progression across the four experiments of the average misclassification error rates.

5.5   Discussion

We first discuss the clustering results. Let us give some additional information about the criteria in question. The *asw* indicates the strength (asw close to 1) or the weakness ($asw < 0.5$) of the obtained partitions, while *wbr* measures the compactness (i.e. within-cluster variability) and the separability (between-clusters variability) of the obtained clusters. A good partition is characterized by a lower within/between ratio. Finally, RI allows us to measure the similarity between the obtained partitions and the true ones ($RI = 1$ for a high level of similarity, and $RI = 0$ for non-similarity).

Figures 3, 4, and 5 show that the clustering based on δ_E gives, for experiments 1 to 4, weaker partitions than the ones based on COR, CORT, or D_k. Indeed, partitions based on δ_E have the lowest values for *asw* and RI and the highest values for *wbr*. Figure 6 shows that the average values of *asw*, *wbr* and RI of the clustering based δ_E decrease from the experiment 1 to 4, showing the inappropriateness of the Euclidean distance for cases with complex variations. Clustering based on COR gives strong partitions with the best values of asw, wbr, and RI, for the first two experiments. However, this quality decreases drastically in experiments 3 and 4 (Figures 3, 4, 5, and 6) showing the limitations of the Pearson correlation coefficient when faced with tendency variations, as explained in Subsection 3.2. Finally, the best clustering and the strongest partitions across the four experiments are given by CORT and D_k, with asw values varying in $[0.8, 1]$, wbr around 0, and RI varying in $[0.83, 1]$. Note that the quality of the clustering based on D_k is very slightly lower than that based on CORT, revealing that gene expression profiles are more naturally differentiated by their behaviors than by their values. This hypothesis is assessed by the higher values of $k*$ (near 6, with a variability of 0) obtained in the adaptive clustering across the four experiments (Table 2).

Let us now discuss the classification results. Figures 7 and 8 show that for experiments 1 and 2, the four metrics are equally efficient, with misclassification error rates around 0. However, for experiments 3 and 4, we note a drastic increase in the error rate for the partitions based on δ_E, a slight increase in the error rate for the partitions based on COR, and a negligible increase for D_k. Table 2 and Figure 9 indicate the distribution of k* in the adaptive classifications. For experiments 1 and 2, a uniform distribution of k* in $[0, 6]$ is noted. This case arises when a good classification can be obtained both with a metric based on values (k* near 0) and with a metric based on behavior (k* near 6). Indeed, Figures 7 and 8 show that the four metrics are equally efficient at classifying genes across the first two experiments. In experiments 3 and 4, k* takes higher values, indicating that the behavior-based metrics (i.e. CORT, D_k) are the most efficient for classifying gene expression profiles, as can easily be seen in Figures 7 and 8. Finally, according to the results of the four experiments, the metrics CORT and D_k can be said to be the most efficient at classifying gene expression profiles.

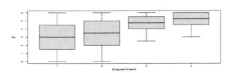

Fig. 9. k* distribution in the adaptive classification

6   Conclusion

The above performances based on simulated data, assessed by the nearly similar results based on a real genes expression profiles developed in [5] allow to conclude that, to cluster or classify cell-cycle gene expression profiles, it is advisable to consider the temporal correlation coefficient as a proximity measure. However, the effectiveness of the learned dissimilarity D_k, which also provides very good partitions and classifications, is worth noting. In general, when faced with data where time should not be warped for proximity evaluation (which is the case of cell-cycle gene expression profiles), the dissimilarity D_k proposed in this paper is recommended.

The adaptive clustering or classification is used to learn the appropriate dissimilarity D_k to use in the next analysis task. The learned dissimilarity D_k can lead to the temporal correlation (for k* near 6), to the Euclidean distance (for k* near 0), or more generally to a metric covering both values and behavior proximities.

References

1. Anagnostopoulos, A., Vlachos, M., Hadjieleftheriou, M., Keogh, E.J., Yu, P.S.: Global Distance-Based Segmentation of Trajectories. In: Proc. of ACM SIGKDD, pp. 34–43 (2006)
2. Bar-Joseph, Z., Gerber, G.K., Gifford, D.K., Jaakkola, T., Simon, I.: Continuous Representations of Time-Series Gene Expression Data. Journal of Computational Biology 10(3), 341–356 (2003)
3. Caiado, J., Crato, N., Pena, D.: A periodogram-based metric for time series classification. Computational Statistics and Data Analysis 50, 2668–2684 (2006)
4. Douzal-Chouakria, A., Nagabhushan, P.N.: Adaptive dissimilarity index for measuring time series proximity. Advances in Data Analysis and Classification Journal 1(5-21) (2007)
5. Douzal-Chouakria, A., Diallo, A., Giroud, F.: Adaptive clustering for time series: application for identifying cell-cycle expressed genes. Computational Statistics and Data Analysis 53(4), 1414–1426 (2009)
6. Džeroski, S., Gjorgjioski, V., Slavkov, I., Struyf, J.: Analysis of time series data with predictive clustering trees. In: Džeroski, S., Struyf, J. (eds.) Knowledge Discovery in Inductive Databases, 5th International Workshop, KDID, Berlin, Germany (2006)
7. Eisen, M.B., Brown, P.O.: DNA arrays for analysis of gene expression. Methods Enzymol. 303, 179–205 (1999)
8. Garcia-Escudero, L.A., Gordaliza, A.: A proposal for robust curve clustering. Journal of Classification 22, 185–201 (2005)
9. Heckman, N.E., Zamar, R.H.: Comparing the shapes of regression functions. Biometrika 22, 135–144 (2000)
10. He, Y., Pan, W., Lin, J.: Cluster analysis using multivariate normal mixture models to detect differential gene expression with microarray data. Computational Statistics and Data Analysis 51(2), 641–658 (2006)
11. Kakizawa, Y., Shumway, R.H., Taniguchi, N.: Discrimination and clustering for multivariate time series. Journal of the American Statistical Association 93, 328–340 (1998)
12. Kaufman, L., Rousseeuw, P.J.: Finding Groups in Data. An Introduction to Cluster Analysis. John Wiley & Sons, New York (1990)
13. Keller, K., Wittfeld, K.: Distances of time series components by means of symbolic dynamics. International Journal of Bifurcation Chaos 14, 693–704 (2004)
14. Keogh, E.J., Pazzani, M.J.: Scaling Up Dynamic Time Warping for Data Mining Applications. In: Proc. of ACM SIGKDD, pp. 285–289 (2000)
15. Kruskall, J.B., Liberman, M.: The symmetric time warping algorithm: From continuous to discrete. In: Time Warps, String Edits and Macromolecules. Addison-Wesley, Reading (1983)
16. Liu, D., Umbach, D.M., Peddada, S.D., Li, L., Crockett, P.W., Weinberg, C.R.: A Random-Periods Model for Expression of Cell-Cycle Genes. Proc. Natl. Acad. Sci. USA 101, 7240–7245 (2004)

17. Liu, X., Lee, S., Casella, G., Peter, G.F.: Assessing agreement of clustering methods with gene expression microarray data. Computational Statistics and Data Analysis 52(12), 5356–5366 (2008)
18. Maharaj, E.A.: Cluster of time series. Journal of Classification 17, 297–314 (2000)
19. Oates, T., Firoiou, L., Cohen, P.R.: Clustering time series with Hidden Markov Models and Dynamic Time Warping. In: Proc. 6th IJCAI 1999, Workshop on Neural, Symbolic and Reinforcement Learning Methods for Sequence Learning, Stockholm, pp. 17–21 (1999)
20. Park, C., Koo, J., Kim, S., Sohn, I., Lee, J.W.: Classification of gene functions using support vector machine for time-course gene expression data. Computational Statistics and Data Analysis 52(5), 2578–2587 (2008)
21. Scrucca, L.: Class prediction and gene selection for DNA microarrays using regularized sliced inverse regression. Computational Statistics and Data Analysis 52(1), 438–451 (2007)
22. Serban, N., Wasserman, L.: CATS: Cluster After Transformation and Smoothing. Journal of the American Statistical Association 100, 990–999 (2004)
23. Shieh, J., Keogh, E.J.: iSAX: Indexing and Mining Terabyte Sized Time Series. In: Proc. of ACM SIGKDD, pp. 623–631 (2008)

Ontology-Driven KDD Process Composition

Claudia Diamantini, Domenico Potena, and Emanuele Storti

Dipartimento di Ingegneria Informatica, Gestionale e dell'Automazione "M. Panti",
Università Politecnica delle Marche - via Brecce Bianche, 60131 Ancona, Italy
{diamantini,potena,storti}@diiga.univpm.it

Abstract. One of the most interesting challenges in Knowledge Discovery in Databases (KDD) field is giving support to users in the composition of tools for forming a valid and useful KDD process. Such an activity implies that users have both to choose tools suitable to their knowledge discovery problem, and to compose them for designing the KDD process. To this end, they need expertise and knowledge about functionalities and properties of all KDD algorithms implemented in available tools. In order to support users in this heavy activity, in this paper we introduce a goal-driven procedure for automatically compose algorithms. The proposed procedure is based on the exploitation of KDDONTO, an ontology formalizing the domain of KDD algorithms, allowing us to generate valid and non-trivial processes.

1 Introduction

Knowledge discovery in databases (KDD) has been defined as the non-trivial extraction of implicit, previously unknown, and potentially useful information from databases [1]. A KDD process is a highly complex, iterative and interactive process, with a goal-driven and domain-dependent nature. Given the huge amount of tools for data manipulation, their various characteristics and different performances, users should have various skills and expertise in order to manage all of them. As a matter of fact, in designing a KDD process they have to choose, to set-up, to compose and to execute the tools most suitable to their problems.

For these reasons, one of the most interesting challenges in KDD field involves the possibility to give support to users both in tool discovery and in process composition. We refer to the former as the activity of searching tools on the basis of the KDD goal to achieve, the characteristics of the dataset at hand, and functional and non-functional properties of the implemented algorithm. The process composition is the activity of linking suitable tools in order to build valid and useful knowledge discovery processes.

We are working on these issues in the ambit of Knowledge Discovery in Databases Virtual Mart (KDDVM) project [2], that is aimed at realizing an open and extensible environment where users can look for implementations, suggestions, evaluations, examples of use of tools implemented as services. In KDDVM each KDD service is represented by three logical layers, with growing abstraction degrees. Algorithm level is the most abstract one, whereas different implementations of the same algorithm are described at the tool level. Finally, different

N. Adams et al. (Eds.): IDA 2009, LNCS 5772, pp. 285–296, 2009.
© Springer-Verlag Berlin Heidelberg 2009

instances of the same tool can be made available on the net as services through several providers. At present, among the available services, we have at disposal a broker service for supporting the discovery of suitable KDD services. Such a service is based on KDDONTO, a domain ontology describing KDD algorithms and their interfaces [3].

In this work we introduce a goal-driven procedure aimed at automatically composing KDD processes; in order to guide the whole composition procedure, algorithm matching functions are defined on the basis of the ontological information contained in KDDONTO. The outcome of the procedure is a subset of all possible workflows of algorithms, which allows to achieve the goal requested by the user and satisfies a set of constraints. The generated processes are then ranked according to both user-defined and built-in criteria, allowing users to choose the most suitable processes w.r.t. their requests, and to let them try more than a single solution. Our composition procedure, working at the algorithm level, allows to produce abstract KDD processes, which are general and reusable, since each instance of algorithm can be replaced with one among the services implementing it. Furthermore, such a generated process can be itself considered as useful, valid and unknown knowledge. In the rest of this section, relevant literature references are discussed. Then, Section 2 presents the KDDONTO and its main concepts and relations. Section 3 introduces algorithm matching functions, which are used as basic step for the process composition procedure, that is then described in detail in Section 4. Finally, Section 5 ends the paper.

1.1   Related Works

In last years researchers in Data Mining and KDD fields have shown more and more interest in techniques for giving support in the design of knowledge discovery processes. To this end several ontologies have been defined, even if they focus only on tools and algorithms for Data Mining, which is one of the phases of the wider and more complex KDD field [4,1]. The first ontology of this kind is *DAMON* (DAta Mining ONtology) [5], that is built for simplifying the development of distributed KDD applications on the Grid, offering to domain experts a taxonomy for discovering tasks, methods and software suitable for a given goal. In [6], the ontology is exploited for selecting algorithms on the basis of the specific application domain they are used for. Finally, *OntoDM* [7] is a general purpose top-level ontology aimed to describe the whole Data Mining domain.

Some research works were also proposed for supporting process composition [8,9,10,11,12]. An early work of this kind is [9], where authors suggests a framework for guiding users in breaking up a complex KDD task into a sequence of manageable subtasks, which are then mapped to appropriate Data Mining techniques. Such an approach was exploited in [11], where the user is supported in iteratively refining a KDD skeleton process, until executable techniques are available to solve low-level tasks. To this end, algorithms and data are modeled into an object oriented schema. In [10] a system is described focusing on setting-up and reusing chains of preprocessing algorithms, which are represented in a relational meta-model.

Although these works help users in choosing the most suitable tools for each KDD phase, no automatic composition procedure is defined. Recent works have dealt with this issue [8,12] for defining effective support on process composition. In detail, in [8] authors define a simple ontology (actually not much more than a taxonomy) of KDD algorithms, that is exploited for designing a KDD process facing with cost-sensitive classification problems. A forward composition, from dataset characteristics towards the goal, is achieved through a systematic enumeration of valid processes, that are ranked on the basis on accuracy achieved on the processed dataset, and on process speed. [12] introduces a KDD ontology representing concrete implementations of algorithms and any piece of knowledge involved in a KDD process (dataset and model), that is exploited for guiding a forward state-space search planning algorithm in the design of a KDD workflow. Such an ontology describes algorithms in very few classes and a poor set of relationships, resulting in a flat knowledge base.

Both in [12] and in [8], ontologies are not rich enough to be extensively used both for deducing hidden relations among algorithms and for supporting relaxed matches among algorithms or complex pruning strategies during planning procedure. In order to overcome the limits of the cited works, in our proposal we define and exploit a formal KDD ontology expressly conceived for supporting composition. Such an ontology is exploited by a backward composition procedure, which composes algorithms not only by exact matches, but also by evaluating similarity between their interfaces, in order to extract unknown and non-trivial processes. Our approach, moreover, is aimed to achieve a higher level of generality, by producing abstract and reusable KDD process. In this work, we use the term *composition* instead of planning in order to emphasize the difference from traditional AI planning, in which execution stops when a single proper solution is found, and also because we ultimately refer to service composition.

2   The KDD ONTOlogy

KDDONTO is an ontology describing the domain of KDD algorithms, conceived for supporting the discovery of KDD algorithms and their composition.

In order to build a KDD ontology, among many methodologies proposed in literature for ontology building, we choose a formal approach based on the goal-oriented step-wise strategy described in [13]; moreover, the quality requirements and formal criteria defined in [14] are taken into account, with the aim to make meaning explicit and not ambiguous.

The key concept of KDDONTO is *algorithm*, because it is the basic component of each process. Other fundamental domain concepts, from which any other concept can be derived, are the following:

- *method*: a methodology, a technique used by an algorithm to extract knowledge from input data;
- *phase*: a phase of a KDD process;
- *task*: the goal at which aims who executes a KDD process;
- *model*: a set of constructs and rules for representing knowledge;

- *dataset*: a set of data in a proper format;
- *parameter*: any information required in input or produced in output by an algorithm;
- *precondition/postcondition*: specific features that an input (or output) must have in order to be used by a method or an algorithm. Such conditions concern format (normalized dataset), type (numeric or literal values), or quality (missing values, balanced dataset) properties of an input/output datum;
- *performance*: an index and a value about the way an algorithm works;
- *optimization function*: the function that an algorithm or a method optimizes with the aim to obtain the best predictive/descriptive model.

Starting from these concepts, top level classes are identified, namely `Algorithm`, `Method`, `Phase`, `Task`, `Data` (which contains `Model`, `Dataset` and `Parameter` as subclasses), `DataFeature` (corresponding to *precondition/postcondition*), `PerformanceIndex` and `PerformanceClass` (for describing *performance* indexes and *performance* values), `ScoreFunction` (corresponding to *optimization function*).

Main relations among the classes are:

- `specifies_phase`, between `Task` and `Phase`;
- `specifies_task`, between `Method` and `Task`;
- `uses`, between `Algorithm` and `Method`;
- `has_input`/`has_output`, a n-ary relation with domain `Algorithm`, `Method` or `Task` and codomain `Data`, and optionally `DataFeature`.
  For each instance of `DataFeature` involved in `has_input`, a value expressing the *precondition strenght* is also provided. Hence, a value equal to 1.0 corresponds to a mandatory precondition, whereas lower values to optional ones; also inverse properties `input_for`/`output_for` are introduced;
- `has_performance`, a n-ary relation with domain `Algorithm`, `Method`, or `Task` and codomain `PerformanceIndex` and `PerformanceClass`.

Subclasses are defined by means of existential restrictions on main classes, that can be considered as fundamental bricks for building the ontology. At first some `Phase` instances are introduced, namely `PREPROCESSING`, `MODELING`, `POSTPROCESSING`. They represent the main phases in a KDD process and are used to start the subclassing as follows:

- `Task` specializes in subclasses, according to the argument of `specifies_phase`, e.g.: `ModelingTask ⊑ Task ⊓ ∃specifies_phase{MODELING}`
- `Method` is detailed in subclasses according to the tasks that each method specifies by means of `specifies_task` relation, e.g.:
  `ClassificationMethod ⊑ Method ⊓ ∃specifies_task{CLASSIFICATION}`
- `Algorithm` specializes in subclasses according to `uses` and `has_output` relations. For example:

```
ClassificationAlgorithm ⊑ Algorithm
                        ⊓ ∃uses.ClassificationMethod
                        ⊓ ∃has_output.ClassificationModel
```

- `Model` is further detailed in subclasses, on the basis of the task which the models are used for, e.g.:
  `ClassificationModel⊑ Model ⊓ ∃output_for{CLASSIFICATION}`

A top-level view of described classes and relations is shown in Figure 1.

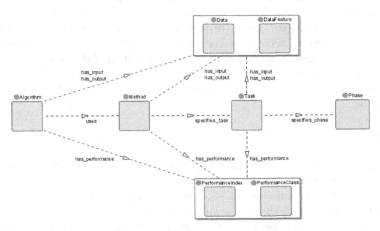

Fig. 1. KDDONTO: main classes and relations

Many other relations are introduced in order to represent information useful to support KDD process composition. Among the most interesting:

- `not_with` links two instances of `Method` that cannot be used in the same process;
- `not_before` links two instances of `Method` such that the first cannot be used in a process before the second;
- `in_module`/`out_module` allow to connect an instance of `Algorithm` to others, which can be executed respectively before or after it. These relations provide suggestions about process composition, representing in an explicit fashion KDD experts' experience about process building;
- `part_of` (and its inverse[1] `has_part`), between an instance of `Model` and an its component (a generic `Data` instance), allows to describe a model in terms of the subcomponents it is made of. These relations are useful for identifying algorithms working on similar models, that is models having common substructures, as discussed in next section.

At present, KDDONTO is represented in OWL-DL, whose logical model is based on Description Logics and is decidable; it is a sublanguage of OWL [15], the de-facto standard language for building ontologies. An implementation of KDDONTO has been obtained after some refinements, whose details are not reported here, and is available at the KDDVM project site[2].

---

[1] We use "inverse" rather than "reciprocal" because both `part_of` and `has_part` are instance-level relations.

[2] http://boole.diiga.univpm.it/kddontology.owl

3   Algorithm Matching

For the purposes of this work, we define a KDD process as a workflow of algorithms that allows to achieve the goal requested by the user. The basic issue in composition is to define the *algorithm matching*, that is to specify under which conditions two or more algorithms[3] can be executed in sequence. Each algorithm takes data with certain features in input, performs some operations and returns data in output, which are then used as input for the next algorithm in the process. Therefore, two algorithms can be matched if the output of the first is *compatible* with the input of the second.

An *exact* match between a set of algorithms $\{A_1,...,A_n\}$ and an algorithm B is defined as:

$$\mathbf{match}_E(\{A_1,...,A_n\},B) \quad \leftrightarrow \quad \forall\, in_B^i\, \exists A_k\, \exists out_{A_k}^j : out_{A_k}^j \equiv_o in_B^i$$

where in_B^i is the ith input of the algorithm B, $out_{A_k}^j$ is the jth output of the algorithm A_k. $\equiv_o$ represents the conceptual equivalence and is defined as follows: let a and b be two parameters, $a \equiv_o b$ if $C_a \sqsubseteq C_b$, i.e. if a and b refer to the concepts C_a and C_b such that C_a is subsumed by C_b (they are the same concept or the former is a subconcept of the latter). In such cases the whole set of algorithms $\{A_1,...,A_n\}$ provide the required data for B, realizing the piece of workflow shown in Figure 2a.

Furthermore, an exact match is *complete* if all the required inputs for an algorithm are provided by a single algorithm, as represented in Figure 2b.

More formally, an *exact complete* match between two algorithms A and B is defined as:

$$\mathbf{match}_{Ec}(A,B) \quad \leftrightarrow \quad \forall\, in_B^i\, \exists out_A^j : out_A^j \equiv_o in_B^i$$

(a)   (b)

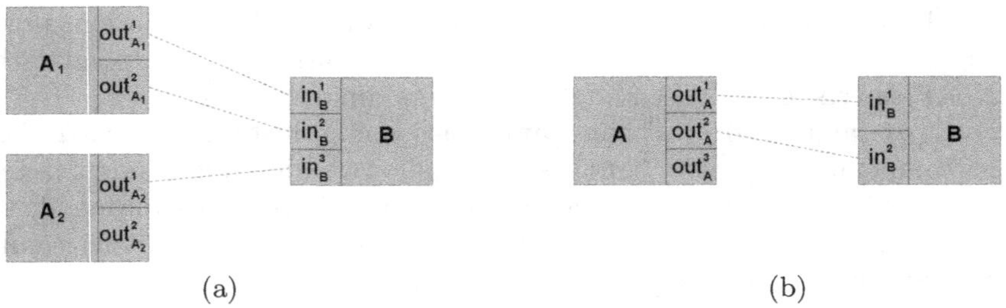

Fig. 2. (a) Exact and (b) exact complete matches (dashed lines represent $\equiv_O$ relation)

By exploiting properties of algorithms, described in the previous section, it is possible to define a match based not only on exact criteria, but also on similarity among data. We can consider compatible two algorithms even if their interfaces

---

[3] Hereafter we use "class" and "concept" as synonyms, and refer to "algorithm" as the `Algorithm` class.

Fig. 3. Approximate match (dashed line represents $\equiv_O$ relation)

are not perfectly equivalent: the relaxation of constraints results in a wider set of possible matches. Hence, an *approximate match* between a set of algorithms $\{A_1,...,A_n\}$ and an algorithm B is defined as:

$$\mathbf{match}_A(\{A_1,...,A_n\},B) \quad \leftrightarrow \quad \forall \ in_B^i \ \exists A_k \ \exists out_{A_k}^j : out_{A_k}^j \equiv_o in_B^i \ \vee$$
$$similar(out_{A_k}^j, in_B^i)$$

where the similarity predicate $similar(x,y)$ is satisfied if x and y are similar concepts, i.e. if there is a path in the ontology graph that links them together. An approximate match is useful not only when an exact match cannot be performed, but also for extracting unknown and non-trivial processes.

The similarity between concepts can be evaluated on the basis of various KD-DONTO relations. The simplest similarity relation is at *hierarchic* level: a specific datum is similar to its siblings, because they share, through an `is-a` relation, the membership to the same class. Moreover, similarity is also at *compositional* level: a datum can be made of simpler data, according to `part_of`/`has_part` relationships, described in Section 2. As shown in Figure 3, a compound datum (e.g. "d") can be used in place of one of its components (e.g. "in_B^1"), because the former is a superset of the latter, containing all the needed information, and other that can be discarded. To give a practical example, a Labeled Vector Quantization model (LVQ) `has_part` a VQ model and a Labeling function: if an algorithm requires VQ model in input, LVQ model can be provided in place of it.

Given two *similar* concepts, we define *ontological distance* as the number of `is-a` or `part_of` relations that are needed to link them in the ontological graph; as only exception, ontological distance from a concept to its subconcepts is considered null. In approximate match, the higher is ontological distance between two concepts, the less they are similar. This allows to assign a score to each match and to define a rank among the generated processes, as described in Subsection 4.3.

In process composition, whatever match is used, it is needed to check the satisfaction of preconditions and postconditions: this means that postconditions of the first algorithm must not be in contrast with preconditions of the second one, as regards the same data.

4   Process Composition Procedure

Based on algorithm matching, in this section a goal-driven procedure for composing KDD processes is described. Our approach is aimed at the generation of all potentially useful, valid and unknown processes satisfying the user requests;

this allows the user to choose among processes with different characteristics and to experiment more than a single solution. We use Jena[4] as a framework for querying the ontology through SPARQL language [16], which is a W3C Recommendation, whereas Pellet[5] is used as reasoner for inferring non-explicit facts. The proposed process composition procedure is formed of the following phases: (I) dataset and goal definition, (II) process building, (III) process ranking.

4.1   Dataset and Goal Definition

Any KDD process is built for achieving a specific KDD goal processing a given dataset. Hence, the first step of our procedure is the description of both the dataset and the goal.

In our framework, the former is described by a set of characteristics (e.g. representation model, size, feature type), which are instances of the `DataFeature` class. The latter is expressed as an instance of the `Task` class, leaving the user to move from complex domain-dependent business goals to one or more well-defined and domain-independent KDD tasks.

The description of both dataset and goal allows us to guide the composition procedure, bounding the number and type of algorithms that can be used at the beginning and at the end of each process.

Moreover, some *process constraints* are provided in this phase for contributing to define a balance between procedure execution speed and composition accuracy. Some of these constraints can be defined by the user; among others: kind of match (only exact or also approximate), maximum ontological distance for each match in a process, maximum number of algorithms in a process, and maximum computational complexity of a process.

Other constraints are predefined and built-in into the procedure for ensuring to produce valid KDD processes. Some examples are the following:

- two algorithms whose methods are linked through `not_with` property cannot coexist in the same process;
- two algorithms whose methods are linked through `not_before` property can coexist in the same process only if the first follows the second;
- more than one FeatureExtraction algorithm cannot coexist in the same process.

4.2   Process Building

Process building is an iterative phase, which starts from the given task and goes backwards adding one or more algorithms to each process and for each iteration. Such algorithms are chosen on the basis of the algorithm matching functionalities defined in the previous section.

The procedure goes on until the first algorithm of each process is compatible with the given dataset, and stops if one of the following conditions come true: no

---

[4] http://jena.sourceforge.net/
[5] http://clarkparsia.com/pellet

Table 1. The composition algorithm

Let $\mathcal{P}$ be the set of processes at each iteration, P_i be the ith process in $\mathcal{P}$, described by the pair $<\mathcal{V}_i, \mathcal{E}_i>$ where $\mathcal{V}_i$ is the set of algorithms in P_i and $\mathcal{E}_i$ is the set of directed edges (A_p, A_q) which connect algorithm A_p to algorithm A_q.
Let $\mathcal{F}$ be the final list of valid generated processes, T be the task, $\mathcal{D}$ be the set of dataset characteristics, $\mathbf{match}_D(\mathcal{D}, P_i)$ be a predicate, which is true if the precondition of the algorithms at the head of P_i are compatible with $\mathcal{D}$.

$\mathcal{P} \leftarrow \varnothing$, $\mathcal{F} \leftarrow \varnothing$;
Find the set $\Gamma = \{A_i: \mathbf{has\_output}(A_i, x) \sqcap \mathbf{output\_for}(x, T)\}$;
foreach $A_i \in \Gamma$ **do**
    initialize $P_i = <A_i, \varnothing>$;
    if ($\mathbf{process\_constraints}(A_i, P_i)$) **then** $\mathcal{P} \leftarrow P_i$;

foreach $P_i \in \mathcal{P}$ **do**
    if ($\mathbf{match}_D(\mathcal{D}, P_i)$) **then** $\mathcal{F} \leftarrow P_i$;
    Define the set $\Delta = \{A_k \in \mathcal{V}_i: \nexists (x, A_k) \in \mathcal{E}_i\}$;
    foreach $A_k \in \Delta$ **do**
        Find the set $\Phi = \{\Phi_1, ..., \Phi_m\}$, where Φ_j is the set of algorithms $\{B_1, ..., B_{m_j}\}$
            such that $\mathbf{match}_E(\Phi_j, A_k) \sqcup \mathbf{match}_A(\Phi_j, A_k)$;
        foreach $\Phi_j \in \Phi$ **do**
            if ($\mathbf{process\_constraints}(\Phi_j, P_i)$) **then**
                define $P' = <\mathcal{V}_i \leftarrow \Phi_j, \mathcal{E}_i \leftarrow \{(B_1, A_k), ..., (B_{m_j}, A_k)\}>$;
                $\mathcal{P} \leftarrow P'$;
    $\mathcal{P} = \mathcal{P} - \{P_i\}$.

given process can be further expanded because no compatible algorithms exist, or one of the process constraints is violated.

The main steps in process building phase are described in Table 1. A process $P_i = <\mathcal{V}_i, \mathcal{E}_i>$ is represented as a directed acyclic graph, where $\mathcal{V}_i$ is the set of nodes, namely algorithms, and $\mathcal{E}_i$ is the set of directed edges linking algorithms together. At first, algorithms A_i, which return as output a model x used for performing the given task T, are found; then, for each of them a process P_i is created. Such a P_i is added to the set $\mathcal{P}$ which contains all the processes that are going to be evaluated in the next step.

Until there is a process P_i in $\mathcal{P}$, algorithms compatible with the one(s) at the head of P_i are extracted. If the process constraints are satisfied, these extracted algorithms are used for forming a new valid process, which is added to the set $\mathcal{P}$. At last, the process P_i is deleted from the set $\mathcal{P}$ because its expansion has ended, and the procedure is iterated. At the beginning of each iteration, P_i is checked against the characteristics of the dataset at hand: if they are compatible, P_i is moved to the set $\mathcal{F}$ of final processes. Note that the process constraints are used as pruning criteria, that ensure to produce useful and valid processes, keeping the complexity of the whole procedure under control.

During the procedure, it may happen that a single algorithm or a set of algorithms can be executed more than one time inside a process. To avoid any possible endless loop, we fix the maximum number of algorithms in a process.

4.3 Process Ranking

In order to support the user in choosing among the generated processes, we define some criteria for ranking them:

- *similarity measurement*: an exact match is more accurate than an approximate one, thus a process can be ranked on the basis of the sum of the ontological distances of each match. The higher the value of the sum, the less the rank of the process;
- *precondition relaxation*: in algorithm matching, preconditions on some data can be relaxed if they have a `condition_strenght` value lower than 1, i.e. a non-mandatory precondition. Relaxing preconditions reduces the process rank, because algorithm execution can lead to lower quality outcomes;
- *use of link modules*: the score of a process in which there are algorithms linked through the properties `in_module` and `out_module` is increased, because these relations state that a specific connection among algorithms was proved to be effective;
- *performance evaluation*: algorithm performances are used to assign a global score to a process. For example, in the case of a computational complexity index, it is possible to determine the whole process complexity as the highest complexity among the algorithms in a process.

4.4 Applicative Example

At present the KDDONTO implementation is formed of 88 classes, 31 relations and more than 150 instances; we describe 15 algorithms of preprocessing, modeling and postprocessing phases, in particular for Feature Extraction, Classification, Clustering, Evaluation and Interpretation tasks.

On this basis, the effectiveness of the composition procedure has been evaluated through a prototype implementation. The following scenario has been assumed: an user wants to perform a *classification* task on a normalized dataset with 2 balanced classes, missing values and both literal and numeric values. The constraints she puts are the following: both exact and approximate matches allowed, maximum number of algorithms for a process equal to 5.

The evaluation has been performed comparing our proposal with other two solutions. As first solution we have defined a procedure using a database for representing information about algorithms, in which no inference is possible. In the other solution, we have exploited a combinatorial approach for composing algorithms, where a Postprocessing algorithm cannot precede Preprocessing or Modeling ones, and a Modeling algorithm cannot precede a Preprocessing algorithm. Resulting processes have been then evaluated by a KDD expert, in order to identify the valid ones, i.e. processes in which the algorithm sequence is both semantically correct w.r.t. all input/output matches and consistent w.r.t. the user goal and requests.

Using a non-ontological approach, we are able to extract 37 processes, that the expert assesses to be all valid. The number of processes considerably increases when the combinatorial approach is exploited, but most of them are invalid and often meaningless, and need to be manually filtered. Finally, our procedure generates a set of 70 processes, which consists of the valid processes extracted through the non-ontological approach and other 33 valid and not explicit processes, composed by using inference and approximate match. Hence, our procedure is able to produce a high number of alternatives, without introducing spurious and semantically incorrect processes.

5 Conclusion

The main contribution of this work is the introduction of a goal-oriented procedure aimed at the automatic composition of algorithms forming valid KDD processes. The proposed procedure is based on the exploitation of KDDONTO, that formalize knowledge about KDD algorithms. The use of such an ontology leads to manifold advantages. Firstly, the resulting processes are valid and semantically correct. Secondly, unlike works in Literature, we are able to generate not only explicit processes formed by directly linkable algorithms, but also implicit, interesting and non-trivial processes where algorithms share similar interfaces. Thirdly, KDDONTO is able to support complex pruning strategies during composition procedure, making also use of inferential mechanism. Finally, processes can be ranked according to both ontological and non-ontological criteria.

Comparing with planning algorithms [12], such an approach allows users to choose more processes suitable w.r.t. their requirements. Moreover, generated processes can be themselves considered as useful, valid and unknown knowledge, valuable both for novice and expert users.

At present we are working on the development of a support service implementing the described process composition procedure, in order to actually integrate it into the KDDVM project. Since abstract KDD processes cannot be directly executed, each of them needs to be substituted with a workflow of services, in which every algorithm is replaced with a service implementing it. As future extensions, we are also working on increasing the number of instances described in KDDONTO and performing more comprehensive tests. Furthermore, we are studying several heuristics to provide an actual ranking of the generated processes.

References

1. Fayyad, U.M., Piatetsky-Shapiro, G., Smyth, P.: From data mining to knowledge discovery: an overview. In: American Association for Artificial Intelligence, Menlo Park, CA, USA, pp. 1–34 (1996)
2. KDDVM project site, http://boole.diiga.univpm.it

3. Diamantini, C., Potena, D.: Semantic Annotation and Services For KDD Tools Sharing and Reuse. In: Proc. of the 8th IEEE International Conference on Data Mining Workshops. 1st Int. Workshop on Semantic Aspects in Data Mining, Pisa, Italy, December 19, pp. 761–770 (2008)

4. CRISP-DM site, http://www.crisp-dm.org

5. Cannataro, M., Comito, C.: A data mining ontology for grid programming. In: Proc. 1st Int. Workshop on Semantics in Peer-to-Peer and Grid Computing, in conjunction with WWW 2003, Budapest, Hungary, pp. 113–134 (2003)

6. Yu-hua, L., Zheng-ding, L., Xiao-lin, S., Kun-mei, W., Rui-xuan, L.: Data mining ontology development for high user usability. Wuhan University Journal of Natural Sciences 11(1), 51–56 (2006)

7. Panov, P., Džeroski, S., Soldatova, L.: OntoDM: An Ontology of Data Mining. In: International Conference on Data Mining Workshops, pp. 752–760. IEEE Computer Society, Los Alamitos (2008)

8. Bernstein, A., Provost, F., Hill, S.: Towards Intelligent Assistance for a Data Mining Process: An Ontology Based Approach for Cost-Sensitive Classification. IEEE Transactions on Knowledge and Data Engineering 17(4), 503–518 (2005)

9. Engels, E.: Planning tasks for knowledge discovery in databases; performing task-oriented user-guidance. In: Proceedings of the 2nd International Conference on Knowledge Discovery in Databases (KDD 1996), Portland, Oregon (August 1996)

10. Morik, K., Scholz, M.: The MiningMart Approach to Knowledge Discovery in Databases. In: Zhong, N., Liu, J. (eds.) Intelligent Technologies for Information Analysis, pp. 47–65. Springer, Heidelberg (2004)

11. Wirth, R., Shearer, C., Grimmer, U., Reinartz, T., Schlösser, J.J., Breitner, C., Engels, R., Lindner, G.: Towards Process-Oriented Tool Support for Knowledge Discovery in Databases. In: Komorowski, J., Żytkow, J.M. (eds.) PKDD 1997. LNCS, vol. 1263, pp. 243–253. Springer, Heidelberg (1997)

12. Žáková, M., Křemen, P., Železný, F., Lavrač, N.: Using Ontological Reasoning and Planning for Data Mining Workflow Composition. In: SoKD: ECML/PKDD 2008 workshop on Third Generation Data Mining: Towards Service-oriented Knowledge Discovery, Antwerp, Belgium (2008)

13. Noy, N., McGuinnes, D.L.: Ontology Development 101: A Guide to Creating Your First Ontology. Stanford University (2002)

14. Gruber, T.: Toward principles for the design of ontologies used for knowledge sharing. Int. J. Hum.-Comput. Stud. 43(5-6), 907–928 (1995)

15. Smith, M.K., Welty, C., McGuinness, D.L.: OWL Web Ontology Language Guide, W3C Recommendation (2004), http://www.w3.org/TR/owl-guide/

16. Prud'hommeaux, E., Seaborne, A.: SPARQL Query Language for RDF, W3C Recommendation (2008), http://www.w3.org/TR/rdf-sparql-query/

Mining Frequent Gradual Itemsets from Large Databases

Lisa Di-Jorio[1], Anne Laurent[1], and Maguelonne Teisseire[2]

[1] LIRMM – Université de Montpellier 2 – CNRS
161 rue Ada, 34392 Montpellier, France
{dijorio,laurent}@lirmm.fr
[2] Cemagref - UMR Tetis
teisseire@teledetection.fr

Abstract. Mining gradual rules plays a crucial role in many real world applications where huge volumes of complex numerical data must be handled, e.g., biological databases, survey databases, data streams or sensor readings. Gradual rules highlight complex order correlations of the form *"The more/less X, then the more/less Y"*. Such rules have been studied since the early 70's, mostly in the fuzzy logic domain, where the main efforts have been focused on how to model and use such rules. However, mining gradual rules remains challenging because of the exponential combination space to explore. In this paper, we tackle the particular problem of handling huge volumes by proposing scalable methods. First, we formally define gradual association rules and we propose an original lattice-based approach. The GRITE algorithm is proposed for extracting gradual itemsets in an efficient manner. An experimental study on large-scale synthetic and real datasets is performed, showing the efficiency and interest of our approach.

1 Introduction

Nowadays, many electronic devices are used to deal with real world applications. Sensors are everywhere and report on phenomena studied by experts for monitoring or science investigation, thus leading to an increasing volume of data containing numerical data attributes. However, even if the problem of mining quantitative attributes has been tackled for many years [1], it remains difficult to extract useful knowledge such as gradual rules. The main reason is that databases are very large, both in the number of tuples and in the number of attributes. Mining such databases is an important and essential task, as experts use such knowledge to take decisions or to understand different behaviours. Many domains are concerned, as for example the biological domain, where most advances are done by analysing genome data. Another major domain is the sensor reading and data stream one, where common behaviours allow for monitoring, intrusion/system failure detection, or behaviour analysis.

In this paper, we address the problem of mining *gradual association rules* from such databases. A gradual rule allows for the modelling of frequent co-variations

N. Adams et al. (Eds.): IDA 2009, LNCS 5772, pp. 297–308, 2009.
© Springer-Verlag Berlin Heidelberg 2009

over a set of objects. Such a rule is built on the following pattern: *"the more/less A_1 and ... the more/less A_n, then the more/less B_1 and ... the more/less B_p"*. This kind of rules has already been addressed in the literature [2,3,4], but there does not exist yet any scalable approach, and mining from more than two ordered attributes remains challenging. Indeed, finding the most representative ordering of all the database objects leads to compute all the possible orders. This problem is linked to order mining, and is more complex than dealing with first ordering objects on a dimension, and then on the following and so on. In this case, order in which items are considered impacts on the sorting operation. Thus, a method independant of the set of items considered is needed. Moreover, this method has to be efficient and scalable.

In this paper, we introduce a formalism integrating two kinds of variations: *increasing* variations (the more), and *decreasing* variations (the less). This is described in Section 3. Moreover, in Section 4, we show that the problem of mining gradual itemsets can be tackled by using a binary matrices-based approach. We also provide algorithms that are directly designed from this framework. In Section 5, our experimental study shows that our method is efficient both in time and memory consumption on large scale synthetic and real databases.

2   Related Work

Gradual rules were designed in the 70's to model system behaviours. They were given by experts, and mainly used as inductive tools into fuzzy controllers [5] (e.g., "the closer the wall, the more the train must apply the brake"). A complete theoretical framework of gradual rules into the fuzzy context is given in [6], with a comparison of fuzzy implication for gradual dependencies. Among them the most used is Rescher-Gaines (RG) implication given by equation 1, where $A(X)$ is the membership degree of X for the fuzzy set A, and $B(Y)$ the membership degree of Y for the fuzzy set B:

$$X \to_{RG} Y = \begin{cases} 1 \ if A(X) \le B(Y) \\ 0 \ else \end{cases} \tag{1}$$

Rescher-Gaines implication ensures that the membership degree of X is constrained by Y's membership degree. Thus, when the value of Y increases, it relaxes the constraint and allows for the increase of X, ensuring that "the more Y is B, the more X is A". However, the use of restrictive fuzzy implication such as RG makes the conjunction hard to implement (see [7] for more details).

In [8], gradual rules are "computed" from a linguistic database. No focus is put on how to mine the rules from huge databases. Authors are interested in the knowledge revealed by such rules, as for example pointing out an empty zone in the database.

According to [2], association rules from data in the presence / absence form can be derived from a *contingency table*. However, as fuzzy sets deal with numerical items, such tables are not suitable anymore. Thus, [2] proposes to model them by the means of a *contingency diagram*. Then, linear regression is applied, revealing correlations between fuzzy items. According to [2], coefficient slope and quality of the linear regression allow to decide of the validity of the gradual rule. The interested reader is referred to [9] for a detailed analysis. However, computing linear regressions is too time consuming to handle huge databases (large number of objects and items).

To the best of our knowledge, [3] is the first using datamining methods through an adaptation of the Apriori algorithm. Gradualness is realised by the use of $\{<,>\}$ operators, leading to a database redefinition. Gradual rules are mined by considering all the pairs of objects. The support is consequently expressed as the number of pairs respecting an order divided by the total number of pairs of the database. The originality of [3] is to take into account rules having compound conditions and conclusions. Moreover, [3] is the first to avoid the use of RG implications. However, the computational complexity of the proposed method remains high, as shown by experiments performed on a small real dataset containing only 6 items.

[10] uses sequential patterns in order to highlight trends over the time. A sequential pattern is a list of ordered itemsets. Traditionally, the order is associated to a time measure. Thus for example, knowledge of the form *"When an engine speed strongly increases, after a very short period of time the truck speed slightly increases for a short period"* is extracted. The aim is different from ours, as [10] considers variations from one timestamp to another one.

Until now, no method allowing for the automatic extraction of gradual itemset has been proposed, although experts are more and more expecting this kind of tool. This paper thus introduces a new efficient algorithm that particularly copes with a large number of attributes.

3   Problem Definition

Gradual association rules can be viewed as an extension of classical association rules, where gradual items are considered. A gradual rule exhibits relations between items such as *"the more/less A, the more/less B"*. This paper is dealing with extracting gradual rules from tables. We consider the classical framework of databases. Let $\mathcal{I} = \{i_1, ..., i_n\}$ be a set of items. A table (or relation) $\mathcal{T}$ can be defined over the schema $\mathcal{I}$. The tuples of $\mathcal{T}$ are denoted by X, and $X[i]$ denotes the value of the attribute i for X.

For example, the relation from Table 1 shows various items about hostels. *Town* and *Pop*(ulation) where an hostel is located are reported, together with its *price* and the *distance* from the town centre. This table contains four tuples: $\{X_{h_1}, X_{h_2}, X_{h_3}, X_{h_4}\}$. A classical analysis of this table as proposed by [1] provides interval based rules, such *"a room located less than one kilometer from the centre will cost between 60 and 200 dollars"*. However, we notice that when the distance from the centre decreases, the price increases. In the same way, the bigger the town, the higher the price. Thus, instead of classifying an item value into an interval or a fuzzy set, it could rather be interesting to study the co-variations from one item to another one, as for example the variation of the town population size and the room price.

Table 1. Touristic sample table

	Hostel	Town	Pop. (10^3)	Dist. from Centre	Price
X_{h_1}	h_1	Paris	2.1	0.3	82
X_{h_2}	h_2	New York	8.0	5	25
X_{h_3}	h_3	New York	8.0	0.2	135
X_{h_4}	h_4	Ocala	0.04	0.1	60

Three kinds of variations have to be considered: increasing variation, decreasing variation, and no variation. Each item will hereafter be considered twice: once to evaluate its increasing strength, and once to evaluate its decreasing strength, using the $\geq$ and $\leq$ operators. This leads to consider new kinds of items, reported here as *gradual items*.

Definition 1. *(Gradual Item) Let $\mathcal{I}$ be a set of items, $i \in \mathcal{I}$ be an item and $* \in \{\geq, \leq\}$ be a comparison operator. A gradual item i^* is defined as an item i associated to an operator $*$.*

Consequently, a gradual itemset is defined as follows:

Definition 2. *(Gradual Itemset) A gradual itemset $s = (i_1^{*1}, ..., i_k^{*k})$ is a non empty set of gradual items. A k-itemset is an itemset containing k gradual items.*

For example, let us consider the rule *"the bigger the town and the nearer from the town centre, then the higher the price"*, formalised by the 3-itemset $s_1 = (Pop^{\geq}\ Dist^{\leq}\ Price^{\geq})$. Such a rule cannot be conveyed by a simple association rule, as it expresses a comparison between the values of items: a hostel does not only have to support an item to increase the frequency, it also needs to satisfy a value variation condition. In this paper, we consider that this variation is measured by comparing the values of different tuples on the same item, using the following definitions:

Definition 3. *(Ordering of two Tuples) Let X and X' be two tuples from $\mathcal{T}$, and $s = (i_1^{*1}, ..., i_k^{*k})$ be a gradual itemset. X preceeds X' if $\forall l \in [1, k]$ $X[i_l] *_l X'[i_l]$ holds, denoted $X \lhd_s X'$.*

Definition 4. *X and X' are comparables according to a given itemset s if $X \lhd_s X'$ or $X' \lhd_s X$. Otherwise, they are incomparables.*

Definition 5. *(List of Ordered Tuples) Let $s = (i_1^{*1}, ..., i_k^{*k})$ be a gradual itemset. A list of tuples $\mathcal{L} =<_L X_1, ..., X_n >_L$ respects s if $\forall p \in [1, n-1], \forall l \in [1, k]$ $X_p[i_l] *_l X_{p+1}[i_l]$ holds.*

Property 1. There is more than one list of tuples respecting s.

Referring back to our previous example from Table 1, we have $X_{h_1}[Pop] \leq X_{h_3}[Pop]$, $X_{h_1}[Dist] \geq X_{h_3}[Dist]$ and $X_{h_1}[Price] \leq X_{h_3}[Price]$. Thus, X_{h_1}

and X_{h_3} are comparable according to s_1. Moreover, from Table 1, s_1 holds for three different sets of hostels: $L_1 =<_L X_{h_4} >_L$, $L_2 =<_L X_{h_1}, X_{h_3} >_L$ and $L_3 =<_L X_{h_2}, X_{h_3} >_L$. Notice that L_1 has only one element, because of the reflexivity property of the used operator. However, as a list of only one element does not bring information in the gradual context, it will be discarded. This point is discussed in Section 4.3. In order to calculate the frequency of an itemset, we consider the more representative set, i.e., the set having the largest size:

Definition 6. *(Frequency) Let $\mathcal{G}_s = \{L_1, ..., L_m\}$ be the set of all the lists respecting a gradual itemset s. Then $Freq(s) = \frac{\max_{1 \leq i \leq m}(|L_i|)}{|T|}$.*

Here, we have $\mathcal{G}_{s_1} = \{L_1, L_2, L_3\}$ and two lists from $\mathcal{G}$ having a maximal size, which is 2. Then, $Freq(s_1) = \frac{2}{4} = 0.5$, meaning that 50% of all of the hostels follow s_1. Notice that we choose to use the most representative set for computing the frequency. One may argue that we could have used the minimal set. However, we adopt an optimistic computation.

Proposition 1. *(Antimonotonicity of gradual itemsets) i) Let s and s' be two gradual itemsets, we have: $s \subseteq s' \Rightarrow Freq(s) \geq Freq(s')$. ii) Let s and s' be two gradual itemsets such that $s \subseteq s'$, then we have: if s' is not frequent then s is not frequent.*

Proof. Let us consider two gradual itemsets s_k and s_{k+1} such that $s_k \subseteq s_{k+1}$, with k and $k + 1$ being the length of these itemsets. Let ml be a maximal list from $\mathcal{G}_{s_k}$. This means that $\forall X, X' \in ml$:

- if $\neg(X \vartriangleleft_{s_{k+1}} X')$ then $Freq(s_k) > Freq(s_{k+1})$
- if $(X \vartriangleleft_{s_{k+1}} X')$, then $Freq(s_k) = Freq(s_{k+1})$

Thus, we have $Freq(s_k) \geq Freq(s_{k+1})$.

A gradual itemset is said to be frequent if its frequency is greater than or equal to a user-defined threshold. The problem of **mining frequent gradual itemsets** is to find the complete set of frequent gradual itemsets in a given table T containing numerical items, with respect to a minimum threshold $minFreq$. In this paper, we propose an efficient Apriori-based algorithm called GRITE, which extracts gradual itemsets.

4   GRITE: Efficient Extraction of Gradual Itemsets

In this section, we present our algorithm, GRITE, for GRadual ITemset Extraction. This algorithm is based on the Apriori algorithm. We explain the principal algorithm features: how to explore the search space, how to join two $(k-1)$-itemsets in order to obtain a k-itemset, and how to compute the frequency of a gradual itemset.

4.1   Gradual Itemset Search Space

Frequent itemset mining has been widely studied since its introduction by [11]. The original algorithm, called Apriori, uses the antimonotonicity property in order to efficiently find all the frequent itemsets. In order to traverse the search space, Apriori uses a prefix tree. We propose the use of an Apriori based method in order to extract gradual itemsets. In order to handle gradual items, we generate two gradual items $i^{\geq}$ and $i^{\leq}$ instead of generating one item i. Moreover, we consider the notion of complementary gradual itemset and the associated properties in order to avoid the consideration of all the combination, as originally proposed in [3].

Definition 7. *(complementary gradual itemset) Let $s = (i_1^{*_1}, ..., i_n^{*_n})$ be a gradual itemset, the complementary gradual itemset of s, denoted $c(s) = (i_1^{*_1^c}, ..., i_n^{*_n^c})$, is defined as $\forall j \in [1, n] *_j^c = c(*_j)$, where $c(\geq) = \leq$ and $c(\leq) = \geq$.*

Proposition 2. $Freq(s) = Freq(c(s))$

Proposition 2 avoids unnecessary computations, as generating only half of the gradual itemsets is sufficient to automatically deduce the other ones. We are using an Apriori-based approach, so we need to define how to join two gradual itemsets of length k in order to obtain a gradual itemset of length $k + 1$. As we associate a binary matrix to each itemset, we need to address the problem of joining using these structure.

4.2   Matrices for Representing Orders

As stated in Proposition 1, we have to keep a track of every list of tuples respecting a gradual itemset. However, the rule extraction process is known as being exponential, which means that we have to be carefull concerning the structures that we use. This is why we propose the use of a bitmap representation, which has already been proved as being efficient in an exponential algorithm [12]. Then, orders are stored in a binary matrix, which is defined as follows:

Definition 8. *(Binary Matrix of Orders) Let s be a gradual itemset, $\mathcal{G}_s$ be the list of object respecting it and $\mathcal{T}_{\mathcal{G}_s}$ be the set of the tuples of $\mathcal{G}_s$. $\mathcal{G}_s$ can be represented by a binary matrix $M_{\mathcal{G}_s} = (m_{a,b})_{a \in \mathcal{T}_{\mathcal{G}_s}, b \in \mathcal{T}_{\mathcal{G}_s}}$, where $m_{a,b} \in \{0, 1\}$.*

If there exists an order relation between a and b, then the bit corresponding to the line of a and the column position of b is set to 1, and to 0 otherwise. For example, let us consider the gradual 1-itemset $s_2 = \{Pop^{\geq}\}$. From Table 1, we have $\mathcal{G}_s = \{<_L X_{h_4}, X_{h_1}, X_{h_2}, X_{h_3} >_L, <_L X_{h_4}, X_{h_1}, X_{h_3}, X_{h_2} >_L\}$, and $\mathcal{T}_{\mathcal{G}_{s_2}} = \{X_{h_1}, X_{h_2}, X_{h_3}, X_{h_4}\}$. This set of orders is modeled by means of a binary matrix of size 4×4, represented by Figure 1a. Figure 1b is the binary matrix for the gradual 1-itemset $s_3 = \{Dist^{\leq}\}$.

Γ	h_1	h_2	h_3	h_4
h_1	0	1	1	0
h_2	0	0	1	0
h_3	0	1	0	0
h_4	1	1	1	0

(a)

Γ	h_1	h_2	h_3	h_4
h_1	0	0	1	1
h_2	1	0	1	1
h_3	0	0	0	1
h_4	0	0	0	0

(b)

Fig. 1. Binary Matrixes M_{s_2} and M_{s_3} for (a) $s_2 = \{Pop^{\geq}\}$, (b) $s_3 = \{Dist^{\leq}\}$

4.3 Candidate Generation

Apriori is a levelwise algorithm: $(k-1)$-itemsets are used to generate k-itemsets. This operation, called *join*, is done an exponential number of times. This operation keeps objects respecting the gradual itemset i in order to compute the frequency of i. In our context, matrices are used for an ordered representation of these objects. Each node of the prefix tree is associated with a matrix. Thus, the joining operation consists in computing all the common orders from two input matrices.

From a binary matrix, common orders are those which bit is set to 1 to each of the input matrices. This is achieved by the AND bitwise operation:

Theorem 1. *Let s'' be the gradual itemset generated using the two gradual itemsets s and s'. The following relation holds: $M_{\mathcal{G}_{s''}} = M_{\mathcal{G}_s}$ AND $M_{\mathcal{G}_{s'}}$*

Theorem 1 allows to efficiently perform the join operation using the levelwise method. Indeed, bitwise operations are among the better performing from a computational point of view, providing thus a scalable algorithm. Figure 2a shows the result of the join operation between M_{s_2} and M_{s_3}.

Γ	h_1	h_2	h_3	h_4
h_1	0	0	1	0
h_2	0	0	1	0
h_3	0	0	0	0
h_4	0	0	0	0

(a)

Γ	h_1	h_2	h_3
h_1	0	0	1
h_2	0	0	1
h_3	0	0	0

(b)

Fig. 2. (a) Binary Matrix M_{s_4} for $s_4 = \{Pop^{\geq}, Dist^{\leq}\}$, (b) Reduced Matrix for s_4

In the end, matrices associated to k-itemsets are obtained. However, some information is not used, for example, isolated lines / columns represent tuples that do not have any relation. These last ones are meaningless in a gradual context, they are deleted. This method allows to gain memory, and time, as deleted tuples are not considered during future joins. On Figure 2a, X_{h_4} is deleted: all bits from the X_{h_4} column and X_{h_4} line are set to 0. Figure 2b represents the final matrix.

Algorithm 1 describes the joining step. Before computing the bitwise AND operation, we initialise a matrix which size corresponds to the number of

---

Algorithm 1. Join

Data: Two matrices $M_{\mathcal{G}_s}$ and $M_{\mathcal{G}_{s'}}$
Result: The matrix $M_{\mathcal{G}_{ss'}}$

1 $M_{ss'} = Initialise(\mathcal{T}_{\mathcal{G}_s} \cap \mathcal{T}_{\mathcal{G}'_s})$
2 $M_{ss'} = M_s \ AND \ M_{s'}$
3 $M_{ss'} \leftarrow DeleteAloneTuples(M_{ss'})$
4 **return** $M_{ss'}$

Algorithm 2. RecursiveCovering

Data: A tuple t
         The memory from previous steps $Memory$
Result: Fill $Memory$

1 $Sons \leftarrow GetSons(t)$ /* all t' set to 1 at line t */
2 **if** $Sons = \emptyset$ **then**
3 $\quad |$ $Memory[node] = 1;$
4 **else**
5 $\quad$ **foreach** $i \in Sons$ **do**
6 $\quad \quad |$ **if** $Memory[i] = -1$ **then**
7 $\quad \quad \quad |$ $RecursiveCovering(i, Memory)$
8 $\quad \quad |$ **end**
9 $\quad$ **end**
10 $\quad$ **foreach** $i \in Sons$ **do**
11 $\quad \quad |$ $Memory[node] = max(Memory[node], Memory[i] + 1)$
12 $\quad$ **end**
13 **end**

---

common tuples from $M_{\mathcal{G}_s}$ and $M_{\mathcal{G}_{s'}}$ (by Proposition 1, an object not respecting gradualness for s or s' does not participate to the support of ss'). This remains to compute the operation $\mathcal{T}_{\mathcal{G}_s} \cap \mathcal{T}_{\mathcal{G}'_s}$. Then, the AND operation is computed (line 2), and in the end, the meaningless tuples are erased by the mean of the function *DeleteAloneTuples* (line 3).

4.4   Frequency Computation

Given a gradual itemset s and its associated matrix $M_{\mathcal{G}_s}$, the frequency of s is the longest list from $M_{\mathcal{G}_s}$. Finding this one could quickly become a bottleneck if we use a naive algorithm (which consist in following all the possible orders). To be efficient, frequency computation must consider each tuple from $M_{\mathcal{G}_s}$ only once. In this section, in order to illustrate our method, we use for the sake of simplicity the Hasse Diagram: if $a \geq b$, a is placed upward of b, and an arrow is drawn from a to b. Usually, redundant relations are not drawn.

In $M_{\mathcal{G}_s}$, a tuple can appear in various levels. For example, tuple i from Figure 3a could be associated to the level 5 or to the level 7. Our aim is to keep level 7

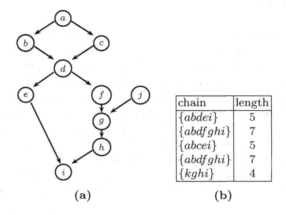

chain	length
{abdei}	5
{abdfghi}	7
{abcei}	5
{abdfghi}	7
{kghi}	4

(a)   (b)

Fig. 3. (a) Hasse Diagram for $\mathcal{M}_s$ and (b) its chains

(longest list). A tuple having more than one father may potentially be assigned with more than one level. When this node is treated for the first time, levels from other fathers are unpredictable. So, we use a *"memory"* conserving knowledge about maximality from other fathers. When many solutions are encountered, we keep the maximal one.

The core of the method is done by the Algorithm 2. Its input is a tuple t, and it updates the memory according to all of its sons (the tuples sets to one on the line t). The following strategy is adopted: a tuple having no relation is considered as a leaf with a level assigned to one. When a "leaf" is encountered, its level is assigned to the value one, and the recursion is broken (line 3). Otherwise, the recursion ensures that when a tuple t is encountered, all the vertices t' in relation with t have been treated. As we are looking for the list having the maximal size, the t' having the hightest level (line 11) is selected. So, the level of t corresponds to the level of t' plus one.

Let us illustrate this method on the Hasse Diagram of Figure 3a. On this diagram, $\{a\}$ is the only tuple having no father. Thus, we will start from it. Notice that it is easy to keep track or efficiently find tuples having no father. Supposing that we treat sons using a lexicographic order, Algorithm 2 is called on a and then on b, d, e, and i. As it does not have any relation, i is considered as a leaf, and has a memory of 1. Each crossed vertex is updated, and we obtain the memory displayed on Table 2a. Then recursion is called on f, leading to change the final level of a. The final results are displayed on Table 2b.

Table 2. Memory obtained after traversing (a) the first chain (b) the entire lattice

(a)

a	b	c	d	e	f	g	h	i	j
-1	-1	-1	-1	2	-1	-1	-1	1	-1

(b)

a	b	c	d	e	f	g	h	i	j
7	6	6	5	2	4	3	2	1	4

5   Experiments

We ran our algorithm on synthetic datasets, in order to measure memory and time performance. It should be recalled here that the issue of scalability is a great challenge as the databases mined by gradual rule methods are very dense and are thus hard to handle. For instance, previous work [2,3] does not provide a scalability study of their method. In this context, we manage to extract gradual rules from databases containing thousands of tuples and hundreds of items (and vice-versa). We used an adapted version of IBM Synthetic Data Generation Code for Associations and Sequential Patterns[1] in order to generate synthetic datasets. Synthetic tests have been done in order to show the efficient behaviour of our algorithm on datasets having a bigger set of objects first (3000 objects and 100 items), and then having a bigger set of items (30 objects, 1500 items).

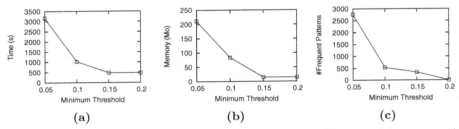

Fig. 4. Performances for 3000 objects and 100 items in (a) time, (b) memory, (c) number of extracted patterns

Figure 4a shows time performance for a database containing a bigger set of objects. Support has been set as low as possible, here up to 0.5%, which means that to each tree node is associated a lattice containing a chain of at least 150 elements. Figure 4b shows that about 200 Mb have been used to extract 2761 patterns (Figure 4c). These experiments show that the first level computation takes time, even if the number of frequent patterns is low. This is due to the ordering operation: for each items, corresponding values have to be ordered. At this stage, all items are frequent, as they all have a numerical value. This is why we output frequent itemsets starting from the second level. Notice that for the 30 objects-sized dataset, this takes less time, as there are fewer objects to order. On Figure 5a, we can see that for a database containing a large set of items, the time-efficiency decreases. This is due to the number of necessary combinations to extract frequent itemsets. On the contrary, this experiment has been more efficient in term of memory (Figure 5b). In fact, less than 30 objects have to be stored per node. With a minimal support set to 0.3, GRITE extracted about 700,000 frequent patterns (Figure 5c).

We also ran GRITE on a real dataset dealing with Alzheimer disease. Researchers in psychology focused on memory and feeling points. Classical tests

---

[1] www.almaden.ibm.com/software/projects/hdb/resources.shtml

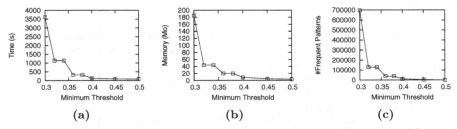

Fig. 5. Performances for 30 objects and 1500 items in (a) time, (b) memory, (c) number of extracted patterns

from the "Diagnostic manual of mental disorders" have been done too. Psychologists ask patients to recall a good moment of their life, as well as a neutral moment and a bad one. Then, they asked them to describe sounds, spatial dispositions and their feelings during these moments. This dataset contains 33 patients and 122 items. Here are some extracted gradual association rules (having 100% confidence):

- *The better the souvenir of persons spatial disposition and the better the souvenir of the day time, then the better the place recall* (87.88%)
- *The more identification during the RI 48 test and the better the spacial position and time of the day souvenir, then the better the MMS test* (81.82%)
- *The older the patient and the worse the souvenir of sound of a good moment, then the worse the MMS results* (81.82%)

The first rule shows that the more a patient has a good memory of persons and day time, then the better they remember the place. The second rule is more useful to experts, as it uses test results and memory information. This links the mental patient state to how they recall neutral moments. Finally, the last rule shows that a decreasing memory of a souvenir is linked with the increasing of the age.

6 Conclusion

Gradual rules have been extensively studied especially in the fuzzy logic domain. However, no efficient method for frequent gradual itemset extraction has been proposed. In this paper, we presented GRITE, an algorithm which takes advantage of a binary representation of lattice structure. As shown in our performance study, GRITE allows to extract gradual itemsets like "the more / the less" from datasets containing more items than objects. Our works is thus directly applicable to numerical databases such as gene databases, or medical databases.

This work raised some interesting perspectives. Firstly, in order to refine the quality of obtained results, we will study how to measure variation strength. From our point of view, a numerical value having a single unit variation has to be treated differently than a large number of unit variations. Secondly, we noticed that the lower the minimal support, the more redundant the rules. Some of them, especially on gene databases, contain noise. The extraction process does need to integrate semantics in order to select the most suitable results.

References

1. Srikant, R., Agrawal, R.: Mining Quantitative Association Rules in Large Relational Tables. In: Proceedings of the 1996 ACM SIGMOD International Conference on Management of Data, pp. 1–12 (1996)
2. Hüllermeier, E.: Association rules for expressing gradual dependencies. In: Elomaa, T., Mannila, H., Toivonen, H. (eds.) PKDD 2002. LNCS (LNAI), vol. 2431, pp. 200–211. Springer, Heidelberg (2002)
3. Berzal, F., Cubero, J.C., Sanchez, D., Vila, M.A., Serrano, J.M.: An alternative approach to discover gradual dependencies. International Journal of Uncertainty, Fuzziness and Knowledge-Based Systems (IJUFKS) 15(5), 559–570 (2007)
4. Dubois, D., Prade, H.: Gradual elements in a fuzzy set. Soft Comput. 12(2), 165–175 (2008)
5. Galichet, S., Dubois, D., Prade, H.: Imprecise specification of ill-known functions using gradual rules. International Journal of Approximate Reasoning 35, 205–222 (2004)
6. Dubois, D., Prade, H.: Gradual inference rules in approximate reasoning. Information Sciences 61(1-2), 103–122 (1992)
7. Jones, H., Dubois, D., Guillaume, S., Charnomordic, B.: A practical inference method with several implicative gradual rules and a fuzzy input: one and two dimensions. In: Fuzzy Systems Conference, 2007. FUZZ-IEEE 2007, IEEE International, pp. 1–6 (2007)
8. Bosc, P., Pivert, O., Ughetto, L.: On data summaries based on gradual rules. In: Proceedings of the 6th International Conference on Computational Intelligence, Theory and Applications, pp. 512–521. Springer, Heidelberg (1999)
9. Hüllermeier, E.: Implication-based fuzzy association rules. In: Siebes, A., De Raedt, L. (eds.) PKDD 2001. LNCS (LNAI), vol. 2168, pp. 241–252. Springer, Heidelberg (2001)
10. Fiot, C., Masseglia, F., Laurent, A., Teisseire, M.: Gradual trends in fuzzy sequential patterns. In: 12th International Conference on Information Processing and Management of Uncertainty in Knowledge-based Systems (2008)
11. Agrawal, R., Srikant, R.: Fast Algorithms for Mining Association Rules. In: 20th International Conference on Very Large Data Bases (VLDB 1994), pp. 487–499 (1994)
12. Ayres, J., Flannick, J., Gehrke, J., Yiu, T.: Sequential pattern mining using a bitmap representation. In: KDD 2002: Proceedings of the eighth ACM SIGKDD International Conference on Knowledge Discovery and Data Mining, pp. 429–435. ACM, New York (2002)

Selecting Computer Architectures by Means of Control-Flow-Graph Mining

Frank Eichinger and Klemens Böhm

Institute for Program Structures and Data Organisation (IPD)
Universität Karlsruhe (TH), Germany
{eichinger,boehm}@ipd.uka.de

Abstract. Deciding which computer architecture provides the best performance for a certain program is an important problem in hardware design and benchmarking. While previous approaches require expensive simulations or program executions, we propose an approach which solely relies on program analysis. We correlate substructures of the control-flow graphs representing the individual functions with the runtime on certain systems. This leads to a prediction framework based on graph mining, classification and classifier fusion. In our evaluation with the SPEC CPU 2000 and 2006 benchmarks, we predict the faster system out of two with high accuracy and achieve significant speedups in execution time.

1   Introduction

The question which computer architecture is best suited for a certain application is of major importance in hardware design and benchmarking. Think of a new scientific tool for which hardware is needed. It is not clear which hardware is most appropriate. Other developments give way to similar questions: With heterogeneous multicore processors, one has to decide at runtime on which processors to execute a certain program. Reconfigurable hardware allows to change the hardware at runtime. These upcoming technologies motivate studying dependencies between program characteristics and computer architectures as well.

To deal with the problem which architecture provides the best performance for a certain application, several approaches have been used, ranging from executions and simulations to analytical models and program analysis. At first sight, it seems feasible to assess the performance of a program by *executions* on the systems in question. But this requires to have access to the machines, and porting the program to them can be expensive. *Simulations* of processor architectures require detailed information on the architectures to choose from and might be very time-consuming. As modern computer architectures have an extreme complexity, *analytical models* describing them are hard to establish and may be unreliable. Some recent approaches make use of *program analysis*. The intuition is that similar programs display a similar runtime behaviour when executed on the same machine. Execution times for a number of programs are known for many systems, e.g., from benchmarks suites. [1, 2] compare similarities of programs based on execution properties such as the CPU instruction mix. These

N. Adams et al. (Eds.): IDA 2009, LNCS 5772, pp. 309–320, 2009.
© Springer-Verlag Berlin Heidelberg 2009

properties are architecture-dependent, but independent of the implementation used. To obtain them, program executions or simulations are necessary.

In this article we investigate another method to find the best computer architecture which does not require any execution or simulation of the application in question. Likewise, we assume that similar applications have a similar execution behaviour – but have consciously decided not to measure any runtime-related characteristics. Instead, we entirely rely on published execution times of benchmark programs. In our approach, we define similarity using structural characteristics of the control-flow graphs (CFGs) [3] of the underlying functions. Our research question is to investigate how well they describe the performance-related characteristics of a program, and if they can be used for performance predictions. In contrast to software metrics like lines of code and statements used, CFGs do not have any potentially distracting characteristics which depend on language specifics, such as the language used or the programming style of the developers. In more detail, we derive structural features from CFGs by means of frequent subgraph mining. The resulting subgraph features characterise a function and can train a classifier which predicts the best architecture for a given application.

The solution just outlined requires a number of contributions at different stages of the analysis process:

Representation of Control-Flow Graphs. To derive subgraph features from CFGs, the nodes of the graphs have to be labelled with information relevant for performance analysis. So far, nodes represent blocks of source code. We have to turn them into concise categorical labels which graph mining algorithms can use. However, such a labelling is not obvious. We propose a labelling scheme with information that is relevant for performance.

Mining Large Graphs. Once we have derived suitable CFGs, mining them is another challenge, due to the size of some of them. We develop an efficient technique consisting of two steps: We first mine a subset of the graphs that are 'easy to mine' and then inspect the remaining graphs. Our technique provides guarantees for the support values achieved.

Classification Framework. We propose a classification setting for our specific context. This is necessary: CFG based information is available at the function level, as we will explain, while we want to choose the best architecture for a program as a whole. We propose a framework that first learns at the function level, before we turn our classification model into a predictor for the architecture where a given program performs best.

Our experimental evaluation is based on the SPEC CPU 2000 and 2006 benchmark suites. The main result is that, for 'relatively similar' computer architectures to choose from, our approach achieves an average prediction accuracy of 69% when choosing between two systems. This also shows the existence of remarkably strong relationships between CFGs and runtime behaviour.

Paper outline: Section 2 presents related work, Section 3 describes CFG representations, and Section 4 says how we mine them. Section 5 describes the prediction framework, Section 6 our results. Section 7 concludes.

2 Related Work

In the areas of computer architecture, high-performance computing and benchmarking, different approaches have been investigated to predict the runtime of applications. Many of them make use of intelligent data-analysis techniques.

As mentioned, analytic models can assess the performance of software on certain machines. For distributed MPI (message passing interface) programs, Kühnemann et al. developed a compiler tool which helps deriving such a model [4]. It builds on source-code analysis and properties of the underlying machines. These properties include the execution times of basic arithmetic and logical operations, which have to be derived for the machines in question. This requires access to the machines or at least a detailed knowledge. The approach then creates a runtime-function model. Another analytical model approach, in the area of superscalar processors, is [5]. Karkhanis et al. use architecture-dependent information such as statistics of branch mispredictions and cache misses to build a performance-prediction model. While predictions are good, the approach requires time-consuming executions to obtain the characteristics used.

The approach which probably is most similar to ours is [1, 2]. Joshi et al. use program characteristics to make statements on the similarities of programs [1]. In contrast to [4, 5], they do not use microarchitecture-dependent measures to characterise programs, but microarchitecture-independent ones, such as the instruction mix and branch probabilities. This limits the approach to a certain instruction-set architecture and a specific compiler. Furthermore, generating the measures requires simulation or execution. Based on [1], Hoste et al. use program-similarity measures and predict performance with programs from the SPEC CPU 2000 benchmark suite [2]. They then normalise the microarchitecture-independent characteristics with techniques such as principal-component analysis. These normalised measures represent a point in the so-called benchmark space for every program. The performance of an unknown application is then predicted as the weighted average of execution times of programs in the neighbourhood. To obtain enough reference points with known performance measures, the authors rely on programs from a benchmark suite. Like our approach, [2] can determine the best platform for an application. Its advantage is that predictions tend to be more accurate than ours. This is achieved by the limitation to a certain instruction-set architecture and by executing or simulating the application in question on an existing platform. Our approach in turn has no such limitation. It uses only measures generated from the source code and does not require any simulation or execution of the program in question.

İpek et al. do not only predict performance for different systems [6] but also contribute to hardware design. The number of design alternatives in computer architecture is huge, and it is hard to develop a good architecture for certain applications. The combination of design parameters is often described as a point in a design space, as is done in [6]. The authors simulate sampled points in design spaces corresponding to the memory hierarchy and to chip multiprocessors, which then serve as input for neural networks. They then use these networks for performance predictions of new computer-architecture designs.

Our approach is also related to work in the field of *graph classification*. One of the first studies, with an application in chemistry, is [7]. The authors propose a graph-classification framework which consists of three steps: (1) search for frequent subgraphs which are then used as binary features indicating if a certain subgraph is included in a graph, (2) a feature-selection strategy to reduce the dimensionality and (3) a model-learning step. Our approach is similar in that frequent subgraphs are generated which serve as features to learn a classification model. However, our application does not require feature selection, but a more complex approach to integrate classifications to a prediction for a program.

3   Control-Flow-Graph Representation

Control-Flow-Graph Generation. Control-flow graphs (CFGs) [3] are a common program representation in compiler technology. They are static in nature and can be derived from source code. They represent all control flows which can possibly occur. The nodes of a CFG stand for *basic blocks* of code, i.e., sequences of statements without any branches. The edges represent the possible control flows, i.e., edges back to previous nodes for loops and different branches for condition statements. This paper studies the usual setting where one CFG describes a single function.

For our work, it is important to define an architecture-independent representation of CFGs. In particular, some compiler optimisations affect the structure of the CFGs, e.g., loop unrolling. This might vary when making optimisations for different architectures. Therefore, we use the *GNU compiler collection (gcc)* to obtain CFGs using the -O0-flag, which prevents the compiler from making any optimisations. However, the *gcc* normalises the source code by using canonical constructs for artefacts which can be expressed in several ways in the programming language. This normalisation is an advantage, as the same algorithms tend to be expressed in the same way, even if the source-code representations vary.

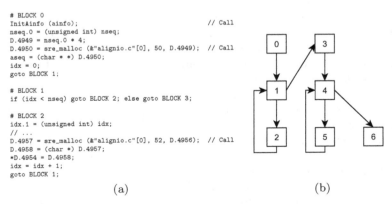

```
# BLOCK 0
InitAinfo (ainfo);                              // Call
nseq.0 = (unsigned int) nseq;
D.4949 = nseq.0 * 4;
D.4950 = sre_malloc (&"alignio.c"[0], 50, D.4949);  // Call
aseq = (char * *) D.4950;
idx = 0;
goto BLOCK 1;

# BLOCK 1
if (idx < nseq) goto BLOCK 2; else goto BLOCK 3;

# BLOCK 2
idx.1 = (unsigned int) idx;
// ...
D.4957 = sre_malloc (&"alignio.c"[0], 52, D.4956);  // Call
D.4958 = (char *) D.4957;
*D.4954 = D.4958;
idx = idx + 1;
goto BLOCK 1;
```

(a)   (b)

Fig. 1. Example control-flow graph (CFG)

Figure 1 is an example of a CFG: (a) is a part of the function `AllocAlignment` from the SPEC program 456.hmmer in a (simplified) intermediate representation derived with the *gcc*. (b) is the CFG derived from the function. In the intermediate representation, `if` and `goto`-statements represent loops.

In addition to the nodes displayed in Figure 1(b), some CFG representations introduce additional entry and exit nodes which do not represent any code. They do not represent any performance-related information. To obtain a more concise graph representation, we do not make use of such nodes.

Node Labelling. To mine CFGs, it seems that one could analyse the pure graph structure ignoring the content of the nodes. However, such an approach would lose a lot of (performance-related) information. To avoid this, we propose the following mapping of source code to node labels:

- *FP* for blocks containing *floating-point operations*.
- *Call* for blocks without *FP* operations but *calls of other functions*.
- *Set* for blocks without *FP* or *Call* operations but *load/store* operations.
- *Int* for blocks containing none of the above (simple *integer* ALU operations).

Table 1. SPEC benchmark programs used

CPU 2000			CPU 2006		
164.gzip	183.equake	255.vortex	400.perlbench	436.cactus-ADM	464.h264ref
175.vpr	186.crafty	256.bzip2	401.bzip2	445.gobmk	470.lbm
176.gcc	188.ammp	300.twolf	403.gcc	454.calculix	481.wrf
177.mesa	197.parser		429.mcf	456.hmmer	482.sphinx3
179.art	253.perlbmk		433.milc	458.sjeng	
181.mcf	254.gap		435.gromacs	462.lib-quantum	

So far, the labelling scheme leaves aside the actual number of statements in a node, which might be important as well. Furthermore, the different labels are quite imbalanced: *FP* is assigned to only 3% of the nodes from CFGs in the SPEC CPU 2000 and 2006 programs used (see Table 1), *Call* is assigned to 19%, *Set* to 61% and *Int* to 17%. Large sets of nodes with the same label, as well as only few different ones, have a negative effect on the performance of graph-mining algorithms (cf. [8]). We therefore propose a more fine-grained labelling scheme: We divide the blocks labelled with *Call* into blocks containing one function call, $Call_1$, and blocks with two or more calls, $Call_{2+}$. As there is a larger variety in the number of load/store operations, and the *Set* class of labels is the largest, we divide it in four labels. Each of them has approximately the same number of corresponding nodes. Blocks with one load/store operation are labelled Set_1, blocks with two with Set_2, those with three to five with Set_{3-5} and those with more than five Set_{6+}. In preliminary experiments, graph mining was one order of magnitude faster with the fine-grained labels, while the accuracy of predictions did not decrease. Figure 2 provides examples of the labelling schemes.

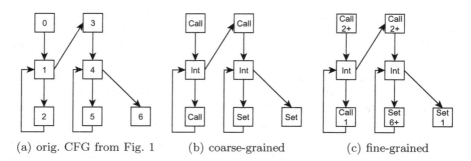

(a) orig. CFG from Fig. 1   (b) coarse-grained   (c) fine-grained

Fig. 2. Examples of CFG node labelling schemes

4   Control-Flow-Graph Mining

For our experiments (see Section 6) we use the C/C++ programs from the SPEC
CPU 2000 and 2006 benchmarks suites listed in Table 1. This results in a set
G_{CFG} of approximately 27,000 CFGs belonging to the 31 programs. The graphs
have an average size of 22 nodes, with high variance. Approximately 30% of the
graphs consist of one node only, while roughly 12% have more than 32 nodes,
including a few with more than 1,000 nodes. As graphs with a single node do not
contain any information which is useful for our scheme, we omit these graphs.
This reduces the size of G_{CFG} to approximately 19,000 graphs.

Though the average graphs in G_{CFG} are not challenging from a state-of-the-
art graph-mining perspective, the large CFGs do lead to scalability problems.
We can mine the entire graph set with, say, the *gSpan* algorithm [9] in a reasonable time, but only with relatively high minimum support values (*minSup*). In
preliminary experiments, this leads to the discovery of very small substructures,
i.e., with a maximum of two nodes only. Further, the *CloseGraph* algorithm [10]
is not helpful in our case. This is because there rarely are closed graphs (with exactly the same support) which offer pruning opportunities. In preliminary experiments, *CloseGraph* even increased the runtime because of the search for closed
structures. – To obtain larger subgraph patterns from our CFG dataset G_{CFG}
by means of frequent subgraph mining, we inspect the larger graphs further.
We observe that they frequently contain nodes with a high degree. This causes
the scalability problems. Node *Int* in Figure 3(a) serves as an illustration. In
many cases, bulky `switch-case` statements which lead to many outgoing edges
in one node and many incoming edges in another one cause these high degrees.

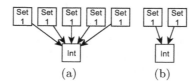

(a)   (b)

Fig. 3. Illustration of CFGs with problematic node degrees

Typically, programs treat many different **case** branches similarly. Therefore, the corresponding nodes frequently have the same labels (Set_1 in Figure 3(a)). The problem with these situations is the number of potential embeddings. As an example, we want to find out the embeddings of the graph in Figure 3(b) in the graph in Figure 3(a). Algorithms like *gSpan* search for all such embeddings (subgraph isomorphisms), which is NP-complete [11]. In the example, there are 20 distinct embeddings. G_{CFG} contains nodes with a degree of 720, which leads to extreme numbers of possible embeddings. To conclude, the graphs in G_{CFG} with high node degrees prohibit mining of G_{CFG} with reasonably low *minSup* values. Early experiments with roughly the same data set, but with the largest graphs excluded, have led to encouraging results in turn. We therefore propose the following mining steps:

1. Mining of all graphs G_{small} smaller than a certain threshold t_{size}, resulting in a set of frequent subgraphs SG. ($G_{\text{small}} := \{g \in G_{\text{CFG}} \,|\, size(g) \leq t_{\text{size}}\}$)
2. Search for subgraph isomorphisms of the subgraphs SG within the large graphs G_{large} which have been omitted in Step 1. ($G_{\text{large}} := G_{\text{CFG}} \setminus G_{\text{small}}$)
3. Unified representation of all graphs in G_{CFG} with feature vectors.

Before we can derive frequent subgraphs (SG) in Step 1, we first systematically identify combinations of the support in G_{small} ($minSup_{\text{small}}$) and the size threshold (t_{size}), which let us mine the data in reasonable time. We do this by means of preliminary mining runs. In general, one wants to have a low $minSup_{\text{small}}$, to facilitate finding large and significant subgraphs, and a high t_{size}. This is to ensure that only few patterns, namely those only included in G_{large}, are missed. With our dataset, we found a t_{size} of 32 (corresponding to G_{large} with a size of 12% of $|G_{\text{CFG}}|$) and a $minSup_{\text{small}}$ of 1.7% to be good values. We assume that similar t_{size} values can be found when mining other CFG datasets, since our numbers are based on a sample of 27,000 CFGs. We then mine G_{small} with the *ParSeMiS* implementation[1] of the *gSpan* algorithm [9], which is a state-of-the-art algorithm in frequent subgraph mining. This results in a set of subgraphs SG which are frequent within G_{small}.

In Step 2 we determine which graphs in G_{large} contain the subgraphs in SG by means of a subgraph-isomorphism test. Although this problem is NP-complete [11], we benefit from properties of our specific dataset. E.g., there are no cliques larger than three nodes, and the average node degree of 3.4 is relatively low. Therefore, this step is less expensive in terms of runtime than Step 1.

In Step 3, we represent every CFG in G_{CFG} with a feature vector. Such a vector contains one bit for every subgraph in SG. A bit states if the respective subgraph is included in the CFG.[2] As the subgraphs in SG have a minimum size of one edge, single nodes are not included. We believe that these nodes provide important information as well. We therefore extend the vector with single nodes

---

[1] http://www2.informatik.uni-erlangen.de/Forschung/Projekte/ParSeMiS/

[2] In preliminary experiments we have used the numbers of embeddings. This has not yielded better results. However, generating boolean features makes the subgraph-isomorphism test in Step 2 much easier.

labelled with the label classes described in Section 3. For these features, we use integers representing their number of occurrences. This allows for a more precise description of the operations contained in a CFG, i.e., in a function. Summing up, we represent every CFG $g \in G_{\mathrm{CFG}}$ with the following vector:

$$g := (sg_1, sg_2, ..., sg_n, FP, Call_1, Call_{2+}, Set_1, Set_2, Set_{3\text{-}5}, Set_{6+}, Int)$$

where $sg_1, sg_2, ..., sg_n \in SG$ are boolean features, $|SG| = n$, and FP, $Call_1$, $Call_{2+}$, Set_1, Set_2, $Set_{3\text{-}5}$, Set_{6+} and Int are integers.

Our technique based on mining G_{small} bears the risk that certain subgraphs may not be found, namely those contained in G_{large}. In the worst case, a subgraph sg is contained in every graph in G_{large}, corresponding to 12% of $|G_{\mathrm{CFG}}|$ in our case, but hardly misses the minimum support when only looking at G_{small}. In other words, sg becomes a part of the result set SG if it has a support of 13.5% in G_{CFG}. More formally, we can guarantee to find all subgraphs with the following minimum support in G_{CFG}:

$$minSup_{\mathrm{guarantee}} = \frac{|G_{\mathrm{large}}|}{|G_{\mathrm{CFG}}|} + \frac{|G_{\mathrm{small}}|}{|G_{\mathrm{CFG}}|} \cdot minSup_{\mathrm{small}}$$

In our dataset, $minSup_{\mathrm{guarantee}}$ is 13.5%. However, we find many more subgraphs as we are mining with a much lower $minSup_{\mathrm{small}}$ in G_{small}. A direct mining of G_{CFG} with a $minSup$ of 13.5% was not possible due to scalability problems – the lowest $minSup$ value possible in preliminary experiments was 20%. Our approach succeeds due to the relatively small fraction of large graphs. It is applicable to other datasets as well, but only if the share of large graphs is similar.

5   Classification Framework

We now describe the subsequent classification process, to predict the best computer architecture for a given program. We formulate this prediction as the selection between a number of architectures. The architectures are the classes in this setting. In the following, we focus on the prediction of the faster one of two architectures. This is due to the limited number of architectures with data available, as we will explain. However, with more training data, we do not see any problems when choosing from an arbitrary number of systems.

For the classification, we are faced with the challenge that our substructure based feature vectors are descriptions at the function level, while we are interested in predictions for a program as a whole. At the same time, SPEC publishes runtimes of several systems – this information is at the level of programs as well. Potentially helpful information on the execution of functions, such as the number of calls and the execution time, is not available. This situation is completely natural, and this is why we propose an approach that is supposed to work without that information. In the following, we develop an approach which does not need any more runtime-related information than the one typically available.

One way to do program-level predictions is to aggregate the information contained in the feature vectors to the program level. Then a classification model could be learned with this data. As one program consists of many functions (typically hundreds to thousands in the benchmarks), such an aggregation would lose potentially important fine-grained information. Further, it would force us to learn a model based on only few tuples (programs). The problem is the limited availability of systems which are evaluated with more than one benchmark suite. E.g., a system evaluated with SPEC CPU 2006 is rarely evaluated with the now outdated CPU 2000 benchmark suite as well. The number of benchmark programs whose execution time for the same machines is known is therefore limited in practice – and deriving this information would be tedious.

Hence, we propose a classification framework containing a simplification which might seem unusual or 'simplistic' at first sight: To learn a classifier at the function level, we assign the fastest architecture for the program as a whole to all feature vectors describing its CFGs (functions). This simplification, caused by a lack of any respective information, clearly does not take the characteristics of the different functions of a program into account. It also ignores the potentially imbalanced distribution of execution times of the individual functions. However, our hope is that the large amount of function-level training data compensates these issues, and we will show this. Once the classification model is learned, we use it to classify functions from programs with unknown runtime behaviour. We aggregate these predictions to the program level with majority vote.

To learn a prediction model, any classification technique can be used in principle. We have carried out preliminary experiments with support vector machines, neural networks and decision trees, and the results were best with the latter. We therefore deploy the C5.0 algorithm, a successor of C4.5 [12]. Our implementation in the SPSS Clementine data-mining suite lets us specify weights for every tuple during the learning process, to emphasise certain tuples. With our approach, we weight every feature vector with two factors:

1. One class might consist of many more tuples than the other one in the learning data set. As this typically leads to an increased number of predictions of the larger class, we increase the weight of the under-represented one. We use the fraction of the number of functions in the larger class divided by the one of the under-represented class as the weight.
2. The difference in runtime of some programs on the two machines considered might be large, while it is marginal with other programs. To give a higher influence to a program with very different execution times, we use the ratio of the execution time of the slower machine to the one of the faster machine as another weight for the feature vectors of a program.

To fuse the classifications on the function level, we use the majority-vote technique [13]. This is standard to combine multiple classifications. In extensive experiments, we have evaluated alternative weights for learning, as well as different combination schemes. In particular, we have examined the two weights mentioned – as well as other ones – as weights for the majority-vote scheme.

Note that, while the graph-mining step of our approach may be time-consuming, it only takes place once, in order to build the classification model. The prediction for a new program is much faster, once the model is built.

Table 2. Systems used for runtime experiments

System 1	System 2	System 3
Bull SAS NovaScale B280	Dell Precision 380	HP Proliant BL465c
Intel Xeon E5335,	Intel Pentium 4 670	AMD Opteron 2220
QuadCore, 2.0 GHz	SingleCore, 3.8 GHz	DualCore, 2.8 GHz
2x4 MB L2, 8 GB RAM	2 MB L2, 2 GB RAM	2x1 MB L2, 16 GB RAM

System 4	System 5	System 6
Intel D975XBX Motherboard	FSC Celsius V830	IBM BladeCenter LS41
Intel Pentium EE 965	AMD Opteron 256	AMD Opteron 8220
DualCore, 3.7 GHz	SingleCore, 3.0 GHz	DualCore, 2.8 GHz
2x2 MB L2, 4 GB RAM	1 MB L2, 2 GB RAM	2x1 MB L2, 32 GB RAM

6   Experiments

In this section we present our experiments and results. For the programs listed in Table 1, the runtimes on a number of systems are published on the SPEC homepage[3]. We make use of this data and use a subset of the systems available, listed in Table 2. Not every system is evaluated with the older CPU 2000 and the more recent CPU 2006 benchmark. Therefore, our selection is motivated by the availability of runtimes for both benchmarks. Although there would have been a few more systems available, we have only used the ones mentioned. They allow us to set up experiments where the balance of systems being fastest with some programs and systems being fastest with other programs is almost equal. This eases the data-mining process. In reality, equality is not necessary when enough data is available. Our experiments cover single-, dual- and quadcore architectures as well as different memory hierarchies and processors, see Table 3.

Table 3. Experiments and results

Exp.	Platforms	Processors	accuracy	speedup reached
1	System 5 vs. System 2	Opteron vs. Pentium 4	71.0%	83.2%
2	System 3 vs. System 4	Opteron vs. Pentium EE	64.5%	58.6%
3	System 6 vs. System 1	Opteron vs. Xeon	71.0%	72.6%

We use the classification framework as described in Section 5, along with 2-fold cross-validation. We use partitions which are stratified with respect to the class and consist of roughly equal numbers of functions. For evaluation we derive the accuracy, i.e., the percentage of programs with correct prediction, and the speedup in terms of execution time. To obtain the latter, we first calculate the total runtime of all programs, each one on the machine predicted. This allows us to derive a percentage, 'speedup reached'. 0% is achieved when the slowest architecture is always selected, 100% if the predictor assigns the fastest architecture to every program. Table 3 contains both measures as well.

---

[3] http://www.spec.org/benchmarks.html

We achieve an accuracy of 69% on average. This indeed corresponds to a speedup. On average, we reach 71% of the speedup that would have been possible in theory. Further, our results show that there is a strong relationship between CFGs and runtime behaviour. Although it would be interesting, we do not compare our results to approaches making use of execution properties, such as [2]. Such a comparison is not possible, since [2] uses other benchmarks and target machines, as well as other evaluation metrics.

To savour our experimental results, one should take several points into account. The programs in the SPEC CPU benchmarks are relatively similar in the sense that they are all compute-intensive (not I/O or memory-intensive). The systems considered are relatively similar as well. All of them are off-the-shelf systems, differing mainly in their configuration. However, it does not affect runtime by much if, say, the number of processors or the size of RAM changes. The programs considered do not use multiple threads and always fit in memory. The only architectural difference of some significance is the instruction set used, i.e., *x86* in the Xeon and Pentium systems and *x86-64* in the Opteron systems.

7 Conclusion and Future Work

In the computer industry, it is important to know which platform provides the best performance for a given program. Most approaches proposed so far require in-depth knowledge of the systems or of runtime-related characteristics. One must obtain them using expensive simulations or executions.

This paper has proposed an approach solely based on the static analysis of programs and on runtime data from benchmark executions, which is available online. It analyses the control flow graphs (CFGs) of the functions. Based on graph-mining results, it correlates programs with similar CFG substructures and assumes that their runtime is similar as well. This leads to our prediction framework for learning at function level and a classifier-fusion technique to derive program-level predictions. Our framework can predict the runtime behaviour of programs on the target platforms. Though our approach to assign the best architecture for a program as a whole to its classes might be unusual and somewhat risky, it is beneficial according to our evaluation. In experiments with the SPEC CPU 2000 and 2006 benchmarks we obtain an accuracy of 69% on average.

From a graph-mining perspective, we propose a technique which can deal with situations when the usual approach does not scale, e.g., because of high node degrees. Our technique leaves aside few graphs in the graph-mining step which are 'problematic'. Then, it maps the results to the graphs we left out before. We provide guarantees on the overall support.

One aspect of our future work is to improve the prediction quality further. We currently investigate the usage of software metrics which provide additional information on a function. We also investigate program-dependence graphs [14]. They feature data dependencies in addition to control-flow information. Such information might help regarding certain aspects of computer architectures, e.g., pipelining and register usage. However, the graphs are much larger than CFGs.

Another aspect is a further investigation from the computer architecture point of view. Rather than correlating properties from source-code representations, e.g., CFG substructures, with architectures as a whole, we are interested in ties with micro-architectural details, such as the cache architecture. Such insights would be of enormous help when designing hardware for specific applications.

Acknowledgments

We thank Dietmar Hauf for much help with all aspects of this study and Wolfgang Karl and David Kramer for their guidance regarding computer architecture.

References

[1] Joshi, A., Phansalkar, A., Eeckhout, L., John, L.: Measuring Benchmark Similarity Using Inherent Program Characteristics. IEEE Trans. Comput. 55(6), 769–782 (2006)

[2] Hoste, K., Phansalkar, A., Eeckhout, L., Georges, A., John, L.K., Bosschere, K.D.: Performance Prediction Based on Inherent Program Similarity. In: Proc. Int. Conf. on Parallel Architectures and Compilation Techniques, PACT (2006)

[3] Allen, F.E.: Control Flow Analysis. In: Proc. Symposium on Compiler Optimization. SIGPLAN Notices, pp. 1–19 (1970)

[4] Kühnemann, M., Rauber, T., Runger, G.: A Source Code Analyzer for Performance Prediction. In: Proc. Int. Symposium on Parallel and Distributed Processing (2004)

[5] Karkhanis, T.S., Smith, J.E.: A First-Order Superscalar Processor Model. SIGARCH Comput. Archit. News 32(2), 338 (2004)

[6] İpek, E., McKee, S.A., Singh, K., Caruana, R., de Supinski, B.R., Schulz, M.: Efficient Architectural Design Space Exploration via Predictive Modeling. ACM Trans. Archit. Code Optim. 4(4), 1–34 (2008)

[7] Deshpande, M., Kuramochi, M., Wale, N.: Frequent Substructure-Based Approaches for Classifying Chemical Compounds. IEEE Trans. Knowl. Data Eng. 17(8), 1036–1050 (2005)

[8] Chakrabarti, D., Faloutsos, C.: Graph Mining: Laws, Generators, and Algorithms. ACM Comput. Surv. 38(1), 2 (2006)

[9] Yan, X., Han, J.: gSpan: Graph-Based Substructure Pattern Mining. In: Proc. Int. Conf. on Data Mining, ICDM (2002)

[10] Yan, X., Han, J.: CloseGraph: Mining Closed Frequent Graph Patterns. In: Proc. Int. Conf. on Knowledge Discovery and Data Mining, KDD (2003)

[11] Garey, M.R., Johnson, D.S.: Computers and Intractability: A Guide to the Theory of NP-Completeness. W. H. Freeman, New York (1979)

[12] Quinlan, J.R.: C4.5: Programs for Machine Learning. Morgan Kaufmann, San Francisco (1993)

[13] Kuncheva, L.I.: Combining Pattern Classifiers: Methods and Algorithms. John Wiley & Sons, Chichester (2004)

[14] Ottenstein, K.J., Ottenstein, L.M.: The Program Dependence Graph in a Software Development Environment. SIGSOFT Softw. Eng. Notes 9(3), 177–184 (1984)

Visualization-Driven Structural and Statistical Analysis of Turbulent Flows

Kenny Gruchalla[1], Mark Rast[2], Elizabeth Bradley[1], John Clyne[3], and Pablo Mininni[4]

[1] Department of Computer Science, University of Colorado, Boulder, Colorado
[2] Laboratory for Atmospheric and Space Physics, Department of Astrophysical and Planetary Sciences, University of Colorado, Boulder, Colorado
[3] Computational and Information Systems Laboratory, National Center for Atmospheric Research, Boulder, Colorado
[4] Departamento de Física, Facultad de Ciencias Exactas y Naturales, Universidad de Buenos Aires, Argentina and Geophysical Turbulence Program, National Center for Atmospheric Research, Boulder, Colorado

Abstract. Knowledge extraction from data volumes of ever increasing size requires ever more flexible tools to facilitate interactive query. Interactivity enables real-time hypothesis testing and scientific discovery, but can generally not be achieved without some level of data reduction. The approach described in this paper combines multi-resolution access, region-of-interest extraction, and structure identification in order to provide interactive spatial and statistical analysis of a terascale data volume. Unique aspects of our approach include the incorporation of both local and global statistics of the flow structures, and iterative refinement facilities, which combine geometry, topology, and statistics to allow the user to effectively tailor the analysis and visualization to the science. Working together, these facilities allow a user to focus the spatial scale and domain of the analysis and perform an appropriately tailored multivariate visualization of the corresponding data. All of these ideas and algorithms are instantiated in a deployed visualization and analysis tool called VAPOR, which is in routine use by scientists internationally. In data from a 1024^3 simulation of a forced turbulent flow, VAPOR allowed us to perform a visual data exploration of the flow properties at interactive speeds, leading to the discovery of novel scientific properties of the flow, in the form of two distinct vortical structure populations. These structures would have been very difficult (if not impossible) to find with statistical overviews or other existing visualization-driven analysis approaches. This kind of intelligent, focused analysis/refinement approach will become even more important as computational science moves towards petascale applications.

1 Challenges to Data Analysis

A critical disparity is growing in the field of computational science: our ability to generate numerical data from scientific computations has in many cases

N. Adams et al. (Eds.): IDA 2009, LNCS 5772, pp. 321–332, 2009.
© Springer-Verlag Berlin Heidelberg 2009

exceeded our ability to analyze those data effectively. Supercomputing systems have now reached *petaflop* performance [1], supporting numerical models of extraordinary complexity, fidelity, and scale. In supercomputing centers, terabyte data sets are now commonplace and petabyte data sets are anticipated within a few years. However, analysis tools and the computational machinery that supports them have not been able to scale to meet the demands of these data. For many computational scientists, this lack of analysis capability is the largest barrier to scientific discovery.

The imbalance of scale between numerical simulation and data analysis is largely due to their contrasting demands on computational resources. Large-scale numerical simulation is typically a *batch* processing operation that proceeds without human interaction on parallel supercomputers. Data analysis, in contrast, is fundamentally an *interactive* process with a human investigator in the loop, posing questions about the data and using the responses to progressively refine those questions[2]. While some data analyses certainly can be performed in batch mode, this is only practical for completely predetermined investigations. Exploratory analysis depends on hypothesis generation and testing, which requires an interactive environment that can provide timely and meaningful feedback to the investigator. Unfortunately, this kind of interactive workflow is not well-suited to batch access on a parallel supercomputer. Another key bottleneck in the analysis process is data storage. If the data exceeds the size of the available random access media, one must manage its storage and exchange across different media. Disk transfer rates are generally inadequate to support interactive processing of large-scale computational data sets.

These dilemma can be addressed by data reduction, and a variety of schemes have been developed to reduce data volumes while maintaining essential properties. Their methods and results depend on the scientific goals of the simulation and analysis. For example, in the investigation of turbulent flows, analysis of strictly statistical or spectral properties can enable significant reduction in data dimensionality, while analysis of local flow dynamics, thermodynamics, or stability does not. In the later cases, the only solution is to reduce the physical volume under analysis. There are two general classes of methods for this. First, one can isolate and extract local sub-regions from the global domain. The success of this strategy depends on locating those regions in the solution that are of particular scientific importance or interest, which is a real challenge to intelligent data analysis. There has been some very interesting work in the IDA community on dimensional reduction for this purpose[3,4]. The visualization-driven approach described in this paper extracts regions of interest using an iterative interactive filtering technique that employs a combination of global and local flow statistics. The second class of data-volume reduction techniques uses a coarsened global approximation of the discrete solution to subsample the data over the entire domain. The obvious challenge here is selecting an appropriate coarsening (method and scale) to maintain accuracy — and confidence in the results.

Both of these data-reduction techniques have been implemented in VAPOR, an open-source desktop visualization-driven analysis application (available at http://www.vapor.ucar.edu). It closely couples advanced visualization with quantitative analysis capabilities, and it handles the complexities of large datasets using a hierarchical data model. It is designed to support a multi-phase analysis process, allowing the investigator to control a speed-and-memory versus a locus-and-quality trade off. This is an ideal context within which to explore intelligent data reduction. Using the ideas described in the previous paragraph, we have extended VAPOR's capabilities to include definition, manipulation, and refinement of feature sub-spaces based on multi-scale statistics of turbulent structures contained within the data. The user can explore the data volume at a coarsened resolution to gain a qualitative understanding and identify regions/structures of interest. Those regions/stuctures can then be investigated at progressively higher resolutions with computationally intensive visualization and analysis performed on progressively smaller sub-domains or structure populations. The base functionality of VAPOR provides data coarsening in the form of multi-resolution access via wavelet decomposition and reconstruction [5]. VAPOR's data coarsening approach coupled with a simple a simple sub-domain selection capability and has a successful track record in the analysis of large-scale simulation data using only modest computing resources [6]. With the addition of intelligent region-of-interest extraction, VAPOR can now provide interactive, scientifically meaningful explorations of tera-scale data volumes.

The following Section describes VAPOR's visualization-driven analysis capabilities; Section 3 demonstrates their power using data from a 1024^3 forced incompressible hydrodynamic simulation.

2   VAPOR: A Desktop Analysis and Visualization Application

Many applications have been developed specifically for the visualization and analysis of large-scale, time-varying numerical data, but all of them have significant limitations in terms of visualization, analysis, and/or scalability. In many of these tools, the emphasis is on the algorithms and the generation of aesthetic images, rather than on scientific discovery [7]. Visualization-oriented applications, like Paraview [8], Visit [9], and Ensight, lack quantitative analysis capabilities, and many of them demand specialized parallel computing resources. High-level, fourth-generation data languages such as ITT's IDL and Mathworks's Matlab are on the opposite end of the analysis-visualization spectrum. They provide a rich set of mathematical utilities for the quantitative analysis of scientific data but only limited visualization capabilities, and they do not scale well to very large data sets.

The goal of the VAPOR project was to address these shortcomings. It provides an integrated suite of advanced visualization capabilities that are specifically tailored to volumetric time-varying, multivariate numerical data. These capabilities, coupled with the intelligent data reduction strategies introduced in the previous sections, allow investigators to rapidly identify scientifically meaningful spatial-temporal regions in large-scale multivariate data. VAPOR's design

— both its functionality and its user interface — is guided by a steering committee of computational physicists to ensure that it truly meets the needs of the end-user community. Scalability is addressed through a multi-phase analysis process that is based on a combination of region-of-interest isolation, feature extraction (with our extensions), and a hierarchical data model. Finally, VAPOR interfaces seamlessly with high-level analysis languages like IDL, allowing its user to perform rigorous quantitative analyses of regions of interest.

2.1   Visualization

VAPOR incorporates a variety of state-of-the-art volume rendering and flow-visualization techniques, including both *direct* and *indirect* volume rendering [10]. Direct volume rendering describes a class of techniques, which generate images directly from volumetric data without any intermediate geometric constructions, while indirect volume rendering constructs geometric *isosurfaces*. To support the visualization of vector fields, VAPOR provides both *sparse* and *dense* particle-tracing methods [11]. The former render the geometry of individual trajectories of particles seeded in a flow field, and can support both steady (time-invariant) and unsteady (time-varying) trajectory integration. Dense particle-tracing methods synthesize textures that represent how the flow convolves input noise.

VAPOR's integrated design allows these volume rendering and flow-visualization techniques to be used in different combinations over a single analysis run, in concert with the intelligent data reduction strategies described in Section 2.3, as the investigator progressively isolates and refines scientifically meaningful regions of the data. The hierarchical data model that supports this is described in the next Section. By utilizing the optimized data-parallel streaming processors of modern *graphics processing units* (GPUs), VAPOR can effectively work with volumes of the order of 1536^3 [6].

2.2   Hierarchical Data Model

The VAPOR data storage model is based on wavelet decomposition [5,12]. Data are stored as a hierarchy of successively coarser wavelet coefficients; each level in this hierarchy represents a halving of the data resolution along each spatial axis, corresponding to an eight-fold reduction in data volume. In this manner, VAPOR maintains a series of useful coarsened approximations of the data, any of which can be accessed on demand during an analysis run, without an undue increase in storage requirements. Wavelet data are organized into a collection of multiple files: one binary file containing the wavelet coefficients for each time step, each variable, and each wavelet transformation level, and a single metadata file that describes the attributes of the field data (e.g., the grid type, the time steps, the spatial resolution, the field names, etc.).

This storage model naturally supports intelligent, interactive data decomposition. It allows VAPOR to operate on any subset of time steps, variables, and wavelet transformation levels, which has a variety of important advantages, including iterative focus and refinement of the analysis effort. An investigator can control the level of interactivity by regulating the fidelity of the data, first browsing a coarsened representation across the global spatial-temporal domain to identify regions or features of interest and then examining those reduced domains in greater detail. The hierarchical data/metadata combination also allows VAPOR to work with very large data collections, as data components can be stored off-line until required, handling *incomplete* data sets smoothly.

2.3   Multivariate Feature Extraction

The VAPOR volume rendering capability forms the basis for the multivariate feature extraction technique we have implemented to isolate structures of interest in large data sets. A multidimensional *transfer function* is used to define a mapping from data values to the color and opacity values in the volume rendering. The opacity component of this function visually separates the volume into opaque features. VAPOR users can construct and refine these functions iteratively, and use them in different thresholding schemes to visually separate the volume into opaque regions.

Once these regions have been visually identified using the transfer function, the individual structures are extracted and tagged using a connected-component labeling algorithm [13], an image-processing technique that assigns groups of ϵ-connected data points[1] a unique feature label. Once the features have been identified in this manner, they can be visualized and analyzed as individual features oppose to a set of opaque voxels. The individual features can be visualized in isolation or as members of sub-groups, and the data points and geometry of each can exported to external analysis packages for further analysis, as described in Section 2.4. This allows any user-defined physical property to be computed on the associated data points contained in each feature, over any field or combination of fields in the data. VAPOR presents the resulting values and distributions to the user as a table of feature-local histograms and property values, as shown in Figure 1. Using this table, the set of features can be culled based on the central moments of their distributions to further focus the study. The entire reduction process—including the transfer function design and feature definition—can be iterated to progressively refine the analysis, providing insight into the multivariate properties of structures across multiple scales.

2.4   Coupled Visual, Quantitative, and Statistical Analysis

Understanding large-scale simulation data is an exploratory process that can be greatly facilitated by combining highly interactive, qualitative visual examination with quantitative numerical analysis. Visualization can be used to motivate analysis through the identification of structures in the data, giving rise to hypotheses that can be validated or rejected through numerical study. Likewise, the analysis can be used to drive the visualization, identifying salient quantitative characteristics of the data through numerical study, and then visualizing their associated geometric shapes and physical properties. VAPOR's design

---

[1] i.e., those that are connected by an ϵ chain.

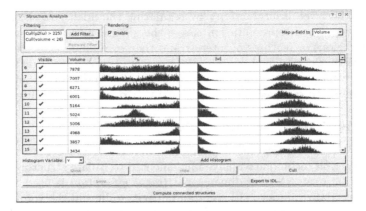

Fig. 1. The VAPOR structure analysis dialog, which displays feature-local histograms of user-selected field distributions

seamlessly combines qualitative visual and quantitative numerical investigation, enabling its users to interactively transition between the two. Its multi-resolution visualization and region-of-interest isolation capabilities, in conjunction with its hierarchical data representation, allow its users to cull data intelligently and pass appropriate subsets to an external quantitative analysis package.

Smooth integration of all of these capabilities required some interesting design decisions. VAPOR performs GPU-accelerated visualization natively, as described in Section 2.1, and hands numerical analysis off to IDL. VAPOR and IDL sessions are run simultaneously; after regions of interest are identified in the former, the associated data volumes are exported via metadata descriptors to the latter for further study. The tight coupling between IDL and VAPOR is accomplished by a library of data-access routines, which allow IDL access to the wavelet-encoded data representation. (This approach is readily generalizable to other analysis packages, complementing and enhancing existing user capabilities.) The qualitative/quantitative tandem is very effective: IDL, as mentioned at the beginning of Section 2, does not scale well to large data sets [12], but VAPOR's ability to focus the study neatly bypasses that problem, and the results of IDL analysis on focused regions can be seamlessly imported back into the VAPOR session for visual investigation. By repeating this process, very large data sets can be interactively explored, visualized, and analyzed without the overhead of reading, writing, and operating on the full data volume.

3   Application to Vortical Structures in Taylor-Green Flow

As an illustration of the power of the ideas described in the previous sections, we use VAPOR to explore data from an incompressible Taylor-Green forced turbulence simulation with a microscale Reynolds number of $R_\lambda \sim 1300$ [14]. The particular structures in this data that are of scientific interest involve vorticity, but the volume contains so many of these structures, of different sizes and strengths, as to pose a truly daunting analysis problem. The small-scale structures are particularly hard to isolate, so that is what we set out to analyze with VAPOR.

3.1   Global Vorticity and Structure Identification

Vortices play important roles in the dynamics and transport properties of fluid flows, but they are surprisingly hard to define, which complicates the task of designing a vortex extraction method. Jiang et al. [15] provide an extensive survey of current techniques. As a working definition, we treat a vortex filament, tube, or sheet as *a connected region with a higher relative amplitude of vorticity than its surrounding* [16]. Many vortex detection and visualization methods use the same definition, and most of them operationalize it by thresholding the magnitude of the vorticity. This is the starting point for our analysis, but VAPOR's capabilities allowed us to add other scientifically meaningful analysis steps and iteratively focus the process. In this particular case, it allowed us to investigate the correlation between vorticity and helicity across multiple scales and discover important structural properties that were previously unknown.

The first step in the process is to threshold the vorticity of the Taylor-Green data using the opacity contribution of the multidimensional tranfer function. The fields in the data include the simulated velocity vector field and two derived fields: a vorticity vector field and a normalized helicity field. Vorticity is defined as the curl of a velocity field, $\boldsymbol{\omega} = \nabla \times \boldsymbol{v}$, characterizing the pointwise rotation of fluid elements. Helicity is a scalar value, $H_n = \frac{\boldsymbol{v} \cdot \boldsymbol{\omega}}{|\boldsymbol{v}||\boldsymbol{\omega}|}$, the cosine of the angle

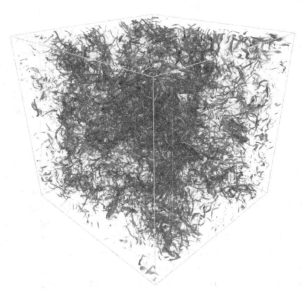

Fig. 2. A volume rendering of areas of strong vorticity in Taylor-Green turbulence isolates tens of thousands of vortical structures

between velocity and vorticity. An initial vorticity threshold was chosen to begin separating the tube-like vortical structures in the data volume. This step isolates tens of thousands of vortical structures, as shown in Figure 2. Using VAPOR's iterative refinement capabilities, we focus the study by further considering the helicity within these structures. A global analysis across the entire data volume, Figure 3, shows that both helicity and its pointwise correlation with vorticity are distributed in a nearly uniform fashion—i.e., that all angles between velocity and vorticity vectors occur with similar frequencies across all values of vorticity. While this is a useful result, it lumps the whole data volume together, possibly obscuring important local differences. Using VAPOR to generate feature-local histograms, we find that different high-vorticity regions do indeed have distinct helicity distributions, Figure 3(c). Three populations of structures are conspicuously evident: those whose helicity distributions span the full range with no distinct peak, those with a peak at high absolute values of helicity (i.e., dominated by nearly aligned or anti-aligned velocity and vorticity vectors), and those whose helicity distributions peak near zero (i.e., dominated by nearly orthogonal velocity and vorticity).

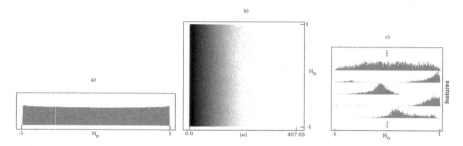

Fig. 3. The relationship between vorticity and helicity in Taylor-Green Turbulence a) the histogram of global normalized helicity, indicating helicity, measured point wise in the domain, has a nearly uniform distribution; b) the scatter plot of vorticity magnitude versus normalized helicity, showing that helicity has a nearly uniform distribution across all values of vorticity; c) a selected subset of the feature-local helicity histograms from features defined by high vorticity that show individual regions of strong vorticity have distinct helicity distributions.

Using VAPOR's intrinsic capabilities, we have thus effectively differentiated the regions of strong vorticity into three structure populations, based on their helicity distributions. In order to investigate the statistics and local dynamics of these structures, we next extend the analysis through a combination of visualization in VAPOR and focused study in the coupled quantitative analysis package. By visualizing individual features in isolation, we find that the wide noisy distributions belong to composite structures that were not well separated into individual components by the original vorticity thresholding, while the other two populations are those of individual tube-like structures. This result allows us to further cull the dataset and focus on the tube-like structures with either high or low helicity magnitude. Both populations have similar geometries, but streamlines seeded in these regions, as shown in Figure 5, reveal that their flow properties are quite different. In the low-helicity tubes, the streamlines twist around the core; in the high-helicity tubes the streamlines more closely follow the writhe of the tube.

Further interactive analysis of these distinct vortex structures can proceed either by examining the statistical properties of the population or the detailed dynamics of any one of them. Looking first at statistics of the population of vortical structures as a whole, we note that, while structures with all values of helicity exist, there seems to be a small deficit of those with high absolute mean value compared to the point-wise helicity distribution (Figure 4a). Moreover, the helicty of any given structure is well defined and symetrically distributed about its mean value (Figure 4b and   4c). The helicity distribution within a great majority of the structures has both small variance and skewness.

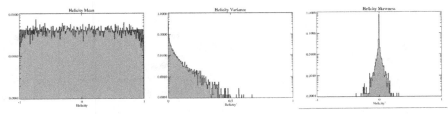

Fig. 4. Distributions of the first three central moments of the feature-local helicity distributions

The detailed dynamics underlying any single vortex structure is also accessible. By exporting planar cross sections through tubes using VAPOR's cross-section capability, average radial profiles of the helicity and vorticity can be constructed (Figure 5c & 5d). Distinct differences between the maximally and minimally helical structures are apparent. The maximally helical structure has one sign helicity throughout, while the minimally helical twisted structure shows a change in the sign of helicity near its border (Figure 5c). This appears to be associated with inward (toward the pinched section midway along the tube) axial flow surrounding the outside of the vortex tube and outward (diverging from a pinched section midway along the tube) axial flow in its core, (Figure 6). A temporal history of these flows would be critical in confirming what looks to be a significant vorticity amplification mechanism in this minimally helical vortex filament. Also critical in future analysis would be the ability to combine the statistical and dynamical analyses presented here to determine how common this mechanism is and whether particular dynamical processes are statistically linked to specific structure populations.

The primary advantage of coupling visual data investigation with a data analysis language is the ability to defer expensive calculations of derived quantities until they are needed and then perform them only over sub-domains of interest. The computational requirements for computing such variables in advance, across the entire domain, is often impratical, overwhelming the available

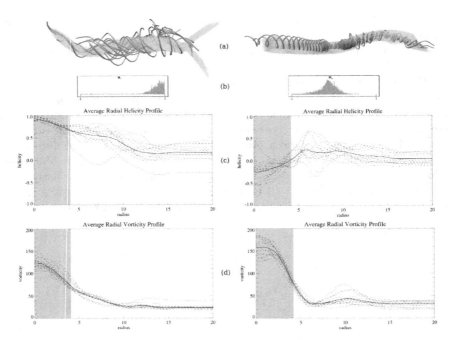

Fig. 5. Local dynamics of two strucutres with different helicity distributions showing: a) streamlines seeded within segmented region; b) the feature-local helicity histogram; c) an average radial helicity profile; d) an average radial vorticity profile. The shaded region of the radial profiles represents the inside of the visualized structure.

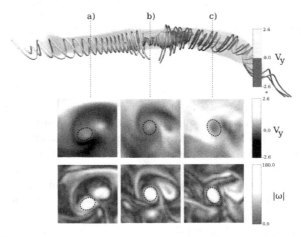

Fig. 6. Top: Streamlines colored by y-component velocity, which approximates the axial velocity. Bottom: vorticity magnitude and Y-component velocity cross-sections taken at positions a, b, & c (extents of the structure are bounded by the dotted line).

analysis resources. Furthermore, some quantities, as was shown by our analysis of the Taylor-Green flow, can only be computed with reference to the location of a flow structure and are therefore not in principle a priori computable. The coupling between VAPOR and IDL facilitates the calculation of derived quantities as needed over sub-regions of the domain, realizing considerable savings in storage space and processing time.

4   Conclusions

VAPOR's tight integration of visualization with traditional techniques like statistics, fourth-generation data languages, and effective information-management strategies meets the challenges that are inherent in visual exploration of complex turbulence data. This will only become more important as data volumes increase. The Taylor-Green flow simulation described in the previous section, which has 1024^3 degrees of freedom, can be readily computed on today's *teraflop* supercomputing platforms. The emergence of *petaflop*-capable machines will enable simulations at vastly greater scales, resulting in substantially larger data volumes. 4096^3 simulations have already been conducted on existing supercomputers [17] and the recent NSF Track 1 Petascale computing solicitation calls for a system capable of executing a homogeneous turbulence simulation with $12,288^3$ degrees of freedom [18]. The interactive analysis model in this paper, with its reliance on progressive data refinement, visual data browsing, and region/structure-of-interest isolation, is intrinsically highly scalable. We have described our experiences with this analysis model in the context of investigating numerically simulated turbulence. However, we believe that these techniques have applicability across a broad spectrum of data-intensive sciences.

Acknowledgements

We wish to thank the National Science Foundation and the National Center for Atmospheric Research for their computational support.

References

1. Barker, K.J., Davis, K., Hoisie, A., Kerbyson, D.J., Lang, M., Pakin, S., Sancho, J.C.: Entering the petaflop era: the architecture and performance of roadrunner. In: 2008 ACM/IEEE conference on Supercomputing, Austin, Texas, pp. 1–11. IEEE Press, Los Alamitos (2008)
2. Keim, D., Ward, M.: Visualization. In: Berthold, M., Hand, D. (eds.) Intelligent Data Analysis: An Introduction, 2nd edn. Springer, Heidelberg (2000)
3. Yang, L.: 3D grand tour for multidimensional data and clusters. In: Hand, D.J., Kok, J.N., R. Berthold, M. (eds.) IDA 1999. LNCS, vol. 1642, pp. 173–184. Springer, Heidelberg (1999)
4. Rehm, F., Klawonn, F., Kruse, R.: Mds-polar: A new approach for dimension reduction to visualize high-dimensional data. In: Famili, A.F., Kok, J.N., Peña, J.M., Siebes, A., Feelders, A. (eds.) IDA 2005. LNCS, vol. 3646, pp. 316–327. Springer, Heidelberg (2005)

5. Clyne, J.: The multiresolution toolkit: Progressive access for regular gridded data, 152–157 (2003)
6. Clyne, J., Mininni, P.D., Norton, A., Rast, M.: Interactive desktop analysis of high resolution simulations: application to turbulent plume dynamics and current sheet formation. New Journal of Physics 9 (2007)
7. Lorensen, B.: On the death of visualization. In: NIH/NSF Fall 2004 Workshop Visualization Research Challenges (2004)
8. Ahrens, J., Brislawn, K., Martin, K., Geveci, B., Law, C.C., Papka, M.: Large-scale data visualization using parallel data streaming. IEEE Computer Graphics and Applications 21, 34–41 (2001)
9. Childs, H., Brugger, E., Bonnell, K., Meredith, J., Miller, M., Whitlock, B., Max, N.: A contract based system for large data visualization. In: Proceedings of IEEE Visualization, pp. 191–198 (2005)
10. Engel, K., Hadwiger, M., Kniss, J.M., Lefohn, A.E., Salama, C.R., Weiskopf, D.: Real-time volume graphics. A K Peters, Ltd., Los Angeles (2006)
11. Weiskopf, D., Erlebacher, G.: Overview of flow visualization. In: Hansen, C., Johnson, C. (eds.) Visualization Handbook. Academic Press, London (2005)
12. Clyne, J., Rast, M.: A prototype discovery environment for analyzing and visualizing terascale turbulent fluid flow simulations. In: Erbacher, R.F., Roberts, J.C., Grohn, M.T., Borner, K. (eds.) Visualization and Data Analysis 2005. SPIE, San Jose, CA, USA, March 2005, vol. 5669, pp. 284–294 (2005)
13. Suzuki, K., Horibia, I., Sugie, N.: Linear-time connected-component labeling based on sequential local operations. Computer Vision and Image Understanding 89, 1–23 (2003)
14. Mininni, P.D., Alexakis, A., Pouquet, A.: Nonlocal interactions in hydrodynamic turbulence at high reynolds numbers: the slow emergence of scaling laws. Physical review. E, Statistical, nonlinear, and soft matter physics 77 (2008)
15. Jiang, M., Machiraju, R., Thompson, D.: Detection and visualization of vortices. In: Hansen, C., Johnson, C. (eds.) Visualization Handbook. Academic Press, London (2005)
16. Wu, J.Z., Ma, H.Y., Zhou, M.D.: Vorticity and Vortex Dynamics, 1st edn. Springer, Heidelberg (2006)
17. Kaneda, Y., Ishihara, T., Yokokawa, M., Itakura, K., Uno, A.: Energy dissipation rate and energy spectrum in high resolution direct numerical simulations of turbulence in a periodic box. Physics of Fluids 15, L21–L24 (2003)
18. Leadership-class system acquisition - creating a petascale computing environment for science and engineering NSF solicitation 06-573

Distributed Algorithm for Computing Formal Concepts Using Map-Reduce Framework*

Petr Krajca[1,2] and Vilem Vychodil[1,2]

[1] T. J. Watson School, State University of New York at Binghamton
[2] Dept. Computer Science, Palacky University, Olomouc
{petr.krajca, vychodil}@binghamton.edu

Abstract. Searching for interesting patterns in binary matrices plays an important role in data mining and, in particular, in formal concept analysis and related disciplines. Several algorithms for computing particular patterns represented by maximal rectangles in binary matrices were proposed but their major drawback is their computational complexity limiting their application on relatively small datasets. In this paper we introduce a scalable distributed algorithm for computing maximal rectangles that uses the map-reduce approach to data processing.

1 Introduction

We introduce a novel distributed algorithm for extracting rectangular patterns in binary object-attribute relational data. Our approach is unique among other approaches in that we employ the map-reduce framework which is traditionally used for searching and querying in large data collections. This paper contains a preliminary study and a proof of concept of how the map-reduce framework can be used for particular data-mining tasks.

In this paper, we focus on extracting rectangular patterns, so called formal concepts, in binary object-attribute relational data. The input data, we are interested in, takes form of a two-dimensional data table with rows corresponding to objects, columns corresponding to attributes (features), and table entries being crosses (or 1's) and blanks (or 0's) indicating presence/absence of attributes (features): a table has $\times$ on the intersection of row corresponding to object x and column corresponding to attribute y iff "object x has attributes y" ("feature y is present in object x"). Given a data table, we wish to find all maximal submatrices full of $\times$'s that are present in the table. These submatrices have a direct interpretation: they represent natural concepts hidden in the data which are the subjects of study of formal concept analysis [3,7] invented by Rudolf Wille [19]. Recently, it has been shown by Belohlavek and Vychodil [1] that maximal rectangles can be used to find optimal factorization of Boolean matrices. In fact, maximal rectangles correspond with optimal solutions to the discrete basis problem discussed by Miettinen et al. [11]. Finding maximal rectangles in data tables is therefore an important task.

---

* Supported by institutional support, research plan MSM 6198959214.

N. Adams et al. (Eds.): IDA 2009, LNCS 5772, pp. 333–344, 2009.
© Springer-Verlag Berlin Heidelberg 2009

The algorithm we propose in this paper may help overcome problems with generating all formal concepts from large data sets. In general, the problem of listing all formal concepts is $\#P$-complete [14]. Fortunately, if the input data is sparse, one can get sets of all formal concepts in a reasonable time. Still, listing all formal concepts can be time and space demanding and, therefore, there is a need to provide scalable distributed algorithms that may help distribute the burden over a large amount of low-cost computation nodes. In this paper we offer a possible solution using the map-reduce framework.

In the sequel we present a survey of notions from the formal concept analysis and principles of the map-reduce framework. Section 2 describes the algorithm. Furthermore, in Section 3, we describe the implementation and provide experimental evaluation of scalability of the proposed algorithm. The paper is concluded by related works and future research directions.

1.1 Formal Concept Analysis

In this section we recall basic notions of the formal concept analysis (FCA). More details can be found in monographs [7] and [3].

Formal concept analysis deals with binary data tables describing relationship between objects and attributes, respectively. The input for FCA is a data table with rows corresponding to objects, columns corresponding to attributes (or features), and table entries being ×'s and blanks, indicating whether an object given by row has or does not have an attribute given by column. An example of such a data table is depicted in Fig. 1 (left). A data table like that in Fig. 1 can be seen as a binary relation $I \subseteq X \times Y$ such that $\langle x, y \rangle \in I$ iff object x has attribute y. In FCA, I is usually called a *formal context* [7]. In this paper, we are going to use a set $X = \{0, 1, \ldots, m\}$ of objects and a set $Y = \{0, 1, \ldots, n\}$ of attributes, respectively. There is no danger of confusing objects with attributes because we do not mix elements from the sets X and Y in any way.

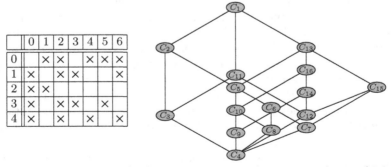

	0	1	2	3	4	5	6
0		×	×		×	×	×
1	×		×	×			×
2	×	×					
3	×		×	×		×	
4	×		×		×		×

Fig. 1. Formal context (left) and the corresponding concept lattice (right)

Each formal context $I \subseteq X \times Y$ induces a couple of operators $^\uparrow$ and $^\downarrow$ defined, for each $A \subseteq X$ and $B \subseteq Y$, as follows:

$$A^\uparrow = \{y \in Y \mid \text{for each } x \in A \colon \langle x, y \rangle \in I\}, \tag{1}$$

$$B^\downarrow = \{x \in X \mid \text{for each } y \in B \colon \langle x, y \rangle \in I\}. \tag{2}$$

Operators $\uparrow \colon 2^X \to 2^Y$ and $\downarrow \colon 2^Y \to 2^X$ defined by (1) and (2) form so-called Galois connection [7]. By definition (1), $A^\uparrow$ is a set of all attributes shared by all objects from A and, by (2), $B^\downarrow$ is a set of all objects sharing all attributes from B.

A pair $\langle A, B \rangle$ where $A \subseteq X$, $B \subseteq Y$, $A^\uparrow = B$, and $B^\downarrow = A$, is called a *formal concept (in $I \subseteq X \times Y$)*. Formal concepts can be seen as particular clusters hidden in the data. Namely, if $\langle A, B \rangle$ is a formal concept, A (called an *extent of $\langle A, B \rangle$*) is the set all objects sharing all attributes from B and, conversely, B (called an *intent of $\langle A, B \rangle$*) is the set of all attributes shared by all objects from A. Note that this approach to "concepts" as entities given by their extent and intent goes back to classical Port Royal logic. From the technical point of view, formal concepts are fixed points of the Galois connection $\langle \uparrow, \downarrow \rangle$ induced by the formal context. Formal concepts in $I \subseteq X \times Y$ correspond to so-called maximal rectangles in I. In a more detail, any $\langle A, B \rangle \in 2^X \times 2^Y$ such that $A \times B \subseteq I$ shall be called a rectangle in I. Rectangle $\langle A, B \rangle$ in I is a maximal one if, for each rectangle $\langle A', B' \rangle$ in I such that $A \times B \subseteq A' \times B'$, we have $A = A'$ and $B = B'$. We have that $\langle A, B \rangle \in 2^X \times 2^Y$ is a maximal rectangle in I iff $A^\uparrow = B$ and $B^\downarrow = A$, i.e. maximal rectangles = formal concepts. Hence, maximal rectangles give us an alternative interpretation of formal concepts.

Let $\mathcal{B}(X, Y, I)$ denote the set of all formal concepts in $I \subseteq X \times Y$. The set $\mathcal{B}(X, Y, I)$ can be equipped with a partial order $\leq$ modeling the subconcept-superconcept hierarchy:

$$\langle A_1, B_1 \rangle \leq \langle A_2, B_2 \rangle \text{ iff } A_1 \subseteq A_2 \text{ (or, equivalently, iff } B_2 \subseteq B_1). \tag{3}$$

If $\langle A_1, B_1 \rangle \leq \langle A_2, B_2 \rangle$ then $\langle A_1, B_1 \rangle$ is called a subconcept of $\langle A_2, B_2 \rangle$. The set $\mathcal{B}(X, Y, I)$ together with $\leq$ form a complete lattice whose structure is described by the Main Theorem of Formal Concept Analysis [7]. The above-described notions are illustrated in the following example.

Example 1. Consider a formal context $I \subseteq X \times Y$ corresponding to the incidence data table from Fig. 1 (left). The concept-forming operators induced by this context have exactly 15 fixpoints $C_1, \ldots, C_{16}$:

$$C_1 = \langle \{0, 1, 2, 3, 4\}, \{\} \rangle, \qquad C_9 = \langle \{4\}, \{0, 2, 4, 6\} \rangle,$$
$$C_2 = \langle \{1, 2, 3, 4\}, \{0\} \rangle, \qquad C_{10} = \langle \{1, 4\}, \{0, 2, 6\} \rangle,$$
$$C_3 = \langle \{2\}, \{0, 1\} \rangle, \qquad C_{11} = \langle \{0, 2\}, \{1\} \rangle,$$
$$C_4 = \langle \{\}, \{0, 1, 2, 3, 4, 5, 6\} \rangle, \qquad C_{12} = \langle \{0\}, \{1, 2, 4, 5, 6\} \rangle,$$
$$C_5 = \langle \{1, 3, 4\}, \{0, 2\} \rangle, \qquad C_{13} = \langle \{0, 1, 3, 4\}, \{2\} \rangle,$$
$$C_6 = \langle \{1, 3\}, \{0, 2, 3\} \rangle, \qquad C_{14} = \langle \{0, 4\}, \{2, 4, 6\} \rangle,$$
$$C_7 = \langle \{3\}, \{0, 2, 3, 5\} \rangle, \qquad C_{15} = \langle \{0, 3\}, \{2, 5\} \rangle,$$
$$C_8 = \langle \{1\}, \{0, 2, 3, 6\} \rangle, \qquad C_{16} = \langle \{0, 1, 4\}, \{2, 6\} \rangle.$$

Hence, $\mathcal{B}(X, Y, I) = \{C_1, \ldots, C_{16}\}$. If we equip $\mathcal{B}(X, Y, I)$ with the partial order (3), we obtain a concept lattice shown in Fig. 1 (right).

The most common algorithms for computing formal concepts include Ganter's algorithm [6], Lindig's algorithm [18], and Berry's [2] algorithm. The algorithm we are going to introduce in Section 2 can be seen as a distributed version of the algorithm proposed in [12,13]. A survey and comparison of algorithms for FCA can be found in [17].

1.2   Processing Data Using Map-Reduce Approach

Distributed computing represents a common approach to processing large data but the complexity of distributed computing usually limits its application only on problems unsolvable in other ways. Common distributed algorithm implementations usually consist of modifications of more or less known algorithms that distribute particular parts of the data to be processed on other computers. This approach is quite comprehensible for programmers but brings several issues. Especially, issues connected with granularity of the task, reliability of the used hardware platform, etc.

A general framework for processing large data in distributed networks consisting of commodity hardware is proposed in [4]. In essence, this framework is based on two basic operations *map* and *reduce* that are applied on the data. These two operations have given the name of the framework—a *map-reduce* framework or (in short, an M/R framework). This approach to data processing has originally been developed by Google for their data centers but has shown to be very practical and later has been adopted by other software companies interested in storing and querying large amounts of data. In the rest of this section we provide a brief overview of the map-reduce framework.

Data in the M/R framework are generally represented in the form of key-value pairs $\langle key, value \rangle$. In the first step of the computation, the framework reads input data and optionally converts it into the desired key-value pairs.

In the second step—the *map phase*, a function f is applied on each pair $\langle k, v \rangle$ and returns a multiset of new key-value pairs, i.e.,

$$f(\langle k, v \rangle) = \{\langle k_1, v_1 \rangle, \ldots, \langle k_n, v_n \rangle\}$$

Notice that there is certain similarity with function `map` present in many programming languages (e.g., LISP, Python, and Scheme). In contrast with the usual `map`, function f may return arbitrary number of results and they are all collected during the *map phase*.

Subsequently, in the *reduce phase*, all pairs generated in the previous step are grouped by their keys and their values are aggregated (reduced) by a function g:

$$g(\{\langle k, v_1 \rangle, \langle k, v_2 \rangle, \ldots, \langle k, v_n \rangle\}) = \langle k, v \rangle.$$

The following example illustrate how the M/R framework can be used to perform a computation which might appear in information retrieval.

Example 2. Let us consider that we want to compute frequencies of letters in a text consisting of three words—*alice, barbara,* and *carol.* Now, consider function

f accepting a word and returning a multiset where each letter in the word is represented as a pair $\langle letter, 1 \rangle$. The map phase will produce the following results:

$$f(alice) = \{\langle a, 1 \rangle, \langle l, 1 \rangle, \langle i, 1 \rangle, \langle c, 1 \rangle, \langle e, 1 \rangle\},$$
$$f(barbara) = \{\langle b, 1 \rangle, \langle a, 1 \rangle, \langle r, 1 \rangle, \langle b, 1 \rangle, \langle a, 1 \rangle, \langle r, 1 \rangle, \langle a, 1 \rangle\},$$
$$f(carol) = \{\langle c, 1 \rangle, \langle a, 1 \rangle, \langle r, 1 \rangle, \langle o, 1 \rangle, \langle l, 1 \rangle\}.$$

In the reduce phase, we group all pairs by the key and function g sums all 1's in the key-value pairs as follows:

$$g(\{\langle a, 1 \rangle, \langle a, 1 \rangle, \langle a, 1 \rangle, \langle a, 1 \rangle, \langle a, 1 \rangle\}) = \langle a, 5 \rangle,$$
$$g(\{\langle b, 1 \rangle, \langle b, 1 \rangle\}) = \langle b, 2 \rangle,$$
$$g(\{\langle c, 1 \rangle, \langle c, 1 \rangle\}) = \langle c, 2 \rangle,$$
$$g(\{\langle e, 1 \rangle\}) = \langle e, 1 \rangle,$$
$$g(\{\langle l, 1 \rangle, \langle l, 1 \rangle\}) = \langle l, 2 \rangle,$$
$$g(\{\langle i, 1 \rangle\}) = \langle i, 1 \rangle,$$
$$g(\{\langle o, 1 \rangle\}) = \langle o, 1 \rangle,$$
$$g(\{\langle r, 1 \rangle, \langle r, 1 \rangle, \langle r, 1 \rangle\}) = \langle r, 3 \rangle.$$

One can see from the previous illustrative example and the sketch of the algorithm that since the functions f and g are applied only on particular pairs, it is possible to easily distribute the computation through several computers. For the sake of completeness, we should stress that functions f and g used to map and aggregate values must always be implemented by procedures (i.e., computer functions) that do not have any side effects. That means, the results of f and g depend only on given arguments, i.e., the procedures computing results of f and g behave as maps.

The actual implementation details of M/R processing are far more complex but their detailed description is out of the scope of this paper. The basic outline we have provided should be sufficient for understanding of our algorithm for computing formal concepts.

2   The Algorithm

In this section we present an overview of a parallel algorithm PCbO we have proposed in [12]. Then, we introduce a distributed variant of PCbO based on the M/R framework.

2.1   Overview

The distributed algorithm is an adaptation of Kuznetsov's Close-by-One (CbO, see [15,16]) and its parallel variant PCbO [12]. The CbO can be formalized by a recursive procedure GENERATEFROM($\langle A, B \rangle, y$), which lists all formal concepts using a depth-first search through the space of all formal concepts. The procedure accepts a formal concept $\langle A, B \rangle$ (an initial formal concept) and an attribute $y \in Y$ (first attribute to be processed) as its arguments. The procedure recursively descends through the space of formal concepts, beginning with the formal concept $\langle A, B \rangle$.

When invoked with $\langle A, B \rangle$ and $y \in Y$, GENERATEFROM first processes $\langle A, B \rangle$ (e.g., prints it on the screen or stores it in a data structure) and then it checks its halting condition. Computation stops either when $\langle A, B \rangle$ equals $\langle Y^\downarrow, Y \rangle$ (the least formal concept has been reached) or $y > n$ (there are no more remaining attributes to be processed). Otherwise, the procedure goes through all attributes $j \in Y$ such that $j \geq y$ which are not present in the intent B. For each $j \in Y$ having these properties, a new couple $\langle C, D \rangle \in 2^X \times 2^Y$ such that

$$\langle C, D \rangle = \langle A \cap \{j\}^\downarrow, (A \cap \{j\}^\downarrow)^\uparrow \rangle \tag{4}$$

is computed. The pair $\langle C, D \rangle$ is always a formal concept [12] such that $B \subset D$. After obtaining $\langle C, D \rangle$, the algorithm checks whether it should continue with $\langle C, D \rangle$ by recursively calling GENERATEFROM or whether $\langle C, D \rangle$ should be "skipped". The test (so-called canonicity test) is based on comparing $B \cap Y_j = D \cap Y_j$ where $Y_j \subseteq Y$ is defined as follows:

$$Y_j = \{y \in Y \mid y < j\}. \tag{5}$$

The role of the canonicity test is to prevent computing the same formal concept multiple times. GENERATEFROM computes formal concepts in a unique order which ensures that each formal concept is processed exactly once. The proof is elaborated in [13].

Remark 1. Recursive invocations of GENERATEFROM form a tree. The tree corresponding to data from Fig. 1 is depicted in Fig. 2. The root corresponds to the first invocation GENERATEFROM($\langle \emptyset^\downarrow, \emptyset^{\downarrow\uparrow} \rangle, 0$). Each node labeled $\langle C_i, k \rangle$ corresponds to an invocation of GENERATEFROM, where C_i is a formal concept, see Example 1. The nodes denoted by black squares represent concepts which are

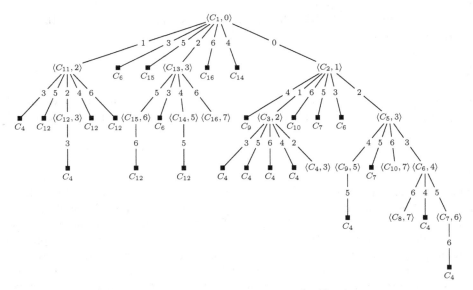

Fig. 2. Example of a call tree for GENERATEFROM($\langle \emptyset^\downarrow, \emptyset^{\downarrow\uparrow} \rangle, 0$) with data from Fig. 1

computed but not processed because the canonicity test fails. The edges in the tree are labeled by the number of attributes that are used to compute new concepts, cf. (4). More details can be found in [13].

2.2   Adaptation for M/R Framework

The procedure GENERATEFROM outlined in Section 2.1 uses a depth-first search strategy to generate new formal concepts. Since GENERATEFROM depends only on its arguments, the strategy in which formal concepts are generated does not play any significant role and can be replaced by another strategy. In our distributed algorithm, we are going to use the breadth-first search strategy. Conversion of the algorithm from the depth-first to the breadth-first search is necessary for the adaptation of CbO algorithm to the M/R framework. Moreover, the original GENERATEFROM will be split into two functions: a map function called MAPCONCEPTS and a reduction function REDUCECONCEPTS. The map function will take care of generating new formal concepts and the reduce function will take care of performing the canonicity tests. The breadth-first search strategy is beneficial since it allows us to compute formal concepts by *levels*, where each level is computed from the previous one by consecutive applications of MAPCONCEPTS and REDUCECONCEPTS on formal concepts that were computed in the previous level.

Remark 2. The adaptation for the map-reduce framework has several aspects. First, instead of using a single recursive procedure GENERATEFROM, we employ two functions which are not recursive (i.e., they do not invoke themselves) but serve as a mapping and a reduction functions used by the framework. Second, the arguments to MAPCONCEPTS and REDUCECONCEPTS, which somehow encode the arguments of the original GENERATEFROM, must be presented as key/value pairs to ensure compatibility with the map-reduce framework. These issues will be addressed in the sequel.

Recall that the M/R framework assumes that all values processed by MAPCONCEPTS and REDUCECONCEPTS are in the form of $\langle key, value \rangle$. In order to ensure the input/output compatibility with the M/R framework, we are going to encode the arguments for MAPCONCEPTS and REDUCECONCEPTS as follows: We consider pairs $\langle key, value \rangle$ such that

- *key* is a tuple $\langle B, y \rangle$ where B is an intent of a concept $\langle A, B \rangle \in \mathcal{B}(X, Y, I)$ and $y \in Y$ is an attribute;
- *value* is a new concept $\langle C, D \rangle \in \mathcal{B}(X, Y, I)$.

The exact meaning of the key/value pair during the computation will become apparent later. The way in which MAPCONCEPTS and REDUCECONCEPTS compute all formal concepts can be summarized by the following steps:

(1) Initially, the first formal concept $\langle \emptyset^{\downarrow}, \emptyset^{\downarrow\uparrow} \rangle$ is computed and MAPCONCEPTS is called with the initial key/value pair $\langle \langle \emptyset^{\downarrow\uparrow}, 0 \rangle, \langle \emptyset^{\downarrow}, \emptyset^{\downarrow\uparrow} \rangle \rangle$ which produces

a multiset of new key/value pairs representing new concepts. The multiset is further reduced by REDUCECONCEPTS by removing key/value pairs with concepts that do not pass the canonicity test. These two steps represents the first iteration.

(2) The MAPCONCEPTS function is applied on each key/value pair from the previous nth iteration and the result is reduced by REDUCECONCEPTS; the returned key/value pairs containing formal concepts are stored as result of the $(n + 1)$th iteration.

(3) If the $(n + 1)$th iteration produces any new concepts, the computation continues with the step (2) for the next iteration; otherwise the computation is stopped.

In the sequel, we provide a detailed description of MAPCONCEPTS and REDUCECONCEPTS.

2.3   Details on MAPCONCEPTS and REDUCECONCEPTS

The mapping function is described in Algorithm 1. It accepts an encoded formal concept $\langle A, B \rangle$ and iterates over all attributes that are equal or greater than y (lines 2–7). If the attribute is not present in the intent B (line 3), it computes a new formal concept $\langle C, D \rangle$ by extending intent B with an attribute j (lines 4 and 5). This corresponds to getting a formal concept of the form (4).

Algorithm 1: MAPCONCEPTS

Input: Pair $\langle key, value \rangle$ where key is $\langle B_0, y \rangle$ and $value$ is $\langle A, B \rangle$.

1   set $result$ to $\emptyset$
2   **for** $j = y$ **to** $|Y|$ **do**
3   **if** $j \in B$ **then continue**;
4   set C to $A \cap \{j\}^{\downarrow}$
5   set D to $C^{\uparrow}$
6   set $result$ **to** $result \cup \{\langle\langle B, j \rangle, \langle C, D \rangle\rangle\}$
7   **end**
8   **return** $result$

Algorithm 1 computes new formal concepts that are derived from $\langle A, B \rangle$. Notice that some of the new concepts obtained this way may be the same. In general, a single concept may result by computing (4) multiple times during the entire computation. To identify redundantly computed formal concepts and to remove them, we use the same canonicity test as the ordinary CbO and its parallel variant PCbO. In our case, the canonicity test will appear in the reduction function. Also note that the value of B_0 is not used by MAPCONCEPTS.

The REDUCECONCEPTS function accepts an encoded tuple $\langle\langle B, j \rangle, \langle C, D \rangle\rangle$ and returns a value as follows:

$$\text{REDUCECONCEPTS}(\langle\langle B, j \rangle, \langle C, D \rangle\rangle) = \begin{cases} \langle\langle B, j + 1 \rangle, \langle C, D \rangle\rangle, & \text{if } B \cap Y_j = D \cap Y_j, \\ \text{void-value}, & \text{otherwise.} \end{cases}$$

Therefore, if the canonicity test is satisfied then the input pair $\langle\langle B, j\rangle, \langle C, D\rangle\rangle$ is reduced to $\langle\langle B, j+1\rangle, \langle C, D\rangle\rangle$ which will be used in the next iteration.

Remark 3. (a) During the computation, each value $\langle B, j\rangle$ of key appears at most once. Thus, we can assume that REDUCECONCEPTS accepts only one argument instead of a set of arguments with the same key.
(b) Practical implementations of M/R frameworks have support for dealing with *void-values*. In practice, the *void-values* are not included into the results.
(c) The process of computing formal concepts may be seen as building a tree (e.g., as depicted in Fig. 2) by its levels as it is shown in Fig. 3.

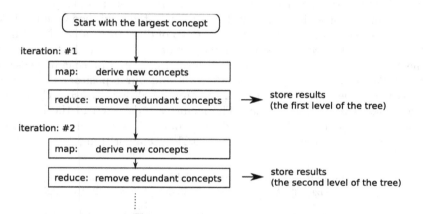

Fig. 3. Particular iterations of the computation

Example 3. This example shows how the algorithm processes input data. As the input dataset we use formal context described in the Fig. 1. We start with the first iteration and create the initial value $\langle\langle\emptyset, 0\rangle, C_1\rangle$ and continue with the first iteration. In the first iteration, map function is applied only on the initial value and generates tuples: $\langle\langle\emptyset, 0\rangle, C_2\rangle$, $\langle\langle\emptyset, 1\rangle, C_{11}\rangle$, $\langle\langle\emptyset, 2\rangle, C_{13}\rangle$, $\langle\langle\emptyset, 3\rangle, C_6\rangle$, $\langle\langle\emptyset, 4\rangle, C_{14}\rangle$, $\langle\langle\emptyset, 5\rangle, C_{15}\rangle$, $\langle\langle\emptyset, 6\rangle, C_{16}\rangle$. In the sequel, i.e., the reduction phase, only tuples passing the canonicity test are retained. That is, tuples $\langle\langle\emptyset, 1\rangle, C_2\rangle$, $\langle\langle\emptyset, 2\rangle, C_{11}\rangle$, $\langle\langle\emptyset, 3\rangle, C_{13}\rangle$ are stored as results of the first iteration and used as input values for the second iteration. The second iteration takes values from the first one and applies map function on them. The map function applied on $\langle\langle\emptyset, 1\rangle, C_2\rangle$ generates tuples: $\langle\langle\{0\}, 1\rangle, C_3\rangle$, $\langle\langle\{0\}, 2\rangle, C_5\rangle$, $\langle\langle\{0\}, 3\rangle, C_6\rangle$, $\langle\langle\{0\}, 4\rangle, C_9\rangle$, $\langle\langle\{0\}, 5\rangle, C_7\rangle$, $\langle\langle\{0\}, 6\rangle, C_{10}\rangle$. Similarly, the map function applied on $\langle\langle\emptyset, 2\rangle, C_{11}\rangle$ returns: $\langle\langle\{1\}, 2\rangle, C_{12}\rangle$, $\langle\langle\{1\}, 3\rangle, C_4\rangle$, $\langle\langle\{1\}, 4\rangle, C_{12}\rangle$, $\langle\langle\{1\}, 5\rangle, C_{12}\rangle$, $\langle\langle\{1\}, 6\rangle, C_{12}\rangle$, and application of map function on $\langle\langle\emptyset, 3\rangle, C_{13}\rangle$ produces: $\langle\langle\{2\}, 3\rangle, C_6\rangle$, $\langle\langle\{2\}, 4\rangle, C_{14}\rangle$, $\langle\langle\{2\}, 5\rangle, C_{15}\rangle$, $\langle\langle\{2\}, 6\rangle, C_{16}\rangle$. From these values only $\langle\langle\{0\}, 2\rangle, C_3\rangle$, $\langle\langle\{0\}, 3\rangle, C_5\rangle$, $\langle\langle\{1\}, 3\rangle, C_{12}\rangle$, $\langle\langle\{2\}, 5\rangle, C_{14}\rangle$, $\langle\langle\{2\}, 6\rangle, C_{15}\rangle$, $\langle\langle\{2\}, 7\rangle, C_{16}\rangle$ pass the canonicity test and thus are stored as results of the second iteration and used in the following iteration. The computation continues in much the same way with the next iteration and stops if the reduction phase does not return any value. One can see that the computation directly corresponds to the tree depicted in Fig. 2. Each iteration represents one level of the tree whereas map function computes descendant nodes and the reduce function determines which nodes should be used in further computation.

3   Implementation and Experiments

We have implemented our algorithm as a Java application using Hadoop Core framework [8] providing infrastructure for map-reduce computations along with distributed file system for storing input data and results. For our experiments, we have used cluster consisting of 15 idle desktop computers equipped with Intel Core 2 Duo (3.0 GHz) processors, 2 GB RAM, 36 GB of disk space, and GNU/Linux.

We present here two sets of preliminary experiments. The first one focuses on the total time needed to compute all formal concepts present in real-world datasets. We selected three datasets from [9] and our repository with various properties and measured time it took to compute all formal using cluster consisting of one, five, and ten nodes. The results are depicted in the Fig. 4. For the comparison, we have also included the running time it takes to compute all formal concepts by the usual Ganter's NextClosure algorithm [7].

dataset	mushroom	debian tags	anon. web
size	8124 × 119	14315 × 475	32710 × 295
density	19 %	< 1 %	1 %
our (1 node)	1259	1379	2935
our (5 nodes)	436	426	830
our (10 nodes)	397	366	577
NextClosure	743	1793	10115

Fig. 4. Time needed to compute all formal concepts for various datasets (in seconds)

While evaluating the distributed algorithm, besides the overall performance, we have to take into account an important feature called scalability. In other words, the ability to decrease the time of the computation by utilizing more computers. In the second set of experiments we focus on this feature. To represent scalability we are using *relative speedup*, a ratio $S = \frac{T_1}{T_n}$, where T_1 is the time of the computation running only one computer and T_n is the time of the computation running on n computers. The theoretical maximal speedup is equal to the number of computers. The real speedup is always smaller than the theoretical one due to many factors including especially communication overhead between particular nodes, network throughput, etc.

Fig. 5 (left) shows scalability of our algorithm for selected datasets. One can see that for small numbers of nodes the speedup is almost linear but with the increasing number the speedup is not so significant. In case of the *mushrooms* dataset, there is even a decline of the speedup, i.e., with increasing number

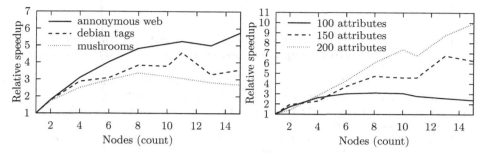

Fig. 5. Relative speedup for real world datasets (left) and for contexts with various counts of attributes (right)

of nodes, the speedup is no longer increasing. This is caused by the overhead of distributed computation that cannot be counterweighted by utilizing more computers. This means, the size of the cluster has to be adequate to the size of the input data. This fact also supports Fig. 5 (right), depicting scalability of the algorithm on randomly generated contexts consisting of 10000 objects, 100, 150, and 200 attributes, where density of 1's is 10%. From the figure Fig. 5 (right) follows that with the increasing size of data the scalability grows and for the data table of size 10000×200 the computation on cluster consisting of 15 may be done even $10\times$ faster than on a single computer.

3.1 Related Works, Conclusions, and Future Research

Several parallel algorithms for computing formal concepts have been proposed. For instance, [5], [10], or [12]. In general, parallel algorithms have a disadvantage of requiring a hardware equipped with several processors or processor cores. Despite the shift in hardware development toward to multicore microprocessors, hardware configuration with large amounts (more than ten) of processor cores are still relatively expensive and rare. Contrary to that, distributed algorithms may run on a coupled commodity hardware. Typically, parallel programs have a smaller overhead of the computation management than the distributed ones but the distributed algorithms are more cost-effective (they can be run on ordinary personal computers connected by a network). Although all mentioned algorithms may be modified to their *ad hoc* distributed versions, as far as we know, there are no distributed implementations of these algorithms. The approach introduced in this paper should be seen as a proof of concept of computing formal concepts by isolated nodes. We have shown that the algorithm is scalable. Therefore, there is a potential to apply techniques of formal concept analysis for much larger data sets than previously indicated. Our future research will focus on improving the algorithm by employing more efficient canonicity tests and providing a more efficient implementation. Furthermore, we intend to test the behavior of the algorithm on larger data sets and with larger amount of nodes.

References

1. Belohlavek, R., Vychodil, V.: Discovery of optimal factors in binary data via a novel method of matrix decomposition. Journal of Computer and System Sciences (to appear)
2. Berry, A., Bordat, J.-P., Sigayret, A.: A local approach to concept generation. Annals of Mathematics and Artificial Intelligence 49, 117–136 (2007)
3. Carpineto, C., Romano, G.: Concept data analysis. Theory and applications. J. Wiley, Chichester (2004)
4. Dean, J., Ghemawat, S.: MapReduce: simplified data processing on large clusters. Commun. ACM 51(1), 107–113 (2008)
5. Fu, H., Nguifo, E.M.: A parallel algorithm to generate formal concepts for large data. In: Eklund, P. (ed.) ICFCA 2004. LNCS (LNAI), vol. 2961, pp. 394–401. Springer, Heidelberg (2004)
6. Ganter, B.: Two basic algorithms in concept analysis (Technical Report FB4-Preprint No. 831). TH Darmstadt (1984)
7. Ganter, B., Wille, R.: Formal concept analysis. Mathematical foundations. Springer, Berlin (1999)
8. Hadoop Core Framework, http://hadoop.apache.org/
9. Hettich, S., Bay, S.D.: The UCI KDD Archive University of California, Irvine, School of Information and Computer Sciences (1999)
10. Kengue, J.F.D., Valtchev, P., Djamegni, C.T.: A parallel algorithm for lattice construction. In: Ganter, B., Godin, R. (eds.) ICFCA 2005. LNCS (LNAI), vol. 3403, pp. 249–264. Springer, Heidelberg (2005)
11. Miettinen, P., Mielikäinen, T., Gionis, A., Das, G., Mannila, H.: The discrete basis problem. In: Fürnkranz, J., Scheffer, T., Spiliopoulou, M. (eds.) PKDD 2006. LNCS (LNAI), vol. 4213, pp. 335–346. Springer, Heidelberg (2006)
12. Krajca, P., Outrata, J., Vychodil, V.: Parallel Recursive Algorithm for FCA. In: Belohlavek, R., Kuznetsov, S.O. (eds.) Proc. CLA 2008, vol. 433, pp. 71–82. CEUR WS (2008) ISBN 978-80-244-2111-7
13. Krajca, P., Outrata, J., Vychodil, V.: Parallel Algorithm for Computing Fixpoints of Galois Connections. Annals of Mathematics and Artificial Intelligence (submitted)
14. Kuznetsov, S.: Interpretation on graphs and complexity characteristics of a search for specific patterns. Automatic Documentation and Mathematical Linguistics 24(1), 37–45 (1989)
15. Kuznetsov, S.: A fast algorithm for computing all intersections of objects in a finite semi-lattice (Быстрый алгоритм построения всех пересечений объектов из конечной полурешетки, in Russian). Automatic Documentation and Mathematical Linguistics 27(5), 11–21 (1993)
16. Kuznetsov, S.O.: Learning of simple conceptual graphs from positive and negative examples. In: Żytkow, J.M., Rauch, J. (eds.) PKDD 1999. LNCS (LNAI), vol. 1704, pp. 384–391. Springer, Heidelberg (1999)
17. Kuznetsov, S., Obiedkov, S.: Comparing performance of algorithms for generating concept lattices. J. Exp. Theor. Artif. Int. 14, 189–216 (2002)
18. Lindig, C.: Fast concept analysis. In: Working with Conceptual Structures — Contributions to ICCS 2000, pp. 152–161. Shaker Verlag, Aachen (2000)
19. Wille, R.: Restructuring lattice theory: an approach based on hierarchies of concepts. In: Ordered Sets, Dordrecht, Boston, pp. 445–470 (1982)

Multi-Optimisation Consensus Clustering

Jian Li, Stephen Swift, and Xiaohui Liu

School of Information Systems, Computing and Mathematics,
Brunel University, Uxbridge, Middlesex, UB8 3PH, UK
`{Jian.Li,Stephen.Swift,Xiaohui.Liu}@brunel.ac.uk`

Abstract. Ensemble Clustering has been developed to provide an alternative way of obtaining more stable and accurate clustering results. It aims to avoid the biases of individual clustering algorithms. However, it is still a challenge to develop an efficient and robust method for Ensemble Clustering. Based on an existing ensemble clustering method, Consensus Clustering (CC), this paper introduces an advanced Consensus Clustering algorithm called Multi-Optimisation Consensus Clustering (MOCC), which utilises an optimised Agreement Separation criterion and a Multi-Optimisation framework to improve the performance of CC. Fifteen different data sets are used for evaluating the performance of MOCC. The results reveal that MOCC can generate more accurate clustering results than the original CC algorithm.

Keywords: Consensus Clustering, Simulated Annealing, Multi-optimisation.

1 Introduction

Ensemble Clustering has been suggested to tackle the biases of individual clustering methods [6], [8]. Its aim is to combine the results of a number of clustering algorithms to improve its clustering accurate [4], [5], [6], [7], [8]. The Consensus Clustering (CC) algorithm, developed by Swift et al. [1], is one of the existing ensemble clustering approaches. Hirsch *et al.* [2] compared CC with six different ensemble clustering methods developed by Hornik [3]. The results showed that CC achieved comparable or even better performance than those six ensemble methods.

However, to develop an effective method for ensemble clustering is still a challenge. We take Consensus Clustering as an example. CC seeks for the optimised results by only maximising the agreement fitness, which is calculated by a function that combines the input base clustering arrangements. This function depends fully on the input base clustering, so that the final results would be bad if many of the input base clustering had low accuracies. This leads to the following question: would it be beneficial if the candidate clustering solutions were evaluated not only by the agreement fitness during the evaluation process of CC but also by the internal clustering characteristics of data sets?

Over the past several years, extensive studies have been conducted in the area of multi-objective optimisation and clustering methods. Based on the experience of developing multi-objective optimisation methods, we present an advanced consensus

N. Adams et al. (Eds.): IDA 2009, LNCS 5772, pp. 345–356, 2009.
© Springer-Verlag Berlin Heidelberg 2009

clustering algorithm called Multi-Optimisation Consensus Clustering (MOCC), which evaluates not only the agreement fitness but also the internal clustering characteristics of data sets to enhance the clustering accuracy of CC. First of all, MOCC exploits an optimised agreement separation criterion to produce a Weighted Agreement Matrix (WAM). And then a Probability Based Solution Generator (PBSG) is developed to create candidate solutions. Based on the WAM and PBSG, finally, a Multi-Optimisation structure is adopted to produce final results. We demonstrated MOCC on fifteen different data sets and compared it with the original CC algorithm. The results reveal that MOCC performs clearly better than CC.

This paper is organised as follows: Section 2 describes the related work. Section 3 details the framework of the MOCC algorithm. The data sets are presented in Section 4, followed by the experimental results and discussion in Section 5. Finally in Section 6, we draw conclusions and discuss future work.

2   Related Works

2.1   Consensus Clustering

Consensus Clustering (CC) is an optimisation based ensemble clustering method [1]. The Agreement Matrix, the Simulated Annealing Optimisation Section (SAOS) and the Agreement Fitness Function (AFF) are key components of CC. The Agreement Matrix, A, is built based on a clustering arrangement matrix constructed by the results of a set of individual clustering algorithms. Each element, A_{ij}, indicates how many individual algorithms agree instances i and j to be assigned into the same cluster. The Simulated Annealing (SA) algorithm is applied to generate the final clustering results. The AFF is utilised by SA as the only evaluation criterion for the optimisation search, which causes the final results of CC depending heavily on the accuracy of input results.

Within the AFF, the agreement threshold β is the key parameter, which directly affects the behaviour of the optimisation search. CC recommends β to be the mean of the maximum and minimum values in A. However, in some cases, this agreement threshold does not perform well. In order to solve these limitations in CC, we use multi-optimisation framework for the optimisation search, and exploit a comparative optimal agreement separation value to optimise β. The more detailed description of the CC algorithm can be found in [1].

2.2   Separation Score Evaluation Criterion

The Homogeneity and Separation criterion, which proposed by Shamir and Sharan, is one of the widely-applied cluster validation indices [10]. Homogeneity and Separation are relative to each other. Homogeneity is defined as the average distance between each pattern and the centre of the cluster the pattern belongs to. It reflects the inside compactness of clusters. Separation is defined as the weighted average distance between cluster centres. It reflects the outside distance between clusters. Increasing the Separation score or decreasing the Homogeneity score suggests an improvement in the clustering results [10]. According to above characteristics of Homogeneity and

Separation and the cost of computation, we adopt Separation Score to evaluate the intrinsic clustering characteristics for the optimisation search in MOCC. Equation (1) displays the way of calculating the Separation Score S.

$$S = \frac{\sum_{i \neq j} N_{CL_i} N_{CL_j} D(Cent_i, Cent_j)}{\sum_{i \neq j} N_{CL_i} N_{CL_j}} \tag{1}$$

N_{CL_i} indicates the number of instances in the ith cluster. The $D(Cent_i, Cent_j)$ is the distance function that calculates the distance between two cluster centres $Cent_i$ and $Cent_j$. Euclidean Distance is adopted in the MOCC algorithm.

2.3 Weighted Kappa Index

The Weighted-Kappa (WK) [12] index was employed by Swift et al. [1] and Hirsch et al. [2] to validate the accuracy of clustering results. The WK score denotes the similarity (or agreement) between a candidate clustering solution and the true clustering arrangement. In this paper, we also use this evaluation index to indicate the clustering accuracy.

WK constructs a (2x2) contingency table to measure the similarity between two arrangements [12]. All unique pairs of instances will be counted in this table. The total number of unique pairs for n instances is $n(n-1)/2$.

The value of WK is between 1.0 (which means these two clustering arrangements are the same) and -1.0 (which means the two clustering arrangements are totally different). Two totally random clustering arrangements should have a score of zero. Further details of WK can be found in [12].

2.4 Correlation Index

Maximising the value of the objective function to seek for more accurate clustering results (i.e. the results have higher WK scores) is the basic idea of CC and MOCC. Therefore the correlation of the function value and the WK score is meaningful for CC and MOCC (the correlation definitions of CC and MOCC are described in Section 5.2, and they are different from each other). In this paper, we employ the Pearson Product-Moment Correlation Coefficient (PMCC) [11] to evaluate the correlation. PMCC denotes the degree of the linear correlation between two variables. The range of its values is from -1 to +1; where -1 means these two variables have totally opposite linear correlation, +1 indicates the two variables have the best positive linear correlation [11]. For CC and MOCC, we expect the PMCC value to be as close to +1 as possible.

3 Multi-Optimisation Consensus Clustering (MOCC)

The MOCC algorithm consists of four key components, which are the weighted agreement matrix, the fitness function, the probability based solution generator, and the multi-optimisation section. The details of each component are described as follows.

3.1 The Weighted Agreement Matrix and the Agreement Fitness Function

The weighted agreement matrix is generated by the agreement threshold β and the agreement matrix A. The β is used to evaluate the agreement of clustering each pair of instances. It rewards (or penalizes) a clustering that has an agreement value above (or below) the threshold. MOCC redefines the agreement threshold by equation (2), where the optimised agreement separation is 0.6 (the original value is 0.5). The details of the experiments and analysis of how 0.6 was derived are described in Section 5.1. $Max(A)$ and $Min(A)$ indicate the maximal agreement value and the minimal agreement value respectively in the agreement matrix A.

$$\beta = [Max(A) - Min(A)] \times 0.6 + Min(A) \tag{2}$$

$$A' = A - \beta \tag{3}$$

Equation (3) shows the definition of the weighted agreement matrix A'. It means that the A' is given by subtracting β from each element of A. The leading diagonal elements of A' will never be used, and we set them to be zeros.

The agreement fitness function is shown as equation (5), where $f'(C_i)$ is defined by equation (4).

$$f'(C_i) = \begin{cases} \sum_{k=1}^{S_i-1} \sum_{q=k+1}^{S_i} A'_{C_{ik}C_{iq}}, & S_i > 1 \\ 0, & otherwise \end{cases} \tag{4}$$

$$f(C) = \sum_{i=1}^{m} f'(C_i) \tag{5}$$

The function $f'(C_i)$ calculates the agreement fitness of the ith cluster of the clustering arrangement C. The variable m is the number of clusters in C. Each $A'_{C_{ik}C_{iq}}$ indicates the corresponding weighted agreement value, which related to the instances k and q in the ith cluster C_i, in the weighted agreement matrix A'. The variable S_i denotes the size of the ith cluster (i.e. the number of instances in the ith cluster).

3.2 The Probability Based Solution Generator

In the original CC algorithm, the solution generator has three different operators, which are Move (moving an instance form one cluster to another cluster), Merge (merging two random clusters to be one cluster), and Split (splitting one cluster to be two clusters) [1]. Each of these three operators is randomly chosen to generate a clustering solution (i.e. each operator can be chosen with an equal probability).

The proposed MOCC algorithm also adopts these three operators for the solution generator but with different probabilities. The purpose of the Move operator is to find better solutions by estimating the neighbours of current solutions. The Merge and Split operators are used for avoiding results to be stuck at local maximums. The total probability of these three operators is 1. We set the operation probability of Move to be 80%, Merge to be 10% and Split to be 10%.

The Simulated Annealing algorithm in CC starts from a random clustering solution to seek for the optimal solution. For analysing high-dimensional data sets, it might cost much time to get convergence if the random solutions were very bad. Therefore we adopt a well-known clustering algorithm, K-means, to generate initial solutions for SA in MOCC. The number of clusters for K-means can be roughly inferred from the input clustering matrix. In general, the adoption of K-means aims to keep the quality of initial solutions at a relative good level.

3.3 The Multi-Optimisation Section

MOCC implements SA to generate more accurate clustering results by a multi-optimisation framework. This framework integrates the Agreement Fitness Evaluation (AFE) with the Clustering Quality Evaluation (CQE) to evaluate solutions during the optimisation search of SA. The purpose of the AFE is to seek for the solutions that have the maximal agreement fitness. The CQE uses Separation Score (SS) as a criterion to evaluate the clustering quality. With the same number of clusters, the value of SS is expected to be as high as possible. During the optimisation search of SA, candidate solutions will be accepted if they are eligible for both criteria otherwise will be discarded by a probability based on the annealing temperature.

Table 1. The MOCC algorithm

Input:	The maximal number of clusters, K; A matrix of results of a set of input methods, $Matr$; Original Data set Matrix, X; Initial Temperature, $T0$; Number of Iterations, $Iter$; Cooling Rate, $cool$.
(1)	Construct the Agreement Matrix, A
(2)	Setup the agreement threshold β according to Equation (2)
(3)	Generate the weighted agreement matrix A'
(4)	Use the K-means algorithm to generate the initial clustering arrangement C for SA
(5)	Calculate the Separation score, S, according to Equation (1)
(6)	Calculate the Agreement Fitness function (according to Equation (5)) to get f.
(7)	$Ti= T0$
(8)	For $i = 1$ to $Iter$
(9)	Use Solution Generator to produce a new arrangement C_{new}
(10)	Re-calculate Equation (5) to get a value f_{new}; and re-calculate Equation (1) to get a value S_{new}
(11)	If $f_{new} < f$ or $S_{new} < S$
(12)	Calculate probability p: {If $f_{new} < f$, $p= \exp[(f_{new} -f)/Ti]$;
(13)	Else, $p= \exp[(S_{new} - S)/Ti]$.
(14)	End If }
(15)	If $p > $ random$(0,1)$
(16)	Accept the new clustering arrangement.
(17)	End If
(18)	Else
(19)	Accept the new clustering arrangement.
(20)	End If
(21)	$Ti= cool \times Ti$
(22)	End For
Output:	An optimised clustering arrangement.

The structure of the multi-optimisation framework is described in Table 1 from Step 9 to Step 20. Firstly, a new candidate solution C_{new} is generated by the solution generator. Secondly, the new agreement fitness f_{new} and the new separation score S_{new} are calculated for C_{new}. After that, we compare f_{new} and S_{new} with the previous agreement fitness f and separation score S respectively. If $f_{new} < f$ and $S_{new} > S$, the acceptance probability p of C_{new} is calculated according to Step 12; if $f_{new} > f$ and $S_{new} < S$, p is calculated according to Step 13; if $f_{new} < f$ and $S_{new} < S$, p can be calculated either according to Step 12 or Step 13. Finally, the new candidate solution C_{new} will be accepted based on the p. However, if $f_{new} > f$ and $S_{new} > S$, C_{new} will be accepted unconditionally.

The profile of the MOCC algorithm is described in Table 1. First of all, a weighted agreement matrix A' is generated based on the Agreement Matrix A and the agreement threshold β (Step 1-3). And then an initial clustering arrangement C is produced for SA (Step 4). After that, the agreement fitness f and the separation score S are calculated for C according to equations (5) and (1) respectively (Step 5-6). Finally, SA uses Probability Based Solution Generator to obtain new candidate arrangements, and implements the multi-optimisation framework to seek for an optimal clustering arrangement (Step 7-22).

4   Experimental Data Sets

We evaluate our method, MOCC, against 15 different data sets. Since we not only compare the performance between MOCC and CC but also observe the performance of MOCC and CC between different data sets, the sizes of these data sets have been made to be approximately the same. We chose about 100 instances from each of the original data sets. The value of 100 was chosen as it is not too trivially small, but small enough to allow the large number of experiments run in this paper to complete in a feasible amount of time. For some of the data sets, the instances were randomly chosen from the corresponding original data sets. Some of the data sets were simply formed by the first 100 instances in the original data sets. Because of the stochastic nature of the MOCC and CC methods, we run these methods 10 times, and look at the average performance.

Ten of these data sets, which are Ecoli, Glass, Iris, Lung Cancer, Poker Hand, Soybean, WDBC, Wine, Yeast_2 and Zoo, were downloaded from the UCI Machine Learning Repository database. The information of these experimental data sets is presented in Table 2. The details of the original data sets can be found from the UCI website: (http://archive.ics.uci.edu/ml/datasets.html).

The other five data sets are ASC, Malaria, Normal, VAR and Yeast. The ASC (Amersham Score Card) was introduced and used in [1]. It is a set of multiply repeated control element spots, which using the Amersham Score Card to probe on the Human Gene clone set arrays of Human Genome Mapping Project [1]. This data set has 108 genes/probe elements, which are clustered into 15 clusters. The Malaria microarray data set [9] has fourteen clusters where the genes have the same function within each cluster and different functions between clusters. We choose the first one hundred instances for our experiments. VAR (Vector Auto-Regressive) [14] is a synthetic data set which was obtained from a vector autoregressive process [15]. The total number of

Table 2. Fifteen different data sets

Data Sets	Instances Chosen from the Original Data Sets	Number of Instances	Number of Attributes	Number of Clusters
ASC	1 - 108	108	23	15
Ecoli	CP(1-12); IM(1-12); PP(1-12); and all instances of the rest clusters	100	7	8
Glass	1-16; 71-86; 147-214	100	9	6
Iris	26 - 125	100	4	3
Lung Cancer	1-32; deleted the attributes with "?"	32	54	3
Malaria	1 - 100	100	48	12
Normal	1 - 100	100	20	14
Poker Hand	1 - 100	100	10	8
Soybean	1-30; 71-140	100	35	7
VAR	1 - 100	100	30	14
WDBC	51 - 150	100	30	2
Wine	51 - 150	100	13	3
Yeast	1-20; 68-97; 203-222; 278-297; 330-349	100	17	5
Yeast_2	1 - 100	100	8	8
Zoo	1 - 101	101	16	7

clusters is 60. In each cluster, the number of instances varies from 1 to 60. In this paper, we choose the first one hundred instances of the VAR data set for our experiments. The Normal data set has the same cluster structure as the VAR data set. The difference is that Normal is generated from multivariate normal distribution. The Yeast data set was used in [13]. We choose the first twenty instances from each cluster in the original data set to form the experimental data set. The information of these five data sets is also presented in Table 2.

5 Results and Discussion

5.1 Optimising Agreement Separation for the Agreement Threshold β

According to the definition of the agreement threshold β in equation (2), it is clear that the $Max(A)$ and $Min(A)$ are constants for a specific agreement matrix. The β is therefore only affected by the agreement separation. In order to seek for a comparative optimal agreement separation, we use fifteen data sets to test CC on a number of different agreement separations. The numerical range of the agreement separation is between 0 and 1 (the values 0 and 1 are not included because they have no meaning for β), consequently we analyse a series of agreement separations that varies from 0.1 to 0.9 in steps of 0.1.

The nine agreement separations were tested on fifteen data sets (described in Section 4). We use WK to indicate the clustering accuracy, and the results are illustrated in Fig. 1(a). The solid line with two arrows indicates the range between maximal and minimal WK scores for each agreement separation. The broken line links the mean WK scores for the nine agreement separations. We can see that the agreement separation 0.6 has the highest mean WK score. In addition, PMCC is used to indicate the Fitness-WK correlation for these separations. The mean Fitness-WK correlation values (linked by a broken line) are displayed in Fig. 1(b). It is clear that the agreement separation 0.4 has the highest mean Fitness-WK correlation. By comparing Fig. 1 (a) and (b), we note that the separation 0.4 has a much lower mean WK score than 0.6, 0.7 and 0.8, but the mean Fitness-WK correlation of the separation 0.6 is only slightly lower than the one of 0.4. Therefore, 0.6 seems to be a good agreement separation.

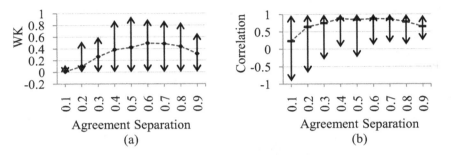

Fig. 1. (a) The WK comparison of nine different agreement separations tested on fifteen data sets; (b) The Fitness-WK correlation comparison of nine different agreement separations tested on fifteen data sets.

In order to validate the inference further, we analyse the distribution of WK scores for each agreement separation. According to the definition of WK, the agreement of two clustering arrangements is almost perfect when the WK score is greater than 0.8 and very poor when the WK score is less than 0.2 [12]. Therefore we count and compare the numbers of WK scores, which are between 0.2 and 0.8, for each separation. Based on Fig. 1(a), the numbers of WK scores between 0.2 and 0.8 for different separations are listed in Table 3. The corresponding mean WK scores (which are different from those in Fig. 1(a)) are illustrated in Fig. 2. It is apparent that the agreement separation 0.6 still has the highest mean WK score, which shows that 0.6 is an ideal separation.

Table 3. Number of WK scores between 0.2 and 0.8 for each of the nine separations

Agreement Separation	0.1	0.2	0.3	0.4	0.5	0.6	0.7	0.8	0.9
Number of WK Scores Between 0.2 and 0.8	0	4	7	9	8	9	7	10	9

5.2   Results and Comparison between CC and MOCC

CC was compared with a number of ensemble clustering methods implemented within the Clue [3] R package in [2]. Therefore, within this paper, we will compare our novel method, MOCC, against CC only, to simplify the number of experiments being

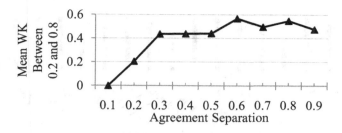

Fig. 2. The comparison of mean WK scores that corresponding to Table 3

evaluated. The experiments were achieved by testing both CC and MOCC on each of the fifteen data sets respectively. Eight individual clustering algorithms were chosen to generate input clustering matrixes for CC and MOCC. The eight clustering algorithms are listed as follows: PAM (Correlation); PAM (Euclidean); Affinity Propagation (Correlation); Affinity Propagation (Euclidean); Hierarchical (Average, Correlation); Hierarchical (Average, Euclidean); Hierarchical (Complete, Correlation); Hierarchical (Complete, Euclidean). The Number of Clusters, which is one of the parameters of these algorithms, is set according to Table 2. For the CC and MOCC algorithms, we do not need to set the number of clusters as an input parameter. We analyse the experimental results and compare the performance between MOCC and CC in two aspects.

Firstly, we look at the accuracy of the clustering results. We still use WK to evaluate the clustering accuracy. Fig. 3 displays the WK score comparison between CC and MOCC. In general, we can say MOCC generated better results than CC (especially for the WDBC and Zoo). We take WDBC as an example, the WK score of CC is zero (it means the result has no agreement related to the true clustering arrangement), but MOCC has a score that is over 0.6 (it means the result has a good agreement with the true clustering arrangement). The reason may be the WDBC has a very small number of clusters (2 clusters) but with a very high number of attributes (30 attributes). For the Poker Hand data set, it is interesting to observe that both MOCC and CC have very low WK scores, which means this data set is not suitable for clustering analysis. In order to highlight the overall difference of WK scores between CC and MOCC, we calculated the mean values of WK, which displayed in Fig. 5 (a). It is clear that MOCC has a better mean WK score than CC.

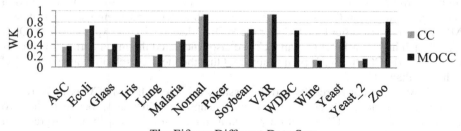

The Fifteen Different Data Sets

Fig. 3. The WK scores comparison of the results between CC and MOCC tested on fifteen different data sets

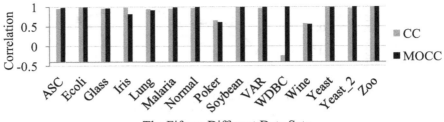

The Fifteen Different Data Sets

Fig. 4. The correlation comparison of the results between CC and MOCC tested on fifteen different data sets

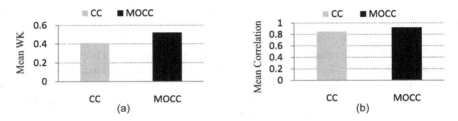

Fig. 5. (a) The comparison between CC and MOCC by the mean WK of the fifteen different data sets; (b) The comparison between CC and MOCC by the mean correlation of the fifteen different data sets.

Secondly, we use PMCC to evaluate the correlation between the WK score and the value of the objective function. The correlation is expected to be as high as possible so that the algorithm can generate more accurate (higher WK scores) results when maximising the value of the objective function. For CC, the objective function value is the agreement fitness value. For MOCC, we define the objective function value to be the sum of the agreement fitness and the separation score. Fig. 4 illustrates the correlation comparison between CC and MOCC. It is clear that MOCC and CC have similar correlation scores for most of the data sets. However, when we compare Fig. 4 with Fig. 3, it is interesting to see that MOCC has better WK scores than CC if it has higher correlation scores, and MOCC still has better WK scores even its correlation scores are lower than CC for some data sets such as Iris, and Lung Cancer. It means MOCC performed better than CC. For the WDBC data set, Fig. 4 explains why MOCC has a much higher WK score than CC. We also calculated the mean correlation values of CC and MOCC. In Fig. 5 (b), it is apparent that the mean correlation value of MOCC is higher than CC's.

Following the above results and discussion, it is clear that MOCC is more robust than CC in general. In addition, the most important difference between the two algorithms is that MOCC is not only seeking for the solutions with high Agreement Fitness, but also impelling the results as close to the true clustering arrangement as possible by the multi-optimisation framework.

6 Conclusions

We have presented a novel Consensus Clustering framework called MOCC, which uses an optimised Agreement Separation and a Multi-Optimisation structure to enhance the ensemble clustering accuracy. Within the Multi-Optimisation structure, a Probability Based Solution Generator is developed to ameliorate the performance of global optimising. The results show that MOCC has better stability of clustering and performs clearly better than original CC.

MOCC combines the Separation Score (SS) index with the Agreement Fitness Evaluation (AFE) for the Multi-Optimisation structure, but the combination is not limited to the SS index. This paper displays the promising preliminary results, and opens up many further research opportunities. For example, to combine other appropriate clustering validity evaluation indexes with the AFE would be a good way forward for further research.

In addition, the purpose of multi-optimisation framework in MOCC is similar to the function of the Multi-Objective Simulated Annealing (MOSA) [16] algorithm, which aims to generate global optimal results by optimising multiple objectives. We will, therefore, further explore these two multi-optimisation structures to see whether we can benefit from MOSA.

References

1. Swift, S., Tucker, A., Vincotti, V., Martin, N., Orengo, C., Liu, X., Kellam, P.: Consensus Clustering and Functional Interpretation of Gene-Expression Data. Genome Biology 5(11), R94.1–R94.16 (2004)
2. Hirsch, M., Swift, S., Liu, X.: Optimal Search Space for Clustering Gene Expression Data Via Consensus. Journal of Computational Biology 14(10), 1327–1341 (2007)
3. Hornik, K.: A Clue for Cluster Ensembles. Journal of Statistical Software 14(12) (2005), http://www.jstatsoft.org/v14/i12/
4. Jain, A.K., Murty, M.N., Flynn, P.J.: Data Clustering: A Review. ACM Computing Surveys 31(3), 264–323 (1999)
5. Lv, T., Huang, S., Zhang, X., Wang, Z.: Combining Multiple Clustering Methods Based on Core Group. In: Second International Conference on Semantics, Knowledge, and Grid (SKG 2006), p. 29 (2006)
6. Strehl, A., Ghosh, J.: Cluster Ensembles — a Knowledge Reuse Framework for Combining Multiple Partitions. J. Mach. Learn. Res. 3, 583–617 (2003)
7. Berkhin, P.: Survey of clustering data mining techniques. Technical Report, Accrue Software (2002)
8. Topchy, A., Jain, A.K., Punch, W.: Clustering Ensembles: Models of Consensus and Weak Partitions. IEEE Transactions on Pattern Analysis and Machine Intelligence 27(12), 1866–1881 (2005)
9. Bozdech, Z., Llinás, M., Pulliam, B.L., Wong, E.D., Zhu, J., DeRisi, J.L.: The Transcriptome of the Intraerythrocytic Developmental Cycle of Plasmodium Falciparum. PLoS Biology 1, 85–100 (2003)
10. Chen, G., Banerjee, N., Jaradat, S.A., Tanaka, T.S., Ko, M.S.H., Zhang, M.Q.: Evaluation and Comparison of Clustering Algorithms in Analyzing ES Cell Gene Expression Data. Statistica Sinica 12, 241–262 (2002)

11. Snedecor, G.W., Cochran, W.G.: Statistical Methods, 7th edn., pp. 175–178. Iowa State Press, Ames (1980)
12. Viera, A.J., Garrett, J.M.: Understanding Interobserver Agreement: the Kappa Statistic. Fam. Med. 37, 360–363 (2005)
13. Yeung, K.Y., Ruzzo, W.L.: Principal Component Analysis for Clustering Gene Expression Data. Bioinformatics 17(9), 763–774 (2001)
14. Lütkepohl, H.: New Introduction to Multiple Time Series Analysis, 1st edn., p. 49 (2007) ISBN:978-3-540-26239-8
15. Sims, C.A.: Macroeconomics and Reality. Econometrica, 1–48 (1980)
16. Bandyopadhyay, S., Saha, S., Maulik, U., Deb, K.: A Simulated Annealing based Multi-objective Optimization Algorithm: AMOSA. IEEE Transactions on Evolutionary Computation 12(3), 269–283 (2008)

Improving Time Series Forecasting by Discovering Frequent Episodes in Sequences

Francisco Martínez-Álvarez[1], Alicia Troncoso[1], and José C. Riquelme[2]

[1] Area of Computer Science, Pablo de Olavide University of Seville, Spain
{fmaralv,ali}@upo.es
[2] Department of Computer Science, University of Seville, Spain
riquelme@lsi.us.es

Abstract. This work aims to improve an existing time series forecasting algorithm –LBF– by the application of frequent episodes techniques as a complementary step to the model. When real-world time series are forecasted, there exist many samples whose values may be specially unexpected. By the combination of frequent episodes and the LBF algorithm, the new procedure does not make better predictions over these outliers but, on the contrary, it is able to predict the apparition of such atypical samples with a great accuracy. In short, this work shows how to detect the occurrence of anomalous samples in time series improving, thus, the general forecasting scheme. Moreover, this hybrid approach has been successfully tested on electricity-related time series.

Keywords: Time series, forecasting, outliers.

1 Introduction

This work provides a new methodology to forecast time series and, in addition, to predict the apparition of outliers. The analysis of temporal data and the forecast of future values of time series are among the most important problems that data analysts face in many fields, ranging from finance and economics, to production operations management or telecommunications. A *forecast* is a prediction of some future events.

The proposed approach is specifically framed in electricity prices time series forecasting, which is a difficult task due to the nonconstant mean and variance and significant outliers typically present in these series.

Thus, the combination of two different techniques are proposed to fulfill this goal. The first one is a general-purpose forecasting algorithm introduced in [9], called LBF. The authors obtained a previous labeling of the elements forming the time series by means of clustering techniques. The forecasting process was performed by using just the information provided by the clustering. Thus, the values of the elements in datasets were discretized and, as a result, the sequence of real values was transformed in a sequence of discrete values or labels. These labels were used to predict the future behavior of the time series, avoiding the

N. Adams et al. (Eds.): IDA 2009, LNCS 5772, pp. 357–368, 2009.
© Springer-Verlag Berlin Heidelberg 2009

use of the real values until the last step of the process. The results returned by the algorithm, however, were not labels but the real values.

The algorithm introduced in [10] is inserted in the general scheme of forecasting with the aim of dealing with the presence of outliers. Concretely, they proposed an algorithm called Q-epiMiner that was, in fact, an improvement of the well-known serial episodes [8]. The main achievement of the Q-epiMiner algorithm was to characterize sequences of similar behavior over all occurrences, as well as providing a tree structure to organize these sequences.

Therefore, the discovery of frequent episodes is used in order to determine possible candidates to be outliers when using the LBF algorithm to forecast time series. The general outline of the new proposed approach is shown in Fig. 1.

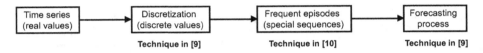

Fig. 1. General outline of the proposed methodology

The rest of the paper is organized as follows. The latest works related to time series forecasting and discovery of outliers are gathered in Section 2. Section 3 presents a brief explanation of the two existing and used algorithms in which the new approach is based on. Section 4 introduces the proposed methodology, showing how the two existing techniques are combined in order to improve the forecasting process. Section 5 shows the results obtained for the electric energy market of Spain for the year 2006, including measures of the quality of them. Finally, Section 6 summarizes the main conclusions achieved and provides clues for future work.

2   Related Work

Regarding to time series forecasting, two mixed models were proposed to obtain the forecasting of the prices for two different prediction horizons in [3]. The first one forecasted electricity prices for each of the 24 hours of the next day using ARIMA models, while the second model computed the predictions by using Bayesian information criteria.

A modification of the nearest neighbors methodology was proposed in [13]. To be precise, the approach presented a simple technique to forecast next-day electricity market prices based on the weighted nearest neighbors methodology.

Li et al. proposed a forecasting system immersed in a grid environment in [6]. In this paper, a fuzzy inference system –adopted due to its transparency and interpretability– and time series methods were proposed for day-ahead electricity price forecasting.

The authors in [12] proposed an artificial neural network-based approach to forecast the energy price in the Spanish market. Thus, a novel training method was presented and applied to the multilayer perceptron in order to improve the forecasting process.

The apparition of outliers in time series has also been widely discussed in the literature. Thus, the authors in [1] proposed a technique to detect outliers in data whose generation was difficult to model. They assumed that the correlation among data close in time is higher than those farther apart.

Another method for detecting outliers was proposed in [7], in which the authors considered two different sources of outliers –additive and innovation– in autorregressive moving-average time series. Concretely, they proposed the application of two different procedures associated to each source simultaneous and coherently.

The occurrence of spike prices (prices significantly higher than the expected value, i. e., outliers) is an usual feature associated with electricity prices-related time series. With the aim of dealing with this peculiarity, the authors in [14] proposed a data mining framework based on both support vector machines and probability classifiers.

The work described in [5] searched for patterns in electricity prices data in order to verify how the outliers may modify the behavior of such prices. To fulfill this goal, they used Box and Jenkins models, Discrete Fourier Transform series smoothing and GARCH approaches.

Also, the work in [4] discussed the use of the fractal theory to forecast the electricity price time series. For this purpose, a forecasting model based on improved fractal was built and solved to forecast short–term electricity price time series.

3   Fundamentals

The proposed approach is a combination of two existing techniques. Thus, this section provides the mathematical fundamentals underlying to both LBF (Subsection 3.1) and Q-epiMiner algorithms (Subsection 3.2). A more detailed explanation can be found in [9] and [10], respectively.

3.1   Time Series Forecasting: The LBF Algorithm

The LBF algorithm was initially presented in [9]. Given the hourly prices recorded in the past, up to day d, the forecasting problem aims to predict the 24 hourly prices corresponding to day d+1.

Let $P_i \in R^{24}$ be a vector composed of the 24 hourly energy prices corresponding to a certain day i

$$P_i = [p_1, p_2, \ldots, p_{24}]. \tag{1}$$

Let L_i be the label of the prices of the day i obtained as a previous step to the forecasting by using a clustering technique. Let S_W^i the subsequence of labels of the prices of the W consecutive days, from day i backward, as follows:

$$S_W^i = [L_{i-W+1}, L_{i-W+2}, \ldots, L_{i-1}, L_i] \tag{2}$$

where the length of the window, W, is a parameter to be determined.

The LBF algorithm first searches for the subsequences of labels which are exactly equals to S_W^d in the data base, providing the equal subsequences set, ES, defined by the equation,

$$ES = \left\{ \text{set of indexes } j \text{ such that } S_W^j = S_W^d \right\} \tag{3}$$

In case of not finding any subsequence in data base equal to S_W^d, the procedure searches for the subsequences of labels which are exactly equals to S_{W-1}^d. That is, the length of the window composed of the subsequence of labels is decreased.

According to the LBF approach, the 24 hourly prices of day $d+1$ are predicted by averaged the prices of the days succeeding those in ES. That is,

$$P_{d+1} = \frac{1}{\text{size}(ES)} \cdot \sum_{j \in ES} P_{j+1} \tag{4}$$

where size(ES) is the number of elements belonging to the set ES. Afterwards, LBF algorithm outputs need to be de-normalized to generate the desired forecasted values.

When the horizon of prediction is greater than one day, the following tasks have to be carried out. First of all, the real values of the predicted sample are linked to the whole dataset. Second, the clustering process is repeated with the enlarged dataset and, finally, the window size is re-calculated and the prediction step is performed.

3.2   Frequent Episodes in Sequences: The Q-epiMiner

The algorithm introduced in [10] analyzes data events or sequences in order to find episodes. Formally, a sequence is an ordered list of events, where an event is identified by the pair $ev = < date, eventType >$. The ordered occurrence of events is called serial episode and represented by:

$$E = [ev_1, ev_2, ..., ev_n] = [< d_1, t_1 >, < d_2, t_2 >, ..., < d_n, t_n >] \tag{5}$$

where n is the number of events forming an episode.

Thus, the algorithm is able to handle with three time constraints, provided as input data: the minimum time span between two events or gap_{min}, the maximum time span between two events or gap_{max} and the maximum time span between the beginning and the end of an episode or $windowSize$. In this way, an episode has to simultaneously satisfy three constraints: the gap_{min} has to be greater or equal to a given threshold, the gap_{max} has to be lesser or equal to a given threshold and the $windowSize$ has to be lesser or equal to a given threshold. The thresholds are set depending on the requirements of the application.

The rules used in the algorithm are computed where the antecedent is a serial episode and the consequent contains only one event type. Then, a list of the positions in the data sequence L_E of a particular episode is built. Concretely, the list contains the time stamps associated with the events comprised in the episode and they are sorted by increasing values.

The next step consists in evaluating the whole sequence of events and the L_E. From this evaluation, the algorithm provides a set of tuples $< ev, L_{ev} >$, where L_{ev} is the list of locations of the event occurrences and ev is an event of the episode E.

Finally, the standard prefix-based strategy [2] is used for the overall enumeration since it fits well with both episodes extraction and the use of the sorted lists. In other words, an episode is used as a prefix and expanded in order to obtain new episodes.

4   Methodology

This section explains the methodology proposed to improve the forecasting process provided by the LBF algorithm. The discovery of frequent episodes is included in the aforementioned algorithm as a crucial step for outliers detection.

Thus, the proposed methodology is divided into two phases clearly differentiated. First of all, the LBF algorithm is trained with the datasets under study. Second, the predictions with the highest error rates made during the training are analyzed by means of frequent episodes techniques. From this analysis, the days likely to be outliers will be determined and not considered in the prediction process.

4.1   Combining the LBF Algorithm with Frequent Episodes

The value of two parameters have to be determined in the LBF process, K and W. With regard to the number of clusters (K), the proposed approach acts exactly equal to what was proposed in [9] (see Section II.C). However, it is important to remark that the length of the window (W) is slightly different calculated from how it was proposed in the original paper. In practice, W is calculated by means of cross-validation.

Concretely, the n−fold cross-validation is used in this work to obtain the optimal value of W. In n−fold cross−validation, the original sample is partitioned into n subsets. Of the n subsets, a single subset is retained as the validation data for testing the model, and the remaining $n-1$ subsets are used as training data. The cross-validation process is then repeated n times (the folds), with each of the n subsets used exactly once as the validation data. The n results from the folds are then averaged (even if some authors prefer a combination of them) to produce a single estimation. The advantage of this method over repeated random sub-sampling is that all observations are used for both training and validation, and each observation is used for validation exactly once.

Twelve folds have been created in this work ($n = 12$) for all datasets, where each fold represents a month. Consequently, the training set consists of one year.

The 12−fold cross−validation is then evaluated. The forecasting errors are calculated in every fold by varying the length of W. These monthly errors are denoted by $e_{month}\{W = j\}$ for $j = 1 \ldots W_{max}$, where $W_{max} = 10$ −no longer sequences were found in datasets. Then, the average errors are calculated for each window size as follows,

$$\mathbf{e}_i = \frac{\sum_{i=1}^{n} e_{month}\{W = i\}}{n} \tag{6}$$

where $n = 12$ and $month = \{Jan, \ldots, Dec\}$.

The W selected is the one that minimizes the average error corresponding to the 24 folds (months) evaluated.

$$W = \arg\min\{\mathbf{e}_i\} \text{ with } i = 1, \ldots, W_{max} \tag{7}$$

This modification has a simple justification. With the former methodology (see Section II.D in [9]), only one test set was evaluated and, consequently, the number of episodes found may be limited and not conclusive. However, with the application of n−fold cross-validation, the number of sequences is increased n times providing thus n training sets instead of one.

It is now −just after the training step and before the prediction process− that the discovery of frequent episodes plays an important role. Hence, the apparition of anomalous days is intended to be predicted. Once the $n-$ fold cross−validation is applied, the number of sequences generated can be calculated as:

$$\#S(W, \overline{FL}) = n\overline{FL} - W + 1 \tag{8}$$

where n is the number of folds, $\overline{FL}$ the average fold length and W the length of the sequence (or window) considered.

Note that the maximum number of dissimilar sequences that can be generated is bounded by:

$$N_{max}(K, W) = K^W \tag{9}$$

which will be a number typically much higher than the sequences found in the training sets ($N_{max} >> \#S(W, \overline{FL})$).

The episodes aimed to be found are those which generate a prediction error greater than the average error in the cross−validation process. For this reason, a set of events that satisfies $e_j > \min\{\mathbf{e}_i\}$, for $i = 1, \ldots, W_{max}$ and $j = 1, \ldots, n\overline{FL}$, is constructed. This set, CS, gathers all the candidates events to be preceded by an episode precursor of outliers.

Nevertheless, not all these candidates have the same probability to be outliers since the associated errors range from values near to the mean error (these candidates should be finally discarded by the approach) to values significantly high. For this reason, each candidate is co-labeled by using clustering techniques, concretely, the K-means algorithm. The decision on how many clusters have to be created is always an open question and many indices might be used. However,

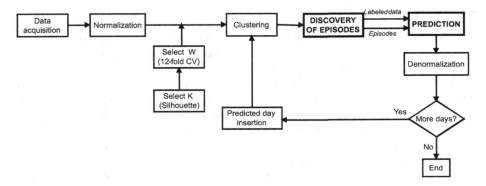

Fig. 2. Illustration of the proposed methodology

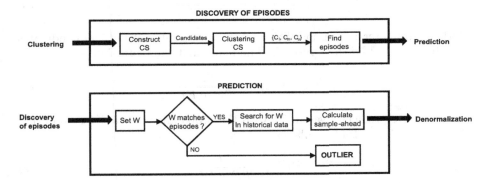

Fig. 3. Detail of both *discovery of episodes* and *prediction* steps

it is worthless to have a large number of clusters and for this reason only three conceptual classes will be created: lower (C_l), central (C_c) and upper (C_u) errors. An a priori reasoning reveals that those candidates belonging to C_u must be more probable to be preceded by sequences generators of outliers than the candidates in C_l or C_c. Results corresponding to each cluster of data will be separately analyzed in Section 5.

The next step consists in computing the episodes (concrete sequences of labels) occurred before the candidates in order to determine the apparition of an outlier. Hence, the approach has to decide which sequences preceding the candidates are the episodes causing errors greater than the expected average error. Then, the sequences that only appear before the candidates are considered frequent episodes and therefore preceding outliers. Fig. 2 illustrates the whole process of prediction when the frequent episodes are included in the LBF algorithm. In addition, the steps corresponding to *discovery of episodes* and *prediction* are further detailed in Fig. 3.

4.2   Parameters of Quality

The parameters used in order to measure the accuracy of the approach are now introduced. Note that in subsequent equations, true positives or TP is the number of candidates that indeed were preceded by episodes that caused errors greater than the average; true negatives or TN is the number of sequences found before a candidate that was properly discarded; false positives or FP is the number of candidates whose preceding episodes were erroneously considered to be causing of errors greater than the average and, finally, false negatives or FN is the number days not considered candidates and eventually preceded by episodes causing errors greater than the average.

According to these definitions, the sensitivity is the probability to detect a frequent episode as precursor of outliers. Its formula is defined as follows:

$$Sensitivity = \frac{TP}{TP + FN} \tag{10}$$

Other parameter is the specificity which is the ratio of outliers candidates properly discarded by the approach. The mathematical expression is:

$$Specificity = \frac{TN}{TN + FP} \tag{11}$$

The positive predictive value (PPV) is the probability that a detected outlier is indeed a real one. Its formula is:

$$PPV = \frac{TP}{TP + FP} \tag{12}$$

Finally, the negative predictive value (NPV) is the probability that a discarded candidate to be outlier was not indeed a real one. Its formula is:

$$NPV = \frac{TN}{TN + FN} \tag{13}$$

5   Results

In order to prove that the LBF works properly over different datasets, the authors in [9] considered several public electricity prices time series. The new approach is applied on the Spanish electricity price time series (OMEL) [11].

This section is structured as follows. First, the training of the LBF is presented, obtaining thus the adequate values for the parameters K and W. The results provided in this step are, then, analyzed by means of frequent episodes techniques intending to find those patterns in the historical data that perform the worst predictions. From this analysis, some days will become candidates to have an anomalous behavior and, consequently, have a higher error prediction. Finally, the validity of considering these days candidates to be outliers is discussed.

5.1   Discovering Frequent Episodes in Time Series

The forecasting process is applied for the year 2006, with a historical data of a length of one year and with a horizon of prediction of one month. Given this situation, every time a month is forecasted the training set changes. When January 2006 is forecasted, the training set comprises the whole year 2005. However, when February 2006 is forecasted the historical data ranges from February 2005 and January 2006, and so on.

The results of January 2006 are now described since the explanation of the remaining eleven months is analogous. As for the LBF, the number of clusters to be generated as well as the length of the window to be searched for are calculated according to the methodology presented in Section 4.1. Thus, this pair of parameters are equal to: $(K, W) \leftarrow (4, 5)$.

Consequently, the number of sequences generated during the training step is $\#S(W, \overline{FL}) = 361$. However, many sequences appeared repeatedly and the final number of different sequences were 43. As the maximum number of possible sequences is $N_{max}(K, W) = 1024$, the aforementioned number of sequences represent the 4.19% of the potential.

With regard to the Q-epiMiner, the parameters are set to $gap_{max} = 1$, $gap_{min} = 1$ and $windowSize = W$ in order to adapt its application to the particular problem tackled in this work. Also note, that the events are the labels generated during the LBF process, the date is the day associated with such label and the type of event is the curve of prices associated to this day.

The CS can be now constructed. For this purpose, the $\mathbf{e}_i$ from (5) have to be considered since the candidates are those days belonging to the training set that obtained an error greater than $\mathbf{e}_i$. The value of the mean error, calculated according to the methodology in Section 4.1 is $\mathbf{e} = 2.23\%$. From the 365 days comprising the training set, 131 had an error greater than 2.23% so the constructed CS contains 131 candidates.

The candidates have to be classified by means of K-means, with $K = 3$ as discussed in Section 4.1. The obtained cutoff values were 3.78% and 5.69% for dividing classes C_l–C_m and C_m–C_u, respectively. From these cutoffs, the candidates were classified as follows: $98 \in C_l$, $25 \in C_m$ and $8 \in C_u$.

Once the candidates are selected and classified, the sequences that generated them are evaluated. From the candidates in C_l, 7 different sequences were found ($\#S_l = 7$); from the candidates in C_m, 4 ($\#S_m = 4$) and from the candidates in C_u, 2 ($\#S_u = 2$). This fact involves that from the 43 sequences found in the training set, only 13 caused errors greater than the average.

Finally, the number of episodes causing outliers are determined. From the sequences C_l, just one appeared exclusively as a precursor of an outliers. With reference to sequences in C_m, three out of four. And both two sequences in C_u were exclusive.

Table 1 summarizes the results for the twelve months of the year 2006.

Table 1. Training parameters, candidates distribution and episodes found for the year 2006

Month	Training K W e			Cutoff C_l–C_m C_m–C_u		Candidates $C_l(\#S_l)$ $C_m(\#S_m)$ $C_u(\#S_u)$			Sequences $\#S(W,\overline{FL})$ N_{max}	
January	4	5	2.23%	3.78%	5.69%	98(7)	25(4)	8(2)	362	1024
February	4	5	4.07%	5.21%	6.87%	87(6)	31(7)	5(2)	362	1024
March	4	5	6.30%	7.03%	7.66%	73(5)	16(3)	8(1)	362	1024
April	4	5	2.79%	3.83%	5.01%	103(9)	30(6)	6(3)	362	1024
May	4	5	7.51%	7.97%	9.43%	65(4)	51(6)	10(4)	362	1024
June	6	4	4.02%	5.42%	6.38%	97(6)	38(5)	4(0)	360	1296
July	5	5	4.98%	5.67%	6.13%	180(8)	27(3)	12(5)	361	3125
August	6	4	5.35%	6.20%	6.94%	101(8)	26(5)	9(4)	360	1296
September	6	4	6.24%	7.30%	8.29%	110(8)	25(5)	5(0)	360	1296
October	6	4	6.38%	7.31%	7.88%	108(7)	23(4)	6(1)	360	1296
November	6	4	8.97%	11.68%	13.57%	120(9)	40(6)	6(3)	360	1296
December	5	5	6.51%	7.93%	8.97%	169(10)	38(9)	10(3)	361	3125

5.2 Quantifying the Improvements Achieved

How the prediction is improved by not considering the days pointed by the episodes precursors to outliers found is shown in this subsection. To evaluate the accuracy of the methodology, different criteria may be taken into consideration. However, two parameters –the mean relative error (MRE) and its standard deviation (σ_{MRE})– are used in order to make a comparison with the results in [9].

Table 2 shows the results of the forecasting process performed by the LBF and the results when the episodes causing outliers were discovered and removed from datasets. Note that the approach improves the forecasting in all the datasets considered but for in April. This fact is due to the absence of episodes found when this month was forecasted.

The greater is the average error, the better works this hybrid methodology since outliers are usually involved in high error rates. Equally remarkable is the reduction in the σ_{MRE} from 0.27 to 0.23. Last but not least, a statistical measure of the accuracy of the proposed methodology is provided. The parameters used are the ones described in Section 4.2 and collected in Table 3. Note that all parameters are referred to the whole year 2006, that is, the numbers gather the twelve sets –months– forecasted.

Note that the number of sequences initially considered was 178 ($\sum_{i=1}^{12}\{\#S_{l_i} + \#S_{m_i} + \#S_{u_i}\} = 178$). From these 178 sequences, 150 were sequences that appeared solely in the subset of candidates in which they were found. Consequently, the system considered 150 episodes to be causing of outliers. From all of them, 145 were indeed episodes that preceded a day with a forecasting error greater than the average during the training. The other five did not cause large errors. That is: $TP = 145$ and $FP = 5$. None of the $178 - 150 = 28$ sequences discarded generated predictions with a high error, so: $TN = 28$. Finally, during the forecasting process there appeared 8 sequences which were not initially considered by the model and that eventually were trigger of outliers.

Table 2. Forecasting for the year 2006 in OMEL time series

Month	LBF		LBF + episodes	
	MRE	σ_{MRE}	MRE	σ_{MRE}
January	7.26%	0.25	6.98%	0.21
February	4.93%	0.19	4.28%	0.16
March	5.88%	0.22	5.07%	0.19
April	3.62%	0.18	3.62%	0.18
May	8.11%	0.21	6.95%	0.19
June	3.76%	0.24	3.67%	0.24
July	4.30%	0.23	4.25%	0.23
August	5.37%	0.34	4.66%	0.27
September	6.41%	0.31	6.40%	0.30
October	7.89%	0.29	7.00%	0.22
November	8.30%	0.40	7.12%	0.29
December	8.02%	0.36	7.61%	0.31
Average	6.15%	0.27	5.63%	0.23

Table 3. Statistical analysis of the method

Parameters	Values
TP	145
TN	28
FP	5
FN	8
Sensitivity	94.77%
Specificity	84.85%
PPV	96.67%
NPV	77.77%

6   Conclusions

The combination of two techniques has been used in order to forecast time series. The initial approach –the LBF– was based on finding similar patterns in time series. However, its application to any kind of time series revealed that there were some samples that cannot be properly forecasted since they showed a stochastic behavior.

The use of frequent episodes techniques is thus applied, not for providing an accurate prediction for these samples, but for indicating that it is reasonably probable that an outlier occurs. The method has been successfully tested on twelve sets of the Spanish electricity price time series.

Future work is directed towards finding not only the days that are going to present an anomalous behavior, but the days whose prediction is going to be specially accurate. In addition, a relaxation for the rule that decides if a given sequence is an episode or not is intended to be created.

Acknowledgments

The financial support from the Spanish Ministry of Science and Technology, project TIN2007-68084-C-02, and from the Junta de Andalucía, project P07-TIC-02611, is acknowledged. The authors also want to acknowledge to Ph. D. Rigotti the source code provided for the statistical analysis of frequent episodes used in this work.

References

1. Basu, S., Meckesheimer, M.: Automatic outlier detection for time series: an application to sensor data. Knowledge and Information Systems 11(2), 137–154 (2007)
2. Esparza, J., Heljanko, K.: Unfoldings: A Partial-Order Approach to Model Checking. Springer, Heidelberg (2008)
3. García-Martos, C., Rodríguez, J., Sánchez, M.J.: Mixed models for short-run forecasting of electricity prices: application for the Spanish market. IEEE Transactions on Power Systems 22(2), 544–552 (2007)
4. Herui, C., Li, Y.: Short-term electricity price forecast based on improved fractal theory. In: Prooceedings of the eighth IEEE International Conference on Computer Engineering and Technology, pp. 347–351 (2009)
5. Jabłońska, M.: Analysis of outliers in electricity spot prices with example of New England and New Zealand markets. PhD thesis, Lappeenranta University, Finland (2008)
6. Li, G., Liu, C.C., Mattson, C., Lawarrée, J.: Day-ahead electricity price forecasting in a grid environment. IEEE Transactions on Power Systems 22(1), 266–274 (2007)
7. Louni, H.: Outlier detection in ARMA models. Journal of Time Series Analysis 29(6), 1057–1065 (2008)
8. Mannila, H., Toivonen, H., Verkamo, A.I.: Discovery of frequent episodes in event sequences. Data Mining and Knowledge Discovery 1, 259–289 (1997)
9. Martínez-Álvarez, F., Troncoso, A., Riquelme, J.C., Ruiz, J.S.A.: LBF: A labeled-based forecasting algorithm and its application to electricity price time series. In: Prooceedings of the eighth IEEE International Conference on Data Mining, pp. 453–461 (2008)
10. Nanni, M., Rigotti, C.: Extracting trees of quantitative serial episodes. In: Džeroski, S., Struyf, J. (eds.) KDID 2006. LNCS, vol. 4747, pp. 170–188. Springer, Heidelberg (2007)
11. Spanish Electricity Price Market Operator, http://www.omel.es
12. Pino, R., Parreno, J., Gómez, A., Priore, P.: Forecasting next-day price of electricity in the Spanish energy market using artificial neural networks. Engineering Applications of Artificial Intelligence 21(1), 53–62 (2008)
13. Troncoso, A., Riquelme, J.C., Riquelme, J.M., Martínez, J.L., Gómez, A.: Electricity market price forecasting based on weighted nearest neighbours techniques. IEEE Transactions on Power Systems 22(3), 1294–1301 (2007)
14. Zhao, J.H., Dong, Z.Y., Li, X., Wong, K.P.: A framework for electricity price spike analysis with advanced data mining methods. IEEE Transactions on Power Systems 22(1), 376–385 (2007)

Measure of Similarity and Compactness in Competitive Space

Nikolay Zagoruiko

Institute of Mathematics of the Siberian Devision
of the Russian Academy of Sciences,
Pr. Koptyg 4, 630090 Novosibirsk, Russia
zag@math.nsc.ru

Abstract. The given work is devoted to measures of similarity which are used at discovering of empirical regularities (knowledge). The function of competitive (rival) similarity (FRiS) is proposed as a similarity measure for classification and pattern recognition applications. This function allows one to design effective algorithms for solving all basic data mining tasks, obtain quantitative estimates of the compactness of patterns and the informativeness of feature spaces, and construct easily interpretable decision rules. The method is suitable for any number of patterns regardless of the nature of their distributions and conditionality of training samples (the ratio of the numbers of objects and features). The usefulness of the FRiS is shown by solving a problems of molecular biology.

Keywords: similarity measure, pattern recognition, compactness, informativeness.

1   Introduction

The measure of similarity plays a key role in the classification of sets of objects (clustering) and in the recognition of the membership of objects in a particular class. The specificity of these tasks is that the measure of similarity used in them is a relative quantity that depends not only on the similarity of an object to a particular class but also on its similarity to other (competitive) classes. Such measures do not satisfy all properties of metric spaces. As a result, the pattern recognition task is solved in a space which we call a competitive space, and the similarity measure in this space is referred to as the **F**unction of **R**ival **S**imilarity (FRiS).

Using the FRiS, it is possible to obtain quantitative estimates of the compactness of patterns, and the value of the compactness can be used as a criterion of informativeness of feature spaces. The FRiS allows one to choose reference objects (stolps), construct effective decision rules, and censor training samples.

The usefulness of the FRiS is shown by solving a problem of molecular biology.

N. Adams et al. (Eds.): IDA 2009, LNCS 5772, pp. 369–380, 2009.
© Springer-Verlag Berlin Heidelberg 2009

2   Similarity Measures in Metric and Competitive Spaces

In metric space the similarity $S(a, b)$ between two objects a and b is usually estimated by a quantity that depends on the distance $R(a, b)$ between these objects. If the maximum pair wise distance between the objects of a set (the diameter of the set) is taken to be 1, then $S(a, b) = 1 - R(a, b)$. A great variety of this kind of similarity measure is described in the literature [1,2]. Both the distance $R(a, b)$ between the objects and the similarity $S(a, b)$ has the symmetry property: the similarity of object a to object b is equal to the similarity of object b to object a. In addition, the similarity between objects a and b does not depend on how much these objects are similar to other objects. This implies that similarity is treated as an absolute category.

However, in determining the membership of object z in one of two patterns A or B, it is important to know not only its similarity to standard a of pattern A, but also its similarity to standard b of competing pattern B. Hence, in pattern recognition, similarity is a relative rather than absolute category. To answer the question "How much z is similar to a?", it is necessary to know the answer to the question "In comparison to what?" An adequate measure of similarity should define the relative value of similarity depending on the features of the competitive environment.

All statistical algorithms use a competition between patterns. If in point z the aprioristic probability of the first pattern is equal P_1, and of the second pattern $- P_2$ for decision-making these sizes are compared and the decision in favor of the first pattern is accepted not on because P_1 exceeds some threshold, but on because $P_1 > P_2$. For cases when laws of distribution of patterns are not known or when the quantity of attributes on orders exceeds quantity of training objects to operate with probabilities it is not possible. Usually the distances $R(z, a_i)$ between control object z and standards (precedents) of patterns A_i, $i = 1, 2...K$ (K – quantity of patterns) are used and the decision is accepted in favor of an pattern, distance to which is smaller. So the algorithm "k the nearest neighbors" (kNN) [3] works for example. Here the relative measure of similarity estimated in a scale of the order is used.

A more complex similarity measure is employed in the RELIEF algorithm [4]. Here the size of relative similarity is measured in stronger absolute scale. The similarity of object z to pattern A in competition with pattern B is determined using the quantity

$$W_{A/B} = \frac{R_B - R_A}{R_{max} - R_{min}}, \tag{1}$$

where R_{max} and R_{min} are the maximum and minimum distances between all pairs of objects. Normalization on size $(R_{max} - R_{min})$ is represented to us unsuccessful. The measure of similarity does not consider local features of distribution of objects near object z. Limits of values of similarity are not limited. If the dispersion of pair distances is small, the size of similarity can be very big, down to infinity.

We formulate the following requirements for the relative measure F of competitive similarity of object z to object a:

1. The similarity measure should depend not on the nature of the distribution of the entire set of objects but on the features of the distribution of objects in immediate proximity to object z.

2. If the measure of similarity of object z to object a is estimated in competition with object b, $b \neq a$, and if a and z are indistinguishable from each other, the measure $F_{za/b}$ should have the maximum value equal to $+1$; if objects z and b are indistinguishable from each other, the measure $F_{za/b}$ should have the maximum negative value (maximum dissimilarity) equal to -1.

3. In the remaining cases, the measure of competitive similarity should have the form of a continuous function and take values between $+1$ and -1.

4. For identical distances R_a and R_b, object z will be equally similar or dissimilar to objects a and b and the similarity functions $F_{za/b}$ and $F_{zb/a}$ should be equal to 0.

The proposed function of competitive similarity FRiS meets all these requirements [5]. As a measure of competitive similarity it is possible to use versions of the function F of the form

$$F_{za/b} = \frac{R_b - R_a}{R_b + R_a} \tag{2}$$

We see that such similarity measures do not keep the positivity property. The presence of both positive and negative values is convenient for the interpretation of similarity. If the object z moves from object a to object b, one can speak of the greater similarity of object z to object a, their moderate similarity, identical similarity to objects a and b, and moderate and then greater negative similarity to a, i.e. one can speak of the dissimilarity of z to a.

We note one more important feature that distinguishes competitive similarity from absolute similarity in metric spaces: the competitive similarity function is *asymmetric*. The similarity of object a to object b for competition of objects b and c it is not equal to the similarity of object b to object a for competition of objects a and c. This is easy to show by an example. Let objects a, b and c are in tops of a triangle with the sides (a,b) = 7, (a,c) = 3 and (b,c) = 9. Similarity between two tops depends on position of the third competing top. In the table 1 the values of similarity of all tops with all are shown at detour of tops on and counter-clockwise.

One can see that object c is more similar to object a, than object a on object c. On the same example it is easy to see, that similarity between pairs tops does

Table 1. The similarity between tops of triangle

	$\longrightarrow$ ca/b	$\longrightarrow$ ab/c	$\longrightarrow$ bc/a	$\longleftarrow$ cb/a	$\longleftarrow$ ba/c	$\longleftarrow$ ac/b
R1	3	7	9	9	7	3
R2	9	3	7	3	9	7
FRiS	0.5	-0.4	-0.125	-0.5	0.125	0.4

not satisfy to property of a triangle. Really: $F_{ab/c} + F_{bc/a} < F_{ca/b}$. Hence the space formed by a competitive measure differs greatly from metric spaces. We shall call this space a *competitive space*.

The absolute similarity measure ($S = 1 - R$) gives a difficult-to-interpret answer to the question: "How much is object z similar to the standard of pattern A?" In the kNN method, the similarity in the order scale answers the question: "Object z is most similar to which pattern's standard?" The competitive similarity in the absolute scale measured by means of the FRiS answers the question: "How much is the similarity of object z to the A greater than the similarity of z to the standard of pattern B?"

3   Compactness in Competitive Space

Almost all recognition algorithms are based on the compactness hypothesis. In the literature on pattern recognition [6], compactness is defined through the notion of a boundary index. Let be an arbitrary set of m objects of pattern A in a discrete space. There exist m_1 objects such that, at distance 1 from them, there is an object that does not belong to pattern A. These m_1 objects are called boundary objects, and the ratio m_1/m is called a boundary index. The smaller the index the higher the compactness of the pattern A. It is assumed that a set that possesses a sufficiently small boundary index can be called a simple pattern. Sometimes, patterns separated from each other by not too complicated boundaries are called simple or compact patterns.

The above definitions of compactness involve fuzzy notions such as small, not too complicated boundary, etc. It would be desirable to obtain a quantitative measure of compactness whose value is directly related to the expected reliability of recognition.

One of such measures is proposed in [7] and consists of calculating the compactness profile. Let for object a training sample all the others ($m - 1$) objects are ordered on their distance up to a. At movement from the most close neighbor a up to its farthest neighbor for each serial position $1, 2, ..., i, ..., (m - 1)$ it is defined: the object which stands on this position, to the same image belongs, as the object a (the "its own" object) or does not belong ("another's" object).

The orders constructed for all m of objects are analyzed. For everyone i-th position on all m orders defines quantity of objects m_i which do not belong to that pattern to which possesses object a. Sizes $V_i = m_i/m$ are forms a profile of compactness as a function $Y = f(V_i)$. The patterns with greater quantity of first serial numbers $V_i = 0$ are more compact. If the patterns strongly overlap each other, the compactness profile is a broken curve oscillating around $Y = 0.5$. If the patterns move away from each other, the initial portion of the curve Y more and more approaches 0, and the final portion approaches the quantity $Y = 1$. The compactness profile provides a qualitative understanding of the situation. Conversion from the function $Y = f(V_i)$ to a quantitative estimate of the compactness can be performed by different methods, but this question is not considered in [7].

We suggest that, to obtain a quantitative estimate of the compactness of patterns, it is possible to use the FRiS-function described above. For this, each object a_i of the training sample of pattern A, $i = l, 2, ... M_A$ is alternatively assigned as the standard of this pattern. For the remaining $(M_A - 1)$ objects a_j, $j \neq i$ of pattern A, it is checked how much object a_i protects object a_j from inclusion in the competing pattern B. For this, an object b of competing pattern that is the nearest to object a_j is determined. The distances from a_j to objects a_i and b are calculated and used to find the value of the competitive similarity $F_{a_j a_i / b}$. This value is added to the counter C_i, $i = 1, 2, ... M_A$. The value of C_i estimates the "protective" properties of object a_i. If the distance from a_j to b is smaller than that to a_i, then C_i decreases by the value of $F_{a_j a_i / b}$. After all objects of pattern A have played the role of standards, it is possible to obtain the average value G_A of the quantities stored in the counters:

$$G_A = \frac{1}{M_A} \sum_{i=1}^{M_A} C_i \tag{3}$$

This quantity can be used as a quantitative estimate of the compactness of pattern A.

Estimates of the compactness G_t of all other K patterns, $t = l, 2, ..., K$ are obtained similarly. The general estimate of the compactness of K patterns in a given feature space can be obtained by arithmetic or geometric averaging of the estimates G_t. If we need to find the most informative attributes for all patterns on the average, then the general criterion of informativeness G' should be as follows:

$$G' = \frac{1}{K} \sum_{t=1}^{K} G_t \tag{4}$$

If we want the compactness of the most noncompact pattern G_t to be the maximum possible, it is necessary to choose a subsystem of attributes for which the following quantity reaches a maximum:

$$G = \sqrt[K]{\prod_{t=1}^{K} G_t} \tag{5}$$

Our experiments with these two criteria of compactness have shown a significant advantage of the second of them.

The described measure of competitive compactness is more, than more density of objects inside of patterns and than further patterns will from each other. This property of compactness precisely corresponds to idea of criterion informativeness of attributes for two normally distributed patterns, offered by Fisher [8]:

$$Q = \frac{|\mu_1 - \mu_2|}{\delta_1^2 + \delta_2^2} \tag{6}$$

Here μ_1 and μ_2 – population means, and δ_1^2 and δ_2^2 – root-mean-square deviations of two patterns. Hence, *the measure of competitive compactness can serve as criterion of informativeness of attributes.* Difference of these criteria consists that the measure of compactness is estimated for any quantity of patterns and with an any kind of their distribution.

Useful recommendations follow from properties of a measure of competitive compactness for the decision of tasks of the data analysis. During training at pattern recognition it is necessary to choose such attributes in which space compactness reaches the maximal size. At a stage of recognition inclusion of each new object in structure of an pattern changes a compactness of all competing patterns. It is necessary to choose such decision which as much as possible increases compactness G. It is necessary to aspire to maximization of compactness both at group recognition and at formation of classes.

4   Censoring of Samples and Construction of Standards

The standard object plays the role of a protector for objects of the pattern. The protective capability C_i of each object a_i, $i = 1, 2, ..., M_A$ of the pattern A was obtained in the estimation of the compactness of this pattern. For an object a_i in the center of a pattern, the value of C_i is larger than that for peripheral objects. For a certain number of objects of pattern A near another object, the value of C_i can be negative. Objects whose nearest neighbors are another object will have the largest negative value. Such objects protects only themselves and will be called individual objects. Usually, it is result from random factors, such as noise in measurements etc. Individual objects significantly worsen the training processes and results, and, hence, it is reasonable to eliminate them from the training sample (to censor). After such censoring a new value of the protective capability C_i is calculated for the remaining objects.

Decision rules based on the FRiS are constructed in the space having the greatest informativeness. For this, in the set of M objects of the training sample, it is necessary to choose a minimum and sufficient subset of m objects that will play the role of standards ("stolps") reliably protecting objects of the patterns. This means that the similarity of each object of a pattern to the nearest own stolp in competition with the nearest another stolp should be higher than the given threshold F^*. The FRiS-Stolp algorithm [5] for choosing stolps includes the following procedures:

1. Object a_i with the maximum protective capability C_i is assigned as the first stolp of pattern A.

2. For the remaining objects $a_j, j \neq i$ of pattern A, the measure $F_{a_j a_i/b}$ of their similarity to stolp a_i in competition with object b of any another pattern that is the nearest to object a_j is calculated.

3. If $F_{a_j a_i/b} > F^*$, the object a_j is considered to be protected by stolp a_i. It is included in the first cluster of pattern A and is eliminated from the further consideration.

4. If not all objects of pattern A are included in the first cluster, for remaining objects a new value of protective capability C_i is calculated, an object with the

maximum value of C_i is chosen as the second stolp, and a second cluster of pattern A is formed.

5. Procedure 3 is repeated until all objects of pattern A become included in clusters.

6. Procedures 1–4 are repeated for all other patterns.

As a result, we have a list of clusters (subclasses of the pattern) with indication of the objects chosen as stolps and the objects protected by these stolps. If the patterns have high compactness, the number of stolps is not large. Ideally, one stolp for a pattern is required. The stolps are located at the centres of gravity of the objects of their clusters. For the same compactness, the number of stolps depends on the value of the threshold F^*: the higher this threshold, the larger the number of stolps required to provide the necessary protection of the objects.

In recognizing control object z it is necessary to calculate the distances R from z to all stolps, choose the two nearest stolps belonging to different patterns, and determine the values of the function of similarity of object z to these stolps. A decision is made in favor of the pattern whose stolp has the greatest similarity to the object.

The algorithm FRiS-Stolp is a part of a complex of programs FRiS-GRAD which is intended for a choice of an informative subset of attributes and construction of decision rules. Below examples of application of this complex for the decision of two tasks of molecular biology are described.

5   Recognition of Two Kinds of Leukemia - ALL and AML

Below, we consider the use of the FRiS approach to recognize two types of leukemia. In the literature, one can find the results of solutions of this problem by various groups of researchers. In particular, the results presented in [9] were the best in the world at the time of their publication. They were obtained with the use of the Support Vector Machine (SVM) methods, whose high efficiency is well-known. This allows a comparison of our results with the best previous results.

The data set being analyzed consists of a matrix of gene expression vectors obtained from DNA micro-arrays [10] for a number of patients with two different types of leukemia (ALL and AML). The training set consists of 38 samples (27 ALL and 11 AML) of bone marrow specimens. The test set includes 34 samples (20 ALL and 14 AML) prepared under different experimental conditions (24 samples of bone marrow specimens and 10 of blood specimens). The number of features is 7129. It corresponds to the normalized gene expression value extracted from the micro-array images.

The results of solution of this problem described in [9] are as follows. The informative subset of attributes was chosen using the RFE method (a version of the Deletion algorithm [11] involving successive elimination of the least informative attributes). The decision rules were based on the SVM method. In the initial space of 7129 attributes, 29 out of 34 control objects were correctly recognized (the results called the "success rate" in [9]). Next, the best subsystems were found whose dimension was a multiple of the power of 2: 4096, 2048,..., 4, and 2. Thirty objects were correctly recognized by the two best attributes chosen according to criteria calculated by the training sample, 31 objects were correctly recognized by the four best attributes, and 33 objects by 128 attributes. In the paper where are subsystems of 2, 8, and 16 attributes which recognize all 34 control objects, but the way how to choose these most efficient subsystems according to results, received during training, wasn't presented in the paper. A Pentium class machine operated for three hours to obtain these results.

Using the same data, we obtained the following results. In the initial feature space without a choice of informative attributes and reference objects (all 38 training objects were considered as stopls), P=28 of 34 control objects were correctly recognized. The informative subset of attributes was produced using the FRiS-GRAD algorithm [12]. This algorithm first estimates each attribute separately, selects a subset of the $n \ll N$ most informative attributes (in this case, $n = 100$), and uses them to construct secondary attributes (granules) in the form of the best pairs and triples of attributes. The choice of the best combinations of granules was performed by an iterative Addition-Deletion procedure [13]. The informativeness of separate attributes and their combinations was estimated by the FRiS compactness criterion. As a result, of the initial number of 7129 attributes 11 attributes were selected and used to construct 10 versions of decision rules with the FRiS-Stolp program (see Table 2). Each rule included 3 or 4 attributes with their weights given after a slash. A correct recognition of 34 out of 34 objects was provided by eight rules, and 33 out of 34 objects were correctly recognized by two rules. Collective decision provided recognition quality P = 33 of 34. Average similarity of training objects with their stolps was $F_{train} \sim 0.6385$, average similarity of test objects was $F_{test} \sim 0.6796$. This fact testifies well correspondence between training and test data sets.

Names and numbers of 11 genes, on expression which it is possible to distinguish two types of a leukemia are following:

"U05259_rnal_at" (2641), "X03934_at" (4049), "X76648_at" (4581), "U20362_at" (2895), "U77665_at" (3716), "U84487_at" (3862), "X68994_at" (4476), "M92439_at" (2358), "U62136_at" (3506), "U79287_at" (3779), "X63469_at" (4388). From Table 2, it is evident that attributes 2641 and 4049 are the most frequent in the rules.

The computing time of the algorithm is equal to $C(N + n^3/6)M^3$, where N is the initial number of attributes, $n \ll N$ is the number of attributes of which granules are composed, and M is the number of objects of the training sample. The computing time depends only slightly on the initial number of attributes N and increases rapidly with increasing number of training objects M. In the task considered, M was insignificant and the above decision was obtained on a Pentium based computer for 50 seconds.

6   Diagnostics of the Cancer of Prostate on Mass-Spectra of Proteins

The goal of this study was to evaluate the ability of proteomic pattern diagnostics to detect and discriminate cancer of prostate from benign conditions in patients with normal or elevated serum PSA (prostate-specific antigen) levels. Information about mass-spectra of the proteins forms [14] received by spectrometer SELDI-MS-TOF is analyzed in this task. Number of attributes (spectral bands) is 15153. Four classes of patients with different levels of PSA are considered. Level PSA defines cancer of prostate stage. 63 healthy patients of a class B have serum PSA level < 1 ng/mL, 26 patients of a class C have PSA levels $4 \div 10$ ng/mL, 43 patients of class D have PSA level > 10 ng/mL and 190 patients of a class A have PSA level > 4 ng/mL.

In the first experiment two groups of classes were formed: the first group consisted of class of healthy patients B, and the second group included two classes of sick patients – classes C and D. On this task algorithm FRiS-GRAD in mode One-Leave-Out (OLO) (132 runs, 131 objects were used for training, 1 object – for testing) was run. As a result 65 attributes (from 15153) were selected as informative ones and were used in decision rules. By them 123 objects from 132 (93.2%) were correctly recognized. Accuracy of recognition of sick patients was 64 of 69 (92,8%) and healthy – 59 of 63 (93.7%).

The same task, but on the other proteomic experimental data, was solved by other authors [15]. In training data set 25 healthy (PSA level < 1 ng/mL) and 31 patients with prostate cancer (PSA level > 4 ng/mL) were presented. Number of attributes was more than 18000. As a test 38 patients and 228 healthy people were classified. In this experiment 213 objects from 266 (80.1%) were correctly recognized. Accuracy of recognition of sick patients was equal 94.7% and healthy – 77.6%.

From these results we draw a conclusion that mass-spectra of proteins contain the information which provides decision rules with good enough level accuracy.

Table 2. Collective of 10 best decision rules

Decision rules	F_{train}	P	F_{test}
2641/2+3862/1+4049/2+4581/1	0.6392	34	0.6744
2641/2+4049/2+4581/1	0.6389	34	0.6739
2641/2+ 2895/1+ 4049/2+4581/1	0.6385	34	0.6811
2641/2+3716/1+4049/2+4581/1	0.6385	34	0.6849
2641/2+4049/2+4388/1+4476/1	0.6385	33	0.6835
2358/1+2641/2+4049/2+4581/1	0.6384	34	0.6805
2641/2+4049/2+4388/1+4581/1	0.6384	34	0.6836
2641/1+4049/1+4388/1	0.6384	33	0.6807
2641/2+3506/1+4049/2+4581/1	0.6384	34	0.6729
2641/2+3779/1+4049/2+4581/1	0.6382	34	0.6809
Collective rule	0.6385	33	0.6796

Then we run similar experiments with other training data sets (see the Table 3). In the second experiment we analyzed data set consisted of two classes – B and C in mode One-Leave-Out. On each step we selected subset of informative attributes on which ten most efficient decision rules were formed. It appeared that it rules, which used 62 attributes, allowed to recognize correctly 62 from 63 healthy people and 25 from 26 sick ones.

Table 3. Results of One-Leave-Out tests

Number of experiment	Classes	B	C	D	Results, %
1	[B–(C+D)]	59/63	64/69		93.2
2	[B–C]	62/63	25/26		97.8
3	[B–D]	59/63		41/43	94.3
4	[C–D]		10/26	33/43	62.3

In analogous experiment 3 for classes B and D in mode OLO 53 informative attributes were selected. Each attribute was a part of some decision rule which consists of from 3 to 5 attributes. Ten decision rules, which used these 53 attributes, allowed to recognize correctly 59 from 63 healthy people and 41 from 43 sick ones.

Attempt to find a rule for division poorly sick (class C) from strongly sick (class D) has not crowned success (experiment 4).

Because of small number of patients we couldn't divide data set on training and test ones. For this reason in next experiments we trained on two classes and for control used objects of the third class. Quality of training and recognition we estimate using the following hypothesis. If we ordered classes of patients according to PSA level class B (level < 1 ng/mL) would be in the beginning of the list, the class C (level 4÷10 ng/mL) would follow its, and class D (level > 10 ng/mL) would be in the end. Patients of a class A (PSA level > 4 ng/mL) would appear among patients of classes C and D. If we constructed efficient rule for discriminate class of healthy patients B from some class of sick patients, for example, class D, according this rule patients of other classes of sick patients (A and C) would be more similar to class D than to class B.

We run many experiments on data sets with different number of objects of different classes in training and test subsets. In the Table 4 results of these experiments are presented, which are confirmed the hypothesis described above. More concretely, on training data set consisted of classes B and D (see experiment number 5) 8 attributes which were a part of 10 best decision rules (3-4 attributes in a rule) were selected. During recognition of objects of class C 25 of 26 (96.2%) were recognized as objects of class D. One object of class C with name C4-10.21.csv was recognized as object of class B.

After training on classes B and D (experiment 6) recognition of class A (190 objects) yielded such results: 137 objects (72.1%) were recognized as sick, and 53 – as healthy. Remind, that in class D only mostly sick (PSA level > 10 ng/mL) patients were, but in a class A as strongly as poorly sick (PSA level > 4 ng/mL)

Table 4. Results of recognition a classes which weren't use for training

Number of experiment	Training	Test	B	C	D	A	Results, %
5	[B D]	C 26	1		25		96.2
6	[B D]	A 190	53		137		72.1
7	[B A]	(C+D) 69	1			68	98.6
8	[B C]	D 43	21	22			51.2
9	[C D]	B 63		16	47		74.6

patients were presented. It is not surprising, that some patients with PSA level < 10 ng/mL were more similar to class B, than to class D.

The training on classes B and A allowed recognizing sick patients (classes C and D) good enough: 68 peoples from 69 have been correctly recognized. When we used intermediate class C in training, obtained results appeared to be less confident. So it is possible to draw a conclusion, that on the data our method well distinguishes sick patients from healthy ones, but not finds attributes on which it would be possible to distinguish groups of patients with different PSA levels.

While we recognized different combination of classes in different modes we selected about 300 attributes from 15153 which were informative in some tasks and formed some decision rules. Most part of these attributes was from the following parts of mass-spectra: [2326, 2330], [3038, 3204], [3233, 3237], [3279, 3281], [6288, 6299], [6324, 6336], [6385, 6396], [8297, 8300].

In addition to these tasks, the methods described above have been successfully used to solve other recognition tasks from the areas of medicine and physics. Common features of these tasks were that there was no information on the distributions of patterns and the dependences of attributes and the number of attributes N was a few orders larger than the number of training objects M.

7 Conclusions

A consideration of relative similarity measures taking into account competitive conditions makes it possible to develop effective decision algorithms for all primary data mining tasks. The function of rival similarity (FRiS) provides quantitative estimates of the compactness of patterns and informativeness of feature spaces and easily interpretable decision rules. The method applies to tasks with any number of patterns regardless of the nature of their distributions and conditionality of training samples (the ratio of M and N). The efficiency of the method allows it to be used for complex real tasks. Quality of solutions of applied problems using this method are does not concede in quality to those obtained by other methods.

Acknowledgements

The author sincerely thanks colleagues V. V. Dyubanov, I. A. Borisova and O. A. Kutnenko for discussions of the problems and participation in the experiments.

This work was supported by the Russian Foundation for Basic Research grant No. 08-01-00040 and International Human Potential Foundation.

References

1. Voronin, Ju.A.: The beginnings of the theory of similarity. Edition by Computer Centre of the Siberian Branch of the Russian Academy of Science, Novosibirsk (1989) (in Russian)
2. Shrejder, J.A.: Equality, similarity and order. "Science", M (1971) (in Russian)
3. Fix, E., Hodges, J.: Discriminatory Analysis: Nonparametric Discrimination: Consistency Properties. Technical report, USAF School of Aviation Med. Randolph Field, TX, Rep. 21-49-004 (1951)
4. Kira, K., Rendell, L.: The Feature Selection Problem: Traditional Methods and a New Algorithm. In: Proc. 10th Nat'l Conf. Artificial Intelligence (AAAI 1992), pp. 129–134 (1992)
5. Zagoruiko, N.G., Borisova, I.A., Dyubanov, V.V., Kutnenko, O.A.: Methods of Recognition Based on the Function of Rival Similarity. Pattern Recognition and Image Analysis 18(1), 1–6 (2008)
6. Braverman, E.M.: Experiences on training the machine to recognition of visual patterns. Automatics and Telemechanics 23(3), 349–365 (1962) (in Russian)
7. Vorontsov, K.V., Koloskov, A.O.: Profiles of compactness and allocation of basic objects in metric algorithms of classification. The Artificial intellect (2006) (in Russian)
8. Fisher, R.: The use of multiple measurements in taxonomic problems. Ann. Eugen. (7), 79–188 (1936)
9. Guy, I., Weston, J., Barnhill, S., Vapnik, V.: Gene Selection for Cancer Classification using Support Vector Machines. Machine Learning 46(1-3), 389–422 (2002)
10. http://www.genome.wi.mit.edu/MPR/data_set_ALL_AML.html
11. Merill, T., Green, O.M.: On the effectiveness of receptions in recognition systems. IEEE Trans. Inform. Theory IT-9, 11–17 (1963)
12. Zagoruiko, N.G., Kutnenko, O.A., Ptitsin, A.A.: Algorithm GRAD for Selection of Informative Genetic Feature. In: Proc. Int. Conf. on Computational Molecular Biology, Moscow, pp. 8–9 (2005)
13. Zagoruiko, N.G., Kutnenko, O.A.: Recognition Methods Based on the AdDel Algorithm. Int. Journal "Pattern Recognition and Image Analysis" 14(2), 198–204 (2004)
14. http://home.ccr.cancer.gov/ncifdaproteomics/ppatterns.asp
15. Adam, B.L., Qu, Y., Davis, J.W., et al.: Serum protein fingerprinting coupled with a pattern- matching algorithm distinguishes prostate cancer from benign prostate hyperplasia and healthy men. Cancer Res. 62, 3609–3614 (2002)

Bayesian Solutions to the Label Switching Problem

Kai Puolamäki and Samuel Kaski

Helsinki Institute for Information Technology HIIT
Department of Information and Computer Science
Helsinki University of Technology
P.O. Box 5400, FI-02015 TKK, Finland

Abstract. The label switching problem, the unidentifiability of the permutation of clusters or more generally latent variables, makes interpretation of results computed with MCMC sampling difficult. We introduce a fully Bayesian treatment of the permutations which performs better than alternatives. The method can even be used to compute summaries of the posterior samples for nonparametric Bayesian methods, for which no good solutions exist so far. Although being approximative in that case, the results are very promising. The summaries are intuitively appealing: A summarized cluster is defined as a set of points for which the likelihood of being in the same cluster is maximized.

1 Introduction

In the recent years there has been a dramatic increase in the use of sampling methods in computing with probabilistic models. The main reason naturally is that Markov Chain Monte Carlo (MCMC) methods make it possible to use complex-structured models, for which variational and other techniques are not feasible. MCMC methods are not without their problems, however.

One of these problems is *label switching* of discrete latent or hidden variables of the probabilistic model, which makes interpretation of the results hard. The problem arises if the prior and the likelihood function, and hence the posterior probability distribution, are invariant under a permutation of the values, "labels," of the discrete latent variable. This leads to non-identifiability of the labels of the latent discrete variable.

As a simple example, consider a Gaussian mixture model with two mixture components, and a data set with two well separated groups A and B. In one MCMC sample the groups A and B may be represented by mixture components 1 and 2, and in another sample by 2 and 1, respectively. It follows that if we try to compute some mixture component-specific quantities as averages over the posterior samples, as we normally would in Bayesian data analysis, we get meaningless results. For instance, the mean position of data points in a given mixture component becomes the mean of the whole data set.

The label switching is inherent to sampling—there is no problem if we are content with a point solution, such as a maximum likelihood or maximum a

N. Adams et al. (Eds.): IDA 2009, LNCS 5772, pp. 381–392, 2009.
© Springer-Verlag Berlin Heidelberg 2009

posteriori solution, or a variational approximation where we can choose an arbitrary labeling.

The non-identifiability poses no problems if the quantities of interest are invariant under permutations of the labels. Problems occur when the individual parameter values need to be compared across samples, as in common measures of convergence of the MCMC simulation, for instance. Another problematic area are label-specific summaries and interpretations. The label switching problem has been discussed extensively in the framework of mixture models. When interpreting a cluster in terms of its typical parameter values, or with the list of data samples mapped to it, switching changes the cluster radically.

There have been many suggestions as to how to deal with the label switching problem. The most straightforward solution is to use a sampler that is inefficient in the sense that the labels are very unlikely to switch. Therefore, a reasonable assumption is that the mixture labels are not permuted across the samples. Many of the samplers, such as the Gibbs sampler for mixture models [1], fall into this category. It turns out that the sampler may perform otherwise adequately even if it is unable to switch labels [2]. This solves the label switching problem in practice, but in an arguably inelegant manner that cannot be proven to always work.

Another solution is to use artificial identifiability constraints to break the symmetry in the likelihood [3]. For example, if the component parameters are denoted by μ_j, a possible constraint is $\mu_j < \mu_{j+1}$, where $j < k$ and k is the number of components. Unfortunately, in the Bayesian context these constraints do not however always perform adequately [4].

[5, 6, 7] re-labeled the mixture components in each sample by using a k-means-type approach. [8, 9] relabeled the points in each sample using label-invariant loss functions, such as functions that compare whether the cluster assignments of the data points are equal in pairs of samples; see also [10]. [11] introduced a probabilistic relabeling method, but not in the context of Bernoulli mixture models as in our work. See [4, 11] for a recent review of attempts to solve the label switching problem in mixture models.

A drawback of the earlier relabeling approaches is that they associate a certain more or less heuristic labeling, or permutation of labels, to each sample to make the samples comparable. As can be seen from the variety of approaches, however, the labeling is not unique. Furthermore, assigning a fixed labeling is slightly inelegant considering that the actual modeling follows the Bayesian approach.

2   Summary of Our Contribution

We propose a probabilistic relabeling that can be implemented in a straightforward way for the MCMC samples from any probability distribution that includes a discrete latent variable. We show that our approach gives a consistent probabilistic labeling. In our examples the probabilistic models are mixture models, but we want to emphasize that the results generalize to all probabilistic models which suffer from the label switching problem.

In a nutshell the idea is to include an additional Bernoulli mixture model that can model the distribution of the discrete latent variable in the original probabilistic model. We will explain below why this is a good solution.

Our contributions in this paper are:

- Fully Bayesian treatment of the label switching problem.
- A straightforward way to obtain mixing matrices that are not affected by label switching (Algorithm 1).
- A principled and probabilistic relabeling of samples in order to compute expectations (Section 4.2).
- An approximation scheme having a polynomial time complexity (instead of the naive $O(k!)$), where k is the number of discrete states in the latent variable, to compute the expectations.
- Experimental proofs of concept, including application to Dirichlet Process Mixture model with varying number of mixture components.

3   Definitions

We introduce the problem first in the general notation and then discuss the case of mixture models in more detail.

3.1   General Derivation

We denote the data by $\mathbf{x} = (x_1, \ldots, x_n)$. We have a probabilistic model that has n instances of a discrete-valued latent variable $\mathbf{z} = (z_1, \ldots, z_n)$ having labels $z_i \in [k]$, where $[k] = \{1, \ldots, k\}$. We can alternatively use binary indicator variables z_{ij} such that $z_{ij} = 1$ if $z_i = j$, $z_{ij} = 0$ otherwise.

We denote by $\phi = (\mathbf{z}, \theta)$ all parameters of the model. Here θ are all other parameters besides the latent variable $\mathbf{z}$. Denote by $\sigma \in S_k$ a *permutation* function of the labels $[k]$. We use $\sigma(\mathbf{z})$ as a shorthand of the application of σ to all the z_i, and by $\sigma(\phi)$ the permutation by σ of labels of the parameters of the model of σ. Invariance under the permutation means that for all permutations $\sigma \in S_k$, the prior and the likelihood satisfy $p(\phi) = p(\sigma(\phi))$ and $p(\mathbf{x} \mid \phi) = p(\mathbf{x} \mid \sigma(\phi))$, respectively; hence the posterior probability distribution $p(\phi \mid \mathbf{x})$ is also invariant under the permutation of the labels as $p(\phi \mid \mathbf{x}) \propto p(\mathbf{x} \mid \phi)p(\phi)$. In the remainder we assume that the model is invariant under the permutation of labels of z_i.

Given the above definitions, we can make th trivial observation that

Observation 1. *For a probabilistic model containing a latent variable z_i and for which the prior probability density and likelihood are invariant under the permutations of the labels of z_i, hence the symmetry needs to be broken before meaningful summaries of the latent variables can be computed.*

$$p(z_{ij} = 1 \mid \mathbf{x}) = \frac{1}{k}.$$

Algorithm 1. Bernoulli Labeling

BernoulliLabeling($\{z_{ij}^t\}$) {Input: $\{z_{ij}^t\}$, the indicator variables z_{ij}^t for all $t \in [T]$, $i \in [n]$ and $j \in [k]$. Output: $\tilde{\beta}$, a $k \times n$ parameter matrix of the Bernoulli mixture model.}
Let $Z(r, i) \leftarrow z_{ij}^t$, where $r = k(t-1) + j$ and $Z \in \{0, 1\}^{Tk \times n}$ for all $i \in [n]$, $j \in [k]$, and $t \in [T]$.
Let $\tilde{\beta} \leftarrow$ BernoulliMixture(Z, k). {Algorithm 3}
return $\tilde{\beta}$.

Algorithm 2. Generalized Bernoulli Labeling

Generalized BernoulliLabeling($\{z_i^t\}$, k) {Input: $\{z_i^t\}$, the cluster indices z_i^t for all $t \in [T]$ and $i \in [n]$; k, cluster of components in Bernoulli Labeling. Output: $\tilde{\beta}$, a $k \times n$ parameter matrix of the Bernoulli mixture model.}
Let Z be an empty matrix with n columns.
for $t = 1$ to T **do**
    Let k^t be the number of non-empty cluster in sample t, and let Y_{ji} be 1 if $z_i^t = j$, 0 otherwise.
    Append the rows of matrix Y to the rows of matrix Z.
end for
Let $\tilde{\beta} \leftarrow$ BernoulliMixture(Z, k).
return $\tilde{\beta}$.

The central contribution of this work is to apply a Bernoulli mixture model to the indicator variables z_{ij}^t as given by Algorithm 1. We discuss the motivation of the algorithm in more detail in Section 4. Briefly put, the idea is to apply a Bernoulli mixture model to the rows of a matrix of indicator variables Z where the rows correspond to mixture components in different samples.

Algorithm 1 uses the Bernoulli mixture model having the likelihood

$$p(Z \mid \tilde{\beta}) = \prod_{r=1}^{R} \sum_{j=1}^{k} \frac{1}{k} \prod_{i=1}^{n} \tilde{\beta}(j, i)^{Z(r,i)} \left(1 - \tilde{\beta}(j, i)\right)^{1 - Z(r,i)}, \qquad (1)$$

where the parameters are given as a mixture matrix $\tilde{\beta} \in [0, 1]^{k \times n}$. The data matrix is $Z \in \{0, 1\}^{R \times n}$, where $R = Tk$. We use Algorithm 3 to maximize the likelihood of Equation (1); the algorithm is a standard EM algorithm. Because EM is guaranteed to find a local but not necessary a global optimum, in our experiments we run Algorithm 3 ten times with different random initializations and pick the solution with the largest likelihood.

Another approach is to take explicitly into account the fact that there is a unique permutation of labels for each sample. The mixing matrix $\tilde{\beta}$ can then be found by optimizing the cost function given by

$$\prod_{t=1}^{T} \sum_{\sigma \in S_k} \frac{1}{k!} \prod_{i=1}^{n} \tilde{\beta}(\sigma(z_i^t), i), \qquad (2)$$

Algorithm 3. Bernoulli Mixture

BernoulliMixture(Z, k) {Input: Z, a $R \times n$ binary matrix; k, the number of mixture components. Output: $\tilde{\beta}$, a $k \times n$ maximum likelihood parameter matrix of the Bernoulli Mixture model.}

Initialize $\tilde{\beta} \in [0,1]^{k \times n}$ at random.

repeat

  {E step:}

  Let $\gamma(r,j) \leftarrow \prod_{i=1}^{n} \tilde{\beta}(j,i)^{Z(r,i)} \left(1 - \tilde{\beta}(j,i)\right)^{1-Z(r,i)}$ for all $r \in [R]$ and $j \in [k]$.

  Let $Z(r) \leftarrow \sum_{j=1}^{k} \gamma(r,j)$ for all $r \in [R]$.

  Let $\gamma(r,j) \leftarrow \gamma(r,j)/Z(r)$ for all $r \in [R]$ and $j \in [k]$.

  {M step:}

  Let $\tilde{\beta}(j,i) \leftarrow \sum_{r=1}^{R} \gamma(r,j)Z(r,i) / \sum_{r=1}^{R} \gamma(r,j)$ for all $j \in [k]$ and $i \in [n]$.

until convergence

return $\tilde{\beta}$.

Algorithm 4. Bernoulli Mixture Permutation

BernoulliMixturePerm($\{z_i^t\}$) {Input: $\{z_i^t\}$, the cluster indices z_i^t for all $t \in [T]$ and $i \in [n]$. Output: $\tilde{\beta}$, $k \times n$ maximum likelihood parameter matrix of the Bernoulli Mixture model.}

Initialize $\tilde{\beta} \in [0,1]^{k \times n}$ in random.

repeat

  {E step:}

  Let $\gamma(t,\sigma) \leftarrow \prod_{i=1}^{n} \tilde{\beta}(\sigma^{-1}(z_i^t),i)$ for all $t \in [T]$ and $\sigma \in S_k$.

  Let $Z(t) \leftarrow \sum_{\sigma \in S_k} \gamma(t,\sigma)$ for all $t \in [T]$.

  Let $\gamma(t,\sigma) \leftarrow \gamma(t,\sigma)/Z(t)$ for all $t \in [T]$ and $\sigma \in S_k$.

  {M step:}

  Let $\tilde{\beta}(j,i) \leftarrow \sum_{t=1}^{T} \sum_{\sigma \in S_k} \gamma(t,\sigma)z_{\sigma(i)}^t/T$ for all $j \in [k]$ and $i \in [n]$.

until convergence

return $\tilde{\beta}$.

---

by using EM algorithm, presented by Algorithm 4. We call this model Bernoulli Mixture Permutation model.

3.2 Mixture Models

Mixture models are a common class of probabilistic models where the label switching is a problem. We use the mixture model as a prototype probabilistic model which suffers from the label switching problem.

In a mixture model with k mixture components, there is a discrete-valued latent variable $\mathbf{z} = (z_1, \ldots, z_n)$. Here $z_i \in [k]$ tells which mixture component the data point x_i comes from. The other parameters θ consist of the mixture probabilities $\pi = (\pi(1), \ldots, \pi(k))$ that satisfy $\sum_{j=1}^{k} \pi(j) = 1$, and the component-specific parameters $\beta = (\beta(1), \ldots, \beta(k))$ that define the likelihood of a data point x_i given a mixture component z_i, according to any parametric likelihood function $p(x_i \mid \beta(z_i))$ such as the multivariate Gaussian. In summary, here $\theta = (\pi, \beta)$.

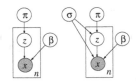

Fig. 1. Graphical representation of mixture model presented in Section 3.2 without (left) and with (right) a permutation sampled from S_k; see the likelihoods of Equations (3) and (4), respectively. The likelihoods and therefore the generative processes of the two models are equivalent.

The likelihood of the mixture model, shown graphically in Figure 1 (left), is given by

$$p(\mathbf{x} \mid \pi, \beta) = \prod_{i=1}^{n} \sum_{z_i=1}^{k} \pi(z_i) p(x_i \mid \beta(z_i)). \tag{3}$$

The likelihood does not change if the labels are permuted by any permutation σ. A special case is where the permutation σ is sampled from S_k uniformly at random, see Figure 1 (right); the likelihood is then

$$p(\mathbf{x} \mid \pi, \beta) = \frac{1}{k!} \sum_{\sigma \in S_k} \prod_{i=1}^{n} \sum_{z_i=1}^{k} \pi(\sigma(z_i)) p(x_i \mid \beta(\sigma(z_i))). \tag{4}$$

See [4] for further discussion.

4   Theoretical Properties

Our idea is, intuitively, to find an assignment of the data items into k mixture components such that if a set of data items co-occurs in the same mixture components in several samples then they should be assigned into the same mixture component.

In this section, we show that the Bernoulli mixture cost function optimized by Algorithm 1 is invariant under the permutation of the labels of the original probabilistic model. We further show that Algorithm 1 exactly reproduces the mixture components of the mixture model described in Section 3.2.

We also provide a principled probabilistic relabeling algorithm of the samples in Section 4.2.

4.1   Properties of the Bernoulli Labeling

Observation 2. *The cost function optimized by Algorithm 1 is invariant under the permutation of labels of the probabilistic model.*

Proof. The Algorithm 1 finds the mixture matrix by maximizing the likelihood given by Equation (1). Any permutation of the labels in the original probabilistic model (in which the discrete variables $\mathbf{z}$ are parameters) corresponds to a permutation of rows of the matrix Z. The likelihood, Equation (1), remains unchanged in any such permutation.   □

The following theorem shows that Algorithm 1 gives consistent results for a mixture model with given fixed parameters π and β.

Theorem 3. *Given a mixture model, parametrized by $\phi = (\mathbf{z}, \theta)$, and data as defined in Section 3.1, Algorithm 1 finds the probability distribution $p(z_i = j \mid \mathbf{x}; \theta)$ in the limit of infinitely many samples, $T \to \infty$.*

Proof. A randomly picked row of matrix Z represents a given component $j \in [k]$ with probability $1/k$. The probability of ones in the ith dimension of component j is given by the distribution $p(z_i = j \mid \mathbf{x}; \theta)$; this distribution can be computed from the mixture model. If we set $\tilde{\beta}(j, i) = p(z_i = j \mid \mathbf{x}; \theta)$, then the Z can be thought of as having been sampled from the Bernoulli mixture model (Equation 1), the probability of each component being $1/k$. Hence, at the limit of infinitely many samples ($T \to \infty$), the maximum of Equation (1) is given by $\tilde{\beta}(j, i) = p(z_i = j \mid \mathbf{x}; \theta)$. Furthermore, as the number of rows of Z approaches infinity, the posterior probability density of the Bernoulli mixture model is essentially a multimodal point estimate with $k!$ modes corresponding to different permutations of the labels, one of the modes being at $\tilde{\beta}$.   □

Theorem 3 is illustrated graphically for a mixture model of Section 3.2 in Figure 2.

Fig. 2. *Left:* Graphical representation of the mixture model presented in Section 3.2, with Bernoulli labeling of Section 4.1. The distribution of the latent variables $\mathbf{z}$ can be equivalently derived either from the mixture model (solid lines), ignoring the dashed lines, or from the Bernoulli mixture model (dashed lines), ignoring the solid lines. *Right:* Graphical representation of mixture model presented in Section 3.2 using Probabilistic Bernoulli Relabeling of Section 4.2: the distribution of the latent variables $\mathbf{z}$ can be equivalently derived either from the mixture model (solid lines), ignoring the dashed lines; or the Bernoulli Mixture model with permutation σ (dashed lines), ignoring parameters π and β.

Observation 2 and Theorem 3 together imply that our approach is completely insensitive to label switching. That is, we can do an arbitrary permutation of labels within each MCMC sample without affecting the results.

It follows that $\tilde{\beta}$ obtained by Algorithm 1 can be used as a principled "point estimate" to summarize the mixture components.

The previous work on relabeling of MCMC samples has focused on finding a single permutation for each sample such that the resulting samples, having permuted labels, can be aggregated that has suffered from the fact that although there usually exists one "most likely" permutation of labels for each sample, the probabilities of *all* possible permutations should be non-vanishing due to the probabilistic nature of the model.

4.2   Probabilistic Bernoulli Relabeling

In this section we consider the problem of computing an expectation of some function $f(\mathbf{z}, \theta)$ of the parameters of the probabilistic model, using the T independently drawn samples at the limit of infinite (or very large) T.

One of the motivations for the model is that it is consistent in the sense that if we had a probabilistic model with a fixed set of parameters θ, but such that the labels in the parameters $\mathbf{z}$ and θ have been relabeled by a permutation function $\sigma \in S_k$, drawn uniformly at random, then the algorithm would asymptotically find correct values.

Now, our task is to compute the posterior distribution for the probability of the permutation for each sample. We can use Algorithm 1 to find the Bernoulli mixing matrix $\tilde{\beta}$ because by Observation 2 the algorithm is unaffected by any permutation σ. Because the number of samples T is very large the posterior distribution is a multimodal point distribution with one of the $k!$ peaks at $\tilde{\beta}$. Figure 2 shows the structure of the model in the case of the mixture model defined in Section 3.2.

Given a fixed $\tilde{\beta}$ we can derive the probability of a permutation for each sample $p(\sigma \mid \mathbf{z}^t, \tilde{\beta})$, and then propose to compute an expectation using *Probabilistic Bernoulli Relabeling* as follows:

$$E_B\left[f(\mathbf{z}, \theta)\right] = \frac{1}{T} \sum_{t=1}^{T} \sum_{\sigma \in S_k} p(\sigma \mid \mathbf{z}^t, \tilde{\beta}) f(\sigma(\mathbf{z}^t), \sigma(\theta^t)), \tag{5}$$

where the posterior probability of a permutation σ for a sample t is given by

$$p(\sigma \mid \mathbf{z}^t, \tilde{\beta}) \propto \sum_{j=1}^{k} \frac{1}{k} \prod_{i=1}^{n} \tilde{\beta}(j, i)^{z^t_{i\sigma(j)}} \left(1 - \tilde{\beta}(j, i)\right)^{1 - z^t_{i\sigma(j)}}, \tag{6}$$

with a normalization defined by $\sum_{\sigma \in S_k} p(\sigma \mid \mathbf{z}^t, \tilde{\beta}) = 1$.

We first note that the expectation defined in Equation (5) reduces to normal expectation in the absence of any label switching.

Observation 4. *If the expectation defined by $f(\mathbf{z}, \theta)$ is invariant under permutation of labels, that is, $f(\mathbf{z}, \theta) = f(\sigma(\mathbf{z}), \sigma(\theta))$ for all $\sigma \in S_k$, then the expectation of Equation (5) reduces to normal expectation of $E[f] = \frac{1}{T} \sum_{t=1}^{T} f(\mathbf{z}^t, \theta^t)$.*

The Probabilistic Bernoulli Relabeling of Equation (5) requires the summation over all $k!$ permutations in S_k. The sum is computable for small enough values of k. For larger values of k, however, the summation can be approximated in polynomial time in k by first finding the most likely permutation by using the

Hungarian algorithm [12]; the time complexity of the Hungarian algorithm is $O(k^3)$. One can then apply Equation (6) to all permutation functions σ that can be reached by at most l swaps from the most likely permutation found by the Hungarian algorithm; the number of these permutation functions is $O(k^{2l})$. All permutations σ which are reachable with more than l permutations can to a good accuracy be approximated with $p(\sigma \mid \mathbf{z}^t, \tilde{\beta}) \approx 0$. As a result, the sum of Equation (5) has only $O(k^{2l})$ non-vanishing terms and the approximate expectation can be therefore be computed in $O(k^3 + k^{2l})$ time.

Finally, we note that although the Figures of Sections 4.1 and 4.2 were given for label switching in the context of the mixture model defined in Section 3.2, the derivations are otherwise general. The method can be applied for any probabilistic model having a non-identifiable discrete latent variable.

5   Experiments

5.1   Mixture Model

We generate an artificial data set by drawing samples from three Gaussian distributions, n samples from each. Each Gaussian has unit variance, and their means are $-x$, 0 and x. We then run a Gibbs sampler for a normal mixture model having $k = 3$ components and conjugate priors (with variance of each component fixed to unity) with parallel tempering as described by [13]. As a consequence of parallel tempering, the sampler switches labels. After 1000 burn-in samples we use the next 1000 samples in our analysis.

Our data analysis task is to use the samples to (i) estimate the means of the mixture components (MEANS), and (ii) to estimate the cluster assignments of the data points (ASSIGN). The error measure in the first task is the difference between the estimated cluster centroids and the "true" cluster centroids at $-x$, 0 and x. The objective measure in the second task is the classification accuracy (sum of probabilities of correct classes) when the true classes (the index of the generating distribution) are known.

The problem is easy for large values x or n; then all methods give equivalent results. For small or moderate values of x and n the methods differ.

Our methods are the Bernoulli mixture model (BM) and Bernoulli mixture model with permutations (BMP). The baseline methods are the identity constraint model (IC), where the samples are permuted such that the means of the mixture components are ordered in an increasing order. The second baseline method [7], denoted by STE, finds permutations using an EM-type approach. We include as a baseline a dummy model DUMB, in which all cluster probabilities are $\frac{1}{3}$.

We chose $n = 5$ and $x = 2$ (tasks MEANS and ASSIGN-2 or $x = \frac{2}{3}$ (task ASSIGN-2/3), and created 100 data sets.

The performance of IC is generally worse than that of the others (Table 1). In task MEANS all algorithms performed comparably. In task ASSIGN-2, BMP was the best, although the differences are very small. In task ASSIGN-2/3 the Bernoulli mixture model (BM) was superior; the reason is that all clusters are

Table 1. Squared prediction errors for the task MEANS (smaller is better); classification accuracy for tasks ASSIGN-2 having $x = 2$ and ASSIGN-2/3 having $x = \frac{2}{3}$ (larger is better), for a data set with $n = 5$. In task ASSIGN-2, BMP outperforms all the other models $(p < 0.05)$. The differences are small, however. In task ASSIGN-2/3, BM outperforms all other models $(p < 10^{-9})$. All tests were one-tailed Wilcoxon Signed Rank Tests.

	BM	BMP	DUMB	IC	STE
MEANS	**0.676**	**0.680**	1.632	1.109	**0.676**
ASSIGN-2	0.598	**0.5995**	0.333	0.575	0.5990
ASSIGN-2/3	**0.442**	0.382	0.333	0.386	0.380

very similar and in many samples one of the clusters remained empty. The BM model will then assign one mixture component to such an empty cluster. The other models suffer from the strong assumption that there must be three clusters (although effectively the number of clusters is smaller).

5.2   Dirichlet Process Mixture

We studied the capability of the Bernoulli mixture to handle varying number of clusters in nonparametric Bayesian settings, by implementing a Dirichlet Process Mixture Model according to [14], with parallel tempering and the hyperparameter α fixed to one. The Gaussian mixture components had a conjugate prior with unit variance. We applied the model to the GALAXY data set [15] consisting of zero-mean relative velocities in 1000 km/sec of 82 galaxies from 6 well-separated conic sections of an unfilled survey of the Corona Borealis region. Multimodality in such surveys is evidence for voids and superclusters in the far universe. The means of the non-empty mixture components are shown in Figure 3. Due to the parallel tempering, the sampling mixes well and there is label switching.

Because the number of clusters varies, out of the introduced algorithms only the generalized Bernoulli labeling (Algorithm 2) is applicable. We ran the algorithm with three numbers of Mixture components, $k = 5$, $k = 6$ and $k = 7$; the

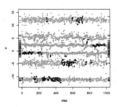

Fig. 3. The cluster means of the Dirichlet Process Mixture on the GALAXY data. Mixture component 2 is highlighted with a darker shade; label switching is evident from the plot. The rug plot on the vertical axis show the data, that is, the relative velocities of the 82 galaxies. There are on average 7.21 non-empty clusters in a sample, the average occupancy of each cluster being 11.57.

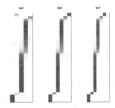

Fig. 4. The mixing matrices $\tilde{\beta}$ for the GALAXY data simulated using the Dirichlet Process Mixture. The y axis corresponds to the 82 galaxies, ordered according to their velocity (lowest velocity at the bottom). The x axis shows the cluster index of the Bernoulli mixture model. Dark shades correspond to a matrix entry of 1, while light shades correspond to zero. Here k is the number of clusters in the Bernoulli mixture model. The mixture components have been ordered for visual clarity.

results are shown in Figure 4. For $k = 7$ mixture components, one of the components turned out to be essentially empty, indicating that the data effectively exhibits six clusters. For $k = 6$ mixture components the mixing matrix looks otherwise similar, except that there is no empty component. For $k = 5$ mixture components, two of the smallest components have been merged to one.

In summary, the Bernoulli Labeling algorithm was capable of extracting the structure of six clusters from the complicated set of samples of the Dirichlet Process Mixture model.

6   Conclusions

We introduced a Bernoulli mixture model for relabeling cluster assignments in mixture models. The model is better motivated than existing solutions to the label switching problem, and outperformed them. The fully Bayesian version requires computation of posteriors for the permutation function which is manageable for models with a fixed number of clusters. For nonparametric Bayesian methods where the number of clusters varies in the MCMC samples, a fully Bayesian method should take into account splits and merges as well, which would be computationally prohibitive.

It turned out that using a Bernoulli mixture without averaging over the posterior of permutations worked very well in solving the label switching problem for nonparametric Bayesian methods, and was rather insensitive to the chosen number of clusters.

In this paper we focused on mixture models, where there is one latent variable per data point, telling which mixture component the point comes from. Furthermore, the simulations were done on one-dimensional data. Both restrictions can naturally be easily removed.

Acknowledgments. SK belongs to Finnish Centre of Excellence in Adaptive Informatics Research and KP to Finnish Centre of Excellence in Algorithmic Data Analysis Research. The work was also supported in part by the PASCAL EU Network of Excellence.

References

[1] Diebolt, J., Robert, C.P.: Estimation of finite mixture distributions through Bayesian sampling. Journal of the Royal Statistical Society. Series B (Methodological) 56(2), 275–363 (1994)

[2] Geweke, J.: Interpretation and inference in mixture models: Simple MCMC works. Computational Statistics & Data Analysis 51, 3529–3550 (2007)

[3] McLachlan, G., Peel, D.: Finite Mixture Models. Wiley Interscience, Hoboken (2000)

[4] Jasra, A., Holmes, C.C., Stephens, D.A.: Markov chain Monte Carlo methods and the label switching problem in Bayesian mixture modeling. Statistical Science 20(1), 50–67 (2005)

[5] Stephens, M.: Bayesian methods for mixtures of normal distributions. PhD thesis, University of Oxford (1997)

[6] Celeux, G.: Bayesian inference for mixtures: The labels-switching problem. In: Payne, R., Green, P. (eds.) Proceedings of XIII Symposium on Computational Statistics (COMPSTAT 1998), Bristol, August 1998, pp. 227–232. Physica-Verlag (1998)

[7] Stephens, M.: Dealing with label switching in mixture models. Journal of the Royal Statistical Society: Series B (Statistical Methodology) 26(4), 795–809 (2000)

[8] Celeux, G., Hurn, M., Robert, C.P.: Computational and inferential difficulties with mixture posterior distributions. Journal of the American Statistical Association 95, 957–970 (2000)

[9] Hurn, M., Justel, A., Robert, C.P.: Estimating mixtures of regressions. Journal of Computational & Graphical Statistics 12(1), 55–79 (2003)

[10] Gerber, G.K., Dowell, R.D., Jaakkola, T.S., Gifford, D.K.: Automated discovery of functional generality of human gene expression programs. PLoS Computational Biology 3(8), 148 (2007)

[11] Jasra, A.: Bayesian Inference for Mixture Models via Monte Carlo Computation. PhD thesis, Imperial College London (2005)

[12] Munkres, J.: Algorithms for the assignment and transportation problems. Journal of the Society for Industrial and Applied Mathematics 5(1), 32–38 (1957)

[13] Liu, J.S.: Monte Carlo Strategies in Scientific Computing. Springer, Heidelberg (2001)

[14] Neal, R.M.: Markov Chain sampling methods for Dirichlet process mixture models. Journal of Computational and Graphical Statistics 9(2), 249–265 (2000)

[15] Postman, M., Huchra, J.P., Geller, M.J.: Probes of large-scale structures in the Corona Borealis region. Astrophysical Journal 92, 1238–1247 (1986)

Efficient Vertical Mining of
Frequent Closures and Generators

Laszlo Szathmary[1], Petko Valtchev[1], Amedeo Napoli[2], and Robert Godin[1]

[1] Dépt. d'Informatique UQAM, C.P. 8888,
Succ. Centre-Ville, Montréal H3C 3P8, Canada
Szathmary.L@gmail.com, {valtchev.petko,godin.robert}@uqam.ca
[2] LORIA UMR 7503, B.P. 239, 54506 Vandœuvre-lès-Nancy Cedex, France
napoli@loria.fr

Abstract. The effective construction of many association rule bases requires the computation of both frequent closed and frequent generator itemsets (FCIs/FGs). However, only few miners address both concerns, typically by applying levelwise breadth-first traversal. As depth-first traversal is known to be superior, we examine here the depth-first FCI/FG-mining. The proposed algorithm, *Touch*, deals with both tasks separately, i.e., uses a well-known vertical method, *Charm*, to extract FCIs and a novel one, *Talky-G*, to extract FGs. The respective outputs are matched in a post-processing step. Experimental results indicate that *Touch* is highly efficient and outperforms its levelwise competitors.

1 Introduction

The discovery of meaningful associations is a key data mining discipline [1]. An association miner typically proceeds in two steps: **(i)** extract all frequent patterns X of a database, and **(ii)** break each X into a *premise* Y, and a *conclusion* $X \setminus Y$ parts to form a rule $Y \to X \setminus Y$. Interestingness measures, such as support and confidence, are applied to prune the set of extracted association rules. However, the number of the remaining rules is usually way too high to be practical. As a remedy, various concise representations of the family of interesting association rules have been proposed [2,3,4], whereas others have been imported from related fields such as concept analysis [5,6]. A good survey can be found in [7].

In this paper we focus on the computation of frequent closed itemsets (FCIs) and frequent generators (FGs), on which are based the minimal non-redundant association rules ($\mathcal{MNR}$) for instance. Following [2], these are rules with the form $P \to Q \setminus P$, where $P \subset Q$, P is a *(minimal) generator* (a.k.a. key-set or free-set) and Q is a *closed itemset*. In other terms, in such rules the premise is minimal and the conclusion is maximal. As shown in [7], $\mathcal{MNR}$ is a *lossless*, *sound*, and *informative* representation of all valid rules. Moreover, further restrictions can be imposed on the rules in $\mathcal{MNR}$, leading to more compact representations such as the *generic basis* or the *proper basis* (see [7] for a complete list).

From a computational point of view, constructing $\mathcal{MNR}$ or its sub-structures requires the family of frequent closed itemsets (FCIs) and their generators (FGs),

N. Adams et al. (Eds.): IDA 2009, LNCS 5772, pp. 393–404, 2009.
© Springer-Verlag Berlin Heidelberg 2009

and possibly the precedence order between FCIs. A few methods for extracting both FCIs and FGs have been published in the mining literature, e.g. *A-Close* [8] or *Titanic* [9]. Generators have been targeted within the concept analysis field as well [10], e.g. by the *Zart* algorithm [11]. Well-known FCI/FG-miners exclusively apply levelwise strategies, although the levelwise itemset miners are known to be outperformed by depth-first methods (e.g. *Charm* [12] and *Closet* [13]) on a broad range of dataset profiles, especially on dense ones. Hence the idea of designing a depth-first FCI/FG-miner. Our method, called *Touch*, tackles the component tasks separately: while the state-of-the-art algorithm *Charm* extracts FCIs, FG-mining is performed by *Talky-G*, an original method following a set inclusion-compliant order in the traversal of the itemset lattice. At a post-processing step of *Touch*, FGs are associated to their respective FCIs, thus providing the necessary starting point for the production of $\mathcal{MNR}$. Experimental results show that *Touch* outperforms two other efficient competitors, *A-Close* [8] and *Zart* [11], especially on dense and highly correlated datasets. Thus, the contributions of our study lay mainly in the design of an efficient method, *Touch*, for constructing the aforementioned rule bases. Additionally, *Talky-G* is a stand-alone algorithm for extracting FGs.

The paper is organized as follows. Section 2 lists the basic concepts of frequent itemset mining and presents the vertical depth-first mining strategy of *Charm*. In Section 3, we introduce a new FG-miner algorithm called *Talky-G*. The *Touch* algorithm that combines the results of *Charm* and *Talky-G* is introduced in Section 4. Finally, conclusions and future work are discussed in Section 5.

2   Background

Consider the following 5×5 sample dataset: $\mathcal{D} = \{(1, \ ACDE), (2, \ ABCDE), (3, \ AB), (4, \ D), (5, \ B)\}$. Throughout the paper, we will refer to this example as **"dataset $\mathcal{D}$"**.

2.1   Basic Concepts from Pattern Mining

We consider a set of *objects* or *transactions* $\mathcal{O} = \{o_1, o_2, \ldots, o_m\}$, a set of *attributes* or *items* $\mathcal{A} = \{a_1, a_2, \ldots, a_n\}$, and a relation $\mathcal{R} \subseteq \mathcal{O} \times \mathcal{A}$. A set of items is called an *itemset*. Each transaction has a unique identifier (*tid*), and a set of transactions is called a *tidset*. The tidset of all transactions sharing a given itemset X is its *image*, denoted $t(X)$. For instance, the image of $\{A, B\}$ in $\mathcal{D}$ is $\{2, 3\}$, i.e., $t(AB) = 23$ in our separator-free set notation. The *length* of an itemset X is $|X|$, whereas an itemset of length i is called an i-itemset. The (absolute) *support* of an itemset X, denoted by $supp(X)$, is the size of its image, i.e. $supp(X) = |t(X)|$. Moreover, X is *frequent*, if its support is not less than a given *minimum support* threshold, min_supp, i.e. $supp(X) \geq min\_supp$. An equivalence relation is induced by t on the power-set of items $\wp(\mathcal{A})$: equivalent itemsets share the same image ($X \cong Z$ iff $t(X) = t(Z)$) [14]. In [12], a subsumption relation is defined as well: X *subsumes* Z, iff $X \supset Z$ and $supp(X) = supp(Z)$.

Consider the equivalence class of X, denoted $[X]$, and its extremal elements w.r.t. set inclusion. $[X]$ has a unique maximum (a *closed* itemset), and a set of minima (*generator* itemsets). The following definition thereof exploits the monotony of support upon set inclusion in $\wp(\mathcal{A})$.

Definition 1. *An itemset X is* closed (generator) *if it has no proper superset (subset) with the same support.*

A *closure* operator underlies the set of closed itemsets; it assigns to X the maximum of $[X]$ (denoted by $\gamma(X)$). Naturally, $X = \gamma(X)$ for closed X. Generators, a.k.a. *key-sets* in database theory, represent a special case of free-sets [15].

By Def. 1, if Z *subsumes* X, then Z cannot be a generator. The following property, which is widely known in the domain, generalizes this observation. It basically states the generator family is a downset within the Boolean lattice $\langle \wp(\mathcal{A}), \subseteq \rangle$.

Property 1. Given $X \subseteq \mathcal{A}$, if X is a generator, then $\forall Y \subseteq X$, Y is a generator, whereas if X is not a generator, $\forall Z \supseteq X$, Z is not a generator.

2.2   Vertical Itemset Mining

Miners from the literature, whether for plain FIs or FCIs, can be roughly split into breadth-first and depth-first ones. Breadth-first algorithms, more specifically the *Apriori*-like [1] ones, apply levelwise traversal of the pattern space exploiting the anti-monotony of the frequent status. Depth-first algorithms, e.g., *Closet* [13], in contrast, organize the search space into a prefix-tree (see Figure 1) thus factoring out the effort to process common prefixes of itemsets. Among them, the *vertical* miners use an encoding of the dataset as a set of pairs (item, tidset), i.e., $\{(i, t(i)) | i \in \mathcal{A}\}$, which reportedly allows the costly database re-scans to be avoided.

Eclat [16] was the first FI-miner to combine the vertical encoding with a depth-first traversal of a tree structure, called IT-tree, whose nodes are $X \times t(X)$ pairs. *Eclat* traverses the IT-tree in a depth-first manner in a pre-order way, from left-to-right [16,17] (see Figure 1).

Charm adapts the computing schema of *Eclat* to the exclusive construction of the FCIs [12]. The key challenges it faces are parsimony in generating the closedness candidates and efficiency of closedness tests on those candidates. To avoid examining the entire IT-tree of the FIs, *Charm* relies on a technique that, given a node $X \times t(X)$, looks for a Z subsuming X by combining X to Y, where $Y \times t(Y)$ is any right sibling node in the tree. Due to the specific traversal discipline, all Z are such that X is a prefix thereof (hence not all X expand to the closure of X).

To certify a candidate Z as closed, it should be checked that no set can subsume Z. Again, the traversal ensures that potential subsumers can only precede Z in the traversal-induced order on $\mathcal{A}$, hence at the moment Z is tested all of them are already processed and the actual closure is known. Thus, the closedness test amounts to a lookup in the working memory for a set Y such that $t(Z) = t(Y)$, absence meaning that Z is the closure of $[Z]$. To avoid extensive

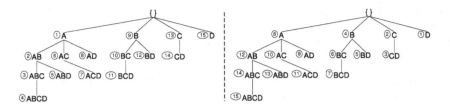

Fig. 1. Left: pre-order traversal with *Eclat*; **Right:** reverse pre-order traversal with *Talky-G*. The direction of traversal is indicated in circles

search through the known part of the FCI family, *Charm* employs a hashing on $t(Z)$ (hashing is discussed in Section 3). For a more detailed presentation of vertical itemset miners, please refer to the report [18].

Charm is known to be one of the fastest FCI-miners, hence we adopt it in our own FCI/FG-miner. A natural question is whether a similar strategy could be defined for FGs. Several generators within a class mean that the pure image-based tests will not work as the existence of a generator with the same image does not disqualify a candidate X. Indeed, beside image equality, the existing generator must be a subset of X in order to invalidate X. Thus, X can only be certified "generator" if no stored generator is a subset thereof with the same image.

Moreover, hidden in the testing principles is a different traversal order: in fact, for the test to be effective, all subsets of a candidate X must be processed before X itself. Only then all generator subsets of X will be known and hence could be used in correctly (in)validating its own generator status. Although such a concern is typically addressed through a breadth-first traversal, the corresponding order could also be achieved with a depth-first one, yet with a different listing order on the items, as discussed in the next section.

3   Talky-G

Talky-G is a vertical FG-miner following a depth-first traversal of the IT-tree and a right-to-left order on sibling nodes.

3.1   Reverse Pre-order Traversal

Talky-G applies an inclusion-compatible traversal: it goes down the IT-tree while listing sibling nodes from right-to-left and not the other way round as in *Eclat* and *Charm*. The resulting order on itemsets is exactly the order on their numerical representations (e.g., E is 1 and ABD is 26 in our dataset $\mathcal{D}$) that is frequently used in combinatorial generation algorithms. This strategy is used in *Next-Closure* [19] under the name of *lectic order*.

The authors of [20] explored that order for mining calling it *reverse pre-order*. They observed that for any itemset X its subsets appear in the IT-tree in nodes that lay either higher on the same branch as $(X, t(X))$ or on branches to the

---

Algorithm 1. (main block of Talky-G):

1) root.itemset ← ∅;   // root *is an IT-node whose itemset is empty*
2) root.tidset ← {all transaction IDs};   // *the empty set is included in every tr.*
3) loop over the vertical representation of the dataset (*attr*) {
4)   if ((*attr*.supp ≥ *min_supp*) and (*attr*.supp < |𝒪|)) {
5)   // |𝒪| *stands for the total number of objects in the database*
6)   root.addChild(*attr*);   // *attr is frequent and generator*
7)   }
8) }
9) loop over the children of root from right-to-left (*child*) {
10)   save(*child*);   // *process the itemset*
11)   extend(*child*);   // *discover the subtree below child*
12) }

Algorithm 2. ("extend(*curr*)" procedure of Talky-G):

Method: extend an IT-node recursively (discover FGs in its subtree)
Input: *curr* – an IT-node whose subtree is to be discovered

1) loop over the right siblings of *curr* from left-to-right (*other*) {
2)   *generator* ← getNextGenerator(*curr, other*);
3)   if (*generator* ≠ null) then *curr*.addChild(*generator*);
4) }
5) loop over the children of *curr* from right-to-left (*child*) {
6)   save(*child*);   // *process the itemset*
7)   extend(*child*);   // *discover the subtree below child*
8) }

---

right of it. Hence, depth-first processing of the branches from right-to-left would perfectly match set inclusion, i.e., all subsets of X will be met before X itself.

While the algorithm in [20] extracts the so-called non-derivable itemsets, our *Talky-G* algorithm uses this traversal to find the set of frequent generators. See Figure 1 for a comparison of *Eclat* and its "reversed" version.

3.2   The Algorithm

Pseudo code. Algorithm 1 provides the main block of *Talky-G*. First, the IT-tree is initialized, which involves the creation of the root node, representing the empty set (of 100% support, by construction). *Talky-G* then transforms the layout of the dataset in vertical format, and inserts below the root node all 1-long frequent itemsets. Such a set is an FG whenever its support is less than 100%. At this point, the dataset is no more needed since larger itemsets can be obtained as unions of smaller ones while for the images intersection must be used.

Algorithm 3. ("getNextGenerator(*curr*, *other*)" function of Talky-G):

Method: create a new frequent generator
Input: two IT-nodes (*curr* and *other*)
Output: a frequent generator or null

1) *cand*.tidset ← *curr*.tidset ∩ *other*.tidset;
2) if (cardinality(*cand*.tidset) < *min_supp*) {   // *test 1*
3)   return null;   // *not frequent*
4) }
5) // *else, if it is frequent*
6) if ((*cand*.tidset = *curr*.tidset) or (*cand*.tidset = *other*.tidset)) {   // *test 2*
7)   return null;   // *not generator*
8) }
9) // *else, if it is a potential generator*
10) *cand*.itemset ← *curr*.itemset ∪ *other*.itemset;
11) if (*cand* has a proper subset with the same support in the hash) {   // *test 3*
12)   return null;   // *not generator*
13) }
14) // *if cand passed all the tests then cand is a frequent generator*
15) return *cand*;

In the core processing, the **extend** procedure is called recursively for each child of the root in a right-to-left order. At the end, the IT-tree contains all FGs. The **addChild** procedure inserts an IT-node under a node. The **save** procedure stores an FG in a dedicated "list" data structure. The **extend** procedure (see Algorithm 2) discovers all FGs in the subtree of a node. First, new FGs are tentatively generated from the right siblings of the current node. Then, certified FGs are added below the current node and later on extended recursively in a right-to-left order.

The **getNextGenerator** function (see Algorithm 3) takes two nodes and returns a new FG, or "null" if no FG can be produced from the input nodes. First, a candidate node is created by taking the union of both itemsets and the intersection of their respective images. The input nodes are thus the candidate's *parents*. Then, the candidate undergoes a frequency test. If successful, the candidate is compared to its parents: if its tidset is equivalent to a parent tidset, then the candidate cannot be a generator. Even with both outcomes positive, an itemset may still not be a generator as a subsumed subset may lay elsewhere in the IT-tree. Due to the traversal strategy in *Talky-G*, all generator subsets of the current candidate are already detected and the algorithm has stored them in a "list" structure (see the **save** procedure). Thus, the ultimate test checks whether the candidate has a proper subset with the same support in this "list". A positive outcome disqualifies the candidate. The test exploits a hash structure that enhances the one used in *Charm* to perform the search for FG subsets efficiently.

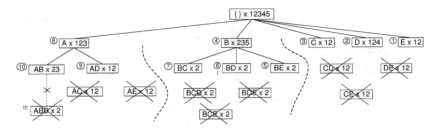

Fig. 2. Execution of *Talky-G* on dataset $\mathcal{D}$ with $min\_supp = 1$ (20%)

Candidates surviving the final test are declared FG and added to the IT-tree. An unsuccessful candidate X is discarded which ultimately prevents any itemset Y having X as a prefix to be generated as candidate and hence substantially reduces the overall search space. When the algorithm stops, all frequent generators (and *only* frequent generators) are inserted in the IT-tree *and* in the "list" of generators.

RUNNING EXAMPLE. The execution of *Talky-G* on dataset $\mathcal{D}$ with $min\_supp = 1$ (20%) is illustrated in Figure 2. Circles beside tree nodes show traversal ranks.

The IT-tree root node is first created and, as there is no full column in the dataset, all items become FGs, thus they are inserted below the root. These nodes are recursively extended in a *right-to-left* order. The rightmost node E has no right sibling, thus it cannot be extended. In contrast, D is extended with E. The result, DE, is discarded since of equal support to its parent E. C is extended with both D and E, but both CD and CE are discarded for this same reason. The processing of the B-branch, in short, yields FGs BC, BD, and BE. As to the 2-long supersets of A, AC and AE fail the second test because of C and E, respectively, while AB and AD succeed. The combination of the latter, ABD, although of strictly smaller support than its parents, fails because of a subsumed FG in the list (BD).

3.3 Fast Subsumption Checking

Recall that in the `getNextGenerator` function, when a new candidate itemset C is created, *Talky-G* checks whether C subsumes a previously found generator. If the test is positive, then clearly C is not a generator. Subsumption might seem expensive here, yet an efficient way to filter non-generators exists.

To that end the hash structure of *Charm* was adapted to the storage of frequent generators. Actually, *Talky-G* hashes the itemsets upon the tidset while storing generators with their support values. Consequently, equivalent itemsets get the same hash value and end up in the same list in the hash structure. In the testing of a candidate Z, the entire list corresponding to its hash value $h(Z)$ is retrieved. Whenever there is a set G in the list such that $supp(Z) = supp(G)$ and $Z \supset G$, Z is discarded, otherwise Z is declared an FG.

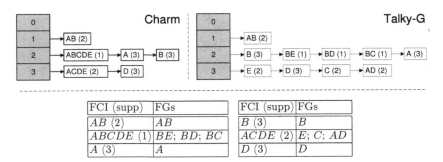

Fig. 3. Top: hash tables for dataset $\mathcal{D}$ with $min\_supp = 1$. **Top left:** hash table of *Charm* containing all FCIs. **Top right:** hash table of *Talky-G* containing all FGs. **Bottom:** output of *Touch* on dataset $\mathcal{D}$ with $min\_supp = 1$.

EXAMPLE. Figure 3 (top right) depicts the hash structure of the IT-tree in Figure 2, which contains all FGs of $\mathcal{D}$. Each entry of the table is a list of (itemset, support) pairs. Here, the size of the hash table is set to four.

Assume we need to test ABD whose absolute support is 1. First, the sum of the tids in its tidset is 2 which, modulo the size of the hash table, is again 2. When traversing the list at position 2 of the table, B is more frequent than ABD while BE, although of the same support, is not a subset of ABD. Yet the next set, BD, is both of identical support and a proper subset of the candidate, hence ABD is discarded.

4   Touch

The algorithm has three steps: **(1)** extracting FCIs, **(2)** extracting FGs, and **(3)** associating FGs to their FCIs.

4.1   The Algorithm

While the above tasks **(1)** and **(2)** are solved by *Charm* and *Talky-G*, respectively, the appropriate associations between the respective outputs of both algorithms, task **(3)**, require some additional effort. Yet as both algorithms provide an additional hash structure (see Figure 3), the problem admits an efficient solution.

The exact method is based on a generalization of the storage strategy for FGs in *Talky-G* to FCIs. Indeed, observe that just as all FGs of the same equivalence class are forced to belong to the same list within the hash structure, their respective closure, whenever hashed to the FGs table would fall into the same list too. Conversely, if hashed against the FCI structure, each FG would fall precisely in the list where its closure lays. In both cases, the same hash value is guaranteed by the shared image.

Yet an effective re-hashing of FCIs or FGs is not necessary: with tables properly sized, i.e. of the *same dimension*, and with *identical hash functions*, the lists from both tables can be directly matched. To that end, FCIs from the list at position n in the closure table should be compared only to FGs from the list at the same n position in the generator table.

Pseudo code. The algorithm *Touch* starts by calling *Charm* and *Talky-G* and taking over their hash structures. Then, *Touch* matches the two hash tables: for each FCI X, it looks up in the hash table of *Talky-G* at the same index position all subsets of X that have the same support.

EXAMPLE. Consider the hash structures of *Charm* and *Talky-G* in Figure 3. Assume the generators of the closed itemset $ACDE$ are sought. As $ACDE$ is stored at position 3 in the hash structure of *Charm*, its generators will also be at position 3 in the hash structure of *Talky-G*. Three members of the corresponding list are subsumed by $ACDE$: E, C, and AD, hence they are the target generators. For the FCI A, the only subsumed FG of the list at index 2 is A, meaning that A is the unique member of its equivalence class $[A]$. The output of *Touch* is shown in Figure 3 (bottom).

4.2   Experimental Results

We evaluated *Touch* against *Zart* [11] and *A-Close* [8]. All the algorithms were implemented in Java in the CORON data mining platform [21].[1] The experiments were carried out on a bi-processor Intel Quad Core Xeon 2.33 GHz machine running under Ubuntu GNU/Linux with 4 GB of RAM. For the experiments we have used the following datasets: T20I6D100K[2], C20D10K, and MUSHROOMS[3]. The T20 is a sparse dataset, constructed according to the properties of market basket data that are typical weakly correlated data. The C20 is a census dataset from the PUMS[4] sample file, while the MUSHROOMS describes mushrooms characteristics. The last two are highly correlated datasets.

Table 1 contains detailed information about the execution of *Touch*. The first three columns correspond to the three main steps of *Touch* namely **(1)** getting FCIs using *Charm*, **(2)** getting FGs using *Talky-G*, and **(3)** associating FGs to their closures. Column 4 indicates the total execution time of the algorithm including input and output. In the sparse dataset T20, almost all frequent itemsets are closed and generators at the same time. It means that most equivalence classes are singletons. It is known that *Charm* is less efficient on sparse datasets. This is due to the fact that *Charm* performs four tests on candidates for reducing the IT-tree. However, in sparse datasets the number of FCIs is almost equivalent to the number of FIs, thus the search space cannot be reduced significantly. *Talky-G* is also less efficient on sparse datasets. However, in dense, highly correlated datasets (C20 and MUSHROOMS), both *Charm* and *Talky-G* are very efficient, even at low minimum support values. Since the number of FCIs and

---

[1] http://coron.loria.fr

[2] http://www.almaden.ibm.com/software/quest/Resources/

[3] http://kdd.ics.uci.edu/

[4] Public Use Microdata Sample

Table 1. Detailed execution times of *Touch* and data-related statistics: number of FCIs, of of FGs, and of FIs (for comparison only, *Touch* does not work with all FIs), ratio of FCIs to FIs, ratio of FGs to FIs.

min_supp	execution time (sec.)				# FCIs	# FGs	(# FIs)	$\frac{\#FCIs}{\#FIs}$	$\frac{\#FGs}{\#FIs}$
	get FCIs *(Charm)*	get FGs *(Talky-G)*	associate FCIs and FGs	total time (with I/O)					
T20I6D100K									
1%	19.07	2.16	0.03	22.76	1,534	1,534	1,534	100.00%	100.00%
0.75%	24.06	2.65	0.05	28.32	4,710	4,710	4,710	100.00%	100.00%
0.5%	35.21	5.01	0.14	42.45	26,208	26,305	26,836	97.66%	98.02%
0.25%	94.59	20.71	0.50	121.60	149,217	149,447	155,163	96.17%	96.32%
C20D10K									
30%	0.20	0.29	0.02	1.06	951	967	5,319	17.88%	18.18%
20%	0.34	0.41	0.03	1.42	2,519	2,671	20,239	12.45%	13.20%
10%	0.71	0.70	0.07	2.27	8,777	9,331	89,883	9.76%	10.38%
5%	1.13	1.06	0.11	3.37	21,213	23,051	352,611	6.02%	6.54%
MUSHROOMS									
30%	0.12	0.21	0.02	0.82	425	544	2,587	16.43%	21.03%
20%	0.19	0.27	0.02	0.98	1,169	1,704	53,337	2.19%	3.19%
10%	0.43	0.46	0.04	1.57	4,850	7,585	600,817	0.81%	1.26%
5%	0.80	0.81	0.08	2.53	12,789	21,391	4,137,547	0.31%	0.52%

Table 2. Response times of *Touch*, compared to *Zart* and *A-Close*

T20I6D100K				C20D10K				MUSHROOMS			
min. supp.	execution time (sec.)			min. supp.	execution time (sec.)			min. supp.	execution time (sec.)		
	Touch	Zart	A-Close		Touch	Zart	A-Close		Touch	Zart	A-Close
1%	22.76	**7.33**	31.25	30%	**1.06**	8.17	15.78	30%	**0.82**	3.65	7.17
0.75%	28.32	**14.96**	39.49	20%	**1.42**	15.84	29.88	20%	**0.98**	10.69	15.28
0.5%	**42.45**	45.52	100.60	10%	**2.27**	36.66	59.41	10%	**1.57**	75.36	36.83
0.25%	**121.60**	159.78	285.41	5%	**3.37**	75.28	94.18	5%	**2.53**	641.54	63.37

FGs is much less than the number of FIs, the two algorithms can take advantage of exploring a much smaller search space. The association of FCIs and FGs is done very efficiently in all cases. That is, the association step gives absolutely no overhead to *Touch*.

Table 2 contains the experimental evaluation of *Touch* against *Zart* and *A-Close*. All times reported are real, wall clock times as obtained from the Unix *time* command between input and output. We have chosen *Zart* and *A-Close* because they represent two efficient algorithms that produce exactly the same output as *Touch*. *Zart* and *A-Close* are both levelwise algorithms. *Zart* is an extension of *Pascal* [14], i.e. first it finds all FIs using pattern-counting inference, then it filters FCIs, and finally the algorithm associates FGs to their closures. *A-Close* reduces the search space to FGs only, then it calculates the closure for each generator. The way *A-Close* computes the closures of generators is quite expensive because of the huge number of intersection operations. *Touch*, just like *A-Close*, reduces the search space to the strict minimum, i.e. it only extracts what it really needs namely the set of FCIs and the set of FGs. Then, *Touch* associates the two sets in a very efficient way. Since *Touch* is based on *Charm* and *Talky-G*, the algorithm is very efficient on dense, highly correlated datasets.

We must admit however that levelwise algorithms are sometimes more suitable for sparse datasets.

5 Conclusions and Future Work

Mining FGs has so far been done largely in a levelwise manner as the breadth-first traversal fits the down-set structure of the FG family. Yet depth-first algorithms have shown superior efficiency in many situations, whence the motivation of our study of depth-first FCI/FG-mining.

As a contribution to this problem, we presented *Touch*, an algorithm that splits the general problem into three tasks: (1) FCI-mining, (2) FG-mining, and (3) association of FGs to their closures (FCIs). While (1) is solved by reusing an existing algorithm, *Charm*, the two others generate innovative solutions. Hence the *Talky-G* vertical FG-miner used in (2) is an original contribution on its own. As all three solutions are highly optimized, the algorithm performs well against comparable levelwise miners. Numerous concise representations of valid association rules can be readily derived from the method's output.

The study led to a range of exciting questions that are currently investigated. Thus, from an algorithmic point of view, it would be interesting to merge steps (1) and (2), e.g. by using the output of *Talky-G* (i.e., the IT-tree of all FGs) as a starting point for the FCI-mining, hence avoiding step (3). A further challenge lays in the computation of the FCI precedence order that underlies some of the association rule bases from the literature. We plan to join *Touch* with our previous algorithm *Snow* [22]. *Snow* allows us to easily compute precedence order using hypergraph theory. Once we have a concept lattice whose nodes are labeled with generators, it is possible to produce all kinds of $\mathcal{MNR}$ rules, including approximate association rules too.

References

1. Agrawal, R., Srikant, R.: Fast Algorithms for Mining Association Rules in Large Databases. In: Proc. of the 20th Intl. Conf. on Very Large Data Bases (VLDB 1994), pp. 487–499. Morgan Kaufmann, San Francisco (1994)
2. Bastide, Y., Pasquier, N., Taouil, R., Stumme, G., Lakhal, L.: Mining Minimal Non-redundant Association Rules Using Frequent Closed Itemsets. In: Palamidessi, C., Moniz Pereira, L., Lloyd, J.W., Dahl, V., Furbach, U., Kerber, M., Lau, K.-K., Sagiv, Y., Stuckey, P.J. (eds.) CL 2000. LNCS (LNAI), vol. 1861, pp. 972–986. Springer, Heidelberg (2000)
3. Kryszkiewicz, M.: Representative Association Rules. In: Wu, X., Kotagiri, R., Korb, K.B. (eds.) PAKDD 1998. LNCS, vol. 1394, pp. 198–209. Springer, Heidelberg (1998)
4. Pasquier, N., Bastide, Y., Taouil, R., Lakhal, L.: Closed Set Based Discovery of Small Covers for Association Rules. In: Proc. 15emes Journees Bases de Donnees Avancees (BDA), pp. 361–381 (1999)
5. Duquenne, V.: Contextual Implications Between Attributes and Some Representational Properties for Finite Lattices. In: Beitraege zur Begriffsanalyse, B.I. Wissenschaftsverlag, Mannheim, pp. 213–239 (1987)

6. Luxenburger, M.: Implications partielles dans un contexte. Mathématiques, Informatique et Sciences Humaines 113, 35–55 (1991)
7. Kryszkiewicz, M.: Concise Representations of Association Rules. In: Proc. of the ESF Exploratory Workshop on Pattern Detection and Discovery, pp. 92–109 (2002)
8. Pasquier, N., Bastide, Y., Taouil, R., Lakhal, L.: Discovering Frequent Closed Itemsets for Association Rules. In: Beeri, C., Bruneman, P. (eds.) ICDT 1999. LNCS, vol. 1540, pp. 398–416. Springer, Heidelberg (1998)
9. Stumme, G., Taouil, R., Bastide, Y., Pasquier, N., Lakhal, L.: Computing Iceberg Concept Lattices with TITANIC. Data and Knowledge Engineering 42(2), 189–222 (2002)
10. Valtchev, P., Missaoui, R., Godin, R.: Formal Concept Analysis for Knowledge Discovery and Data Mining: The New Challenges. In: Proc. of the 2nd Intl. Conf. on Formal Concept Analysis, pp. 352–371. Springer, Heidelberg (2004)
11. Szathmary, L., Napoli, A., Kuznetsov, S.O.: ZART: A Multifunctional Itemset Mining Algorithm. In: Proc. of the 5th Intl. Conf. on Concept Lattices and Their Applications (CLA 2007), pp. 26–37 (2007)
12. Zaki, M.J., Hsiao, C.J.: ChARM: An Efficient Algorithm for Closed Itemset Mining. In: SIAM Intl. Conf. on Data Mining (SDM 2002), pp. 33–43 (2002)
13. Pei, J., Han, J., Mao, R.: CLOSET: An Efficient Algorithm for Mining Frequent Closed Itemsets. In: ACM SIGMOD Workshop on Research Issues in Data Mining and Knowledge Discovery, pp. 21–30 (2000)
14. Bastide, Y., Taouil, R., Pasquier, N., Stumme, G., Lakhal, L.: Mining Frequent Patterns with Counting Inference. SIGKDD Explor. Newsl. 2(2), 66–75 (2000)
15. Boulicaut, J.F., Bykowski, A., Rigotti, C.: Free-Sets: A Condensed Representation of Boolean Data for the Approximation of Frequency Queries. Data Mining and Knowledge Discovery 7(1), 5–22 (2003)
16. Zaki, M.J., Parthasarathy, S., Ogihara, M., Li, W.: New Algorithms for Fast Discovery of Association Rules. In: Proc. of the 3rd Intl. Conf. on Knowledge Discovery in Databases, pp. 283–286 (1997)
17. Zaki, M.J.: Scalable Algorithms for Association Mining. IEEE Transactions on Knowledge and Data Engineering 12(3), 372–390 (2000)
18. Szathmary, L., Valtchev, P., Napoli, A.: Efficient Mining of Frequent Closures with Precedence Links and Associated Generators. Research Report RR-6657, INRIA (2008), http://hal.inria.fr/inria-00322798/en
19. Ganter, B., Wille, R.: Formal Concept Analysis: Mathematical Foundations. Springer, Heidelberg (1999)
20. Calders, T., Goethals, B.: Depth-First Non-Derivable Itemset Mining. In: Proc. of the SIAM Intl. Conf. on Data Mining (SDM 2005), Newport Beach, USA (2005)
21. Szathmary, L.: Symbolic Data Mining Methods with the Coron Platform. PhD Thesis in Computer Science, Univ. Henri Poincaré – Nancy 1, France (2006)
22. Szathmary, L., Valtchev, P., Napoli, A., Godin, R.: Constructing Iceberg Lattices from Frequent Closures Using Generators. In: Boulicaut, J.-F., Berthold, M.R., Horváth, T. (eds.) DS 2008. LNCS (LNAI), vol. 5255, pp. 136–147. Springer, Heidelberg (2008)

Isotonic Classification Trees

Rémon van de Kamp, Ad Feelders, and Nicola Barile

Utrecht University, Department of Information and Computing Sciences,
P.O. Box 80089, 3508TB Utrecht, The Netherlands
{rpkamp,ad,barile}@cs.uu.nl

Abstract. We propose a new algorithm for learning isotonic classification trees. It relabels non-monotone leaf nodes by performing the isotonic regression on the collection of leaf nodes. In case two leaf nodes with a common parent have the same class after relabeling, the tree is pruned in the parent node. Since we consider problems with ordered class labels, all results are evaluated on the basis of L_1 prediction error. We experimentally compare the performance of the new algorithm with standard classification trees.

1 Introduction

In many applications of data analysis it is reasonable to assume that the response variable is increasing (or decreasing) in one or more of the attributes or features. For example, the sale price of a house - all else equal - increases with lot size, and according to economists people tend to buy less of a product if its price increases. Such relations between response and attribute are called monotone. Monotonicity constraints can, for example, also be found in medicine [20,6] and law [12]. Besides being plausible, monotonicity may also be a desirable property of a decision model for reasons of explanation, justification and fairness. Pazzani et al.[17], show that rules learned with monotonicity constraints were significantly more acceptable to medical experts than rules learned without the monotonicity restrictions.

Because the monotonicity constraint is quite common in practice, many data analysis techniques have been adapted to be able to handle such constraints. In this paper we present a new algorithm, called ICT, for learning monotone classification trees for problems with ordered class labels. Our approach differs from earlier monotone tree algorithms such as [5,18,11] in that we adjust the probability estimates in the leaf nodes in case of a violation. This is done in such a way that, subject to the monotonicity constraint, the sum of absolute prediction errors on the training sample is minimized. Another new element of our algorithm is that we can also handle problems where some, but not all, attributes have a monotone relation with the response. The performance of the new algorithm is evaluated through experimental studies on real life data sets.

This paper is organized as follows. In the next section we introduce the basic concepts and notation that will be used throughout the paper. Since the isotonic regression is an important technique for our algorithm, we discuss it shortly in

N. Adams et al. (Eds.): IDA 2009, LNCS 5772, pp. 405–416, 2009.
© Springer-Verlag Berlin Heidelberg 2009

section 3. In section 4 we discuss the main contribution of this paper, the Isotonic Classification Tree (ICT) algorithm. ICT is evaluated in section 5 where we present the results of experiments on real data. Section 6 concludes.

2   Preliminaries

Let $\mathcal{X}$ be a *feature space* $\mathcal{X} = \mathcal{X}_1 \times \mathcal{X}_2 \times \ldots \times \mathcal{X}_p$ consisting of vectors $\mathbf{x} = (x_1, x_2, \ldots, x_p)$ of values on p features or attributes. We assume that each feature takes values x_i in a linearly ordered set $\mathcal{X}_i$. The partial ordering $\preceq$ on $\mathcal{X}$ will be the ordering induced by the order relations of its coordinates $\mathcal{X}_i$: $\mathbf{x} = (x_1, x_2, \ldots, x_p) \preceq \mathbf{x}' = (x_1', x_2', \ldots, x_p')$ if and only if $x_i \leq x_i'$ for all i. Furthermore, let $\mathcal{Y}$ be a finite linearly ordered set of *classes*. Without loss of generality, we assume that $\mathcal{Y} = \{1, 2, \ldots, k\}$ where k is the number of classes.

A *monotone* classification rule is a function $c : \mathcal{X} \to \mathcal{Y}$ for which

$$\mathbf{x} \preceq \mathbf{x}' \Rightarrow c(\mathbf{x}) \leq c(\mathbf{x}') \tag{1}$$

for all instances $\mathbf{x}, \mathbf{x}' \in \mathcal{X}$. A data set $\{\mathbf{x}_i, y_i\}_{i=1}^n$ is *monotone* if for all i, j we have $\mathbf{x}_i \preceq \mathbf{x}_j \Rightarrow y_i \leq y_j$.

The classification rules we consider are univariate binary classification trees. For such trees, at each node a split is made using a test of the form $X_i < d$ for some $d \in \mathcal{X}_i, 1 \leq i \leq p$. Thus, for a binary tree, in each node the associated set $t \subset \mathcal{X}$ is split into the two subsets $t_\ell = \{\mathbf{x} \in t : x_i < d\}$ and $t_r = \{\mathbf{x} \in t : x_i \geq d\}$. The classification rule that is induced by a decision tree T will be denoted by c_T.

For any node or leaf t of T, the subset of the instance space corresponding to that node can be written

$$t = \{\mathbf{x} \in \mathcal{X} : \mathbf{a} \preceq \mathbf{x} \prec \mathbf{b}\} = [\mathbf{a}, \mathbf{b}) \tag{2}$$

for some $\mathbf{a}, \mathbf{b} \in \overline{\mathcal{X}}$ with $\mathbf{a} \preceq \mathbf{b}$. Here $\overline{\mathcal{X}}$ denotes the extension of $\mathcal{X}$ with infinity-elements $-\infty$ and ∞. In some cases we need the infinity elements so we can specify a node as in equation (2).

Below we will call $\min(t) = \mathbf{a}$ the *minimal element* and $\max(t) = \mathbf{b}$ the *maximal element* of t. Together, we call these the *corner elements* of node t. If $\min(t) \prec \max(t')$ then node t contains points that are smaller than some points in node t', hence the monotonicity constraint requires that the label assigned to node t should not be bigger than the label assigned to node t'. Therefore, we call a pair of leaves t, t' non-monotone if $\min(t) \prec \max(t')$ and $c_T(t) > c_T(t')$ [19]. A tree is non-monotone if it contains at least one non-monotone leaf pair.

It is customary to evaluate a classifier on the basis of its error-rate or 0/1 loss. For classification problems with ordered class labels this choice is less obvious. It makes sense to incur a higher cost for those misclassifications that are "far" from the true label, than to those that are "close". One loss function that has this property is L_1 loss:

$$L_1(i, j) = |i - j| \qquad\qquad i, j = 1, \ldots, k \tag{3}$$

where i is the true label, and j the predicted label. We note that this is not the only possible choice. One could also choose L_2 loss for example, or another loss function that has the desired property that misclassifications that are far from the true label incur a higher loss. Nevertheless, L_1 loss is a reasonable candidate, and in this paper we confine our attention to this loss function.

To illustrate the concepts introduced, we discuss a small example. Let $\mathcal{Y} = \{1, 2, 3\}$ and suppose we have a tree with two input attributes with $X_1, X_2 \in [0, 1]^2$. The tree is given in Figure 1 on the left. The corresponding partitioning of the input space is depicted in Figure 1 on the right.

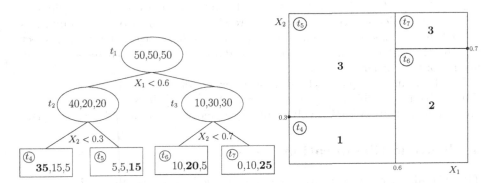

Fig. 1. Left: Classification tree for three-class problem. Numbers in nodes are the counts for class 1, 2 and 3 respectively. In the leaf nodes, the counts of the median class are shown in boldface. The circled labels in the leaf nodes correspond to the labeled regions in the picture on the right. Right: Partitioning of input space corresponding to the tree on the left. The class labels assigned to the different rectangles are shown in boldface. Rectangle t_5 and t_6 form a nonmonotone leaf pair.

To minimize L_1 loss we allocate to the median in leaf nodes, which leads to the class labels as shown in boldface in Figure 1 on the right. Leaf node t_5 and t_6 form a nonmonotone pair, since t_5 has a higher class label but contains points that are smaller than some points in t_6: the lower left corner of t_5 is smaller than the upper right corner of t_6.

3   The Isotonic Regression

In this section we give a general description of the isotonic regression. In section 4 we discuss its application to making trees monotone.

Let $Z = \{z_1, z_2, \ldots, z_m\}$ be a nonempty finite set of constants and let $\preceq$ be a partial order on Z. Any real-valued function f on Z is *isotonic* with respect to $\preceq$ if, for any $z, z' \in Z$, $z \preceq z'$ implies $f(z) \leq f(z')$. We assume that each element z_i of Z is associated with a real number $g(z_i)$; these real numbers typically are estimates of the function values of an unknown isotonic function on Z. Furthermore, each element of Z has associated a positive weight $w(z_i)$ that typically indicates the precision of this estimate. An isotonic function g^* on Z now is an *isotonic regression* of g with respect to the weight function w and the partial order $\preceq$ if and only if it minimizes the sum

$$\sum_{i=1}^{m} w(z_i) \left[f(z_i) - g(z_i) \right]^2 \tag{4}$$

in the class of isotonic functions f on Z. Brunk [8] proved the existence of a unique g^*.

Any real-valued function f on Z is *antitonic* with respect to $\preceq$ if, for any $z, z' \in Z$, $z \preceq z'$ implies $f(z) \geq f(z')$. The antitonic regression of g is defined completely analogous to the isotonic regression as the function that minimizes (4) within the class of antitonic functions. The *isotonic* regression with respect to a partial order is equivalent to the *antitonic* regression with respect to the inverse of that order.

The best time complexity known for an exact solution to the isotonic regression problem for arbitrary partial order is $O(m^4)$ [16]. It is based on a divide-and-conquer strategy that involves solving at most m maximal flow problems.

4   Isotonic Classification Trees

The ICT algorithm can in principle be combined with any standard classification tree algorithm. Here we modify a CART-like algorithm [7] to incorporate the monotonicity constraints. The main principle of ICT is that it makes trees monotone by relabeling its leaf nodes. This is done in such a way that of all monotone trees that can be obtained by relabeling the leaf nodes, the one produced by ICT has lowest absolute error on the training data. The relabeling is not computed directly, but is obtained by first adjusting the probability estimates in the leaf nodes (using the isotonic regression), and then allocating each leaf node to the (smallest) median of its estimated class distribution.

We first discuss the growing of trees in ICT. Then we discuss the adjustment of probability estimates in the leaf nodes, and the corresponding relabeling, which may lead to pruning the tree. We also discuss the incorporation of partial monotonicity constraints in ICT.

4.1   Growing Trees

Let $\tilde{T}$ denote the collection of leaf nodes of tree T, $n(t, j)$ denote the number of observations in t with class label j, and let

$$\hat{P}_j(t) = \frac{n(t, j)}{n(t)}, \qquad t \in \tilde{T}$$

denote the relative frequency of class label j in node t. Furthermore, let

$$\hat{F}_i(t) = \sum_{j \leq i} \hat{P}_j(t), \qquad t \in \tilde{T}$$

denote the unconstrained maximum likelihood estimate of

$$F_i(t) = P(y \leq i \mid \mathbf{x} \in t), \qquad t \in \tilde{T}.$$

Because the median is known to minimize L_1 loss, we allocate to the (smallest) median of the estimated class distribution:

$$c(t) = \min_i : \hat{F}_i(t) \geq 0.5, \tag{5}$$

The standard tree growing algorithm is modified in such a way that it records the corner elements of each node. Since the aim is to minimize L_1 loss, the risk for each node is set to be the mean absolute error for that node:

$$r(t) = \sum_{i:\mathbf{x}_i \in t} \frac{|y_i - c(t)|}{n(t)},$$

where $c(t)$ denotes the class allocated to node t and $n(t)$ denotes the number of observations in node t.

To compute the impurity of a node, we use the gini index combined with absolute error:

$$i(t) = \sum_{j \neq k} |j - k| \hat{P}_j(t) \hat{P}_k(t), \qquad j, k \in \mathcal{Y}$$

As usual, ICT chooses the split that maximizes impurity reduction.

4.2   Making the Tree Monotone

Let $(\tilde{T}, \preceq)$ be the partial order $\preceq$ over $\tilde{T}$ with

$$t \preceq t' \Leftrightarrow \min(t) \prec \max(t'), \qquad t, t' \in \tilde{T},$$

that is, t precedes t' if t contains points that are smaller than some points in t'. Define

$$F_i^*(t), \qquad i = 1, 2, \ldots, k-1; \quad t \in \tilde{T}$$

as the antitonic regression of $g(t) = \hat{F}_i(t)$ with weights $w(t) = n(t)$ and partial order $(\tilde{T}, \preceq)$, for each value $i = 1, 2, \ldots, k-1$. Of course, $F_k^*(t) = 1$ for all $t \in \tilde{T}$. Note that F^* satisfies the stochastic order constraint

$$t \preceq t' \Rightarrow F_i^*(t) \geq F_i^*(t'), \qquad i = 1, \ldots, k. \tag{6}$$

Subsequently, we allocate t to the smallest median of $F^*(t)$:

$$c^*(t) = \min_i : F_i^*(t) \geq 0.5 \tag{7}$$

Because F^* satisfies (6), we have that c^* satisfies the monotonicity constraint [14]

$$t \preceq t' \Rightarrow c^*(t) \leq c^*(t')$$

Furthermore, it can be shown [10,3] that $c^*(t)$ minimizes L_1 loss

$$\sum_{i=1}^{n} |y_i - c(t(\mathbf{x}_i))|$$

within the class of monotone integer-valued functions $c(\cdot)$ on $\tilde{T}$, where $t(\mathbf{x}_i) = t \in \tilde{T} : \mathbf{x}_i \in t$. In other words, of all monotone classifiers on $\tilde{T}$, c^* is among the ones (there may be more than one) that minimize L_1 loss on the training sample.

To illustrate, we continue the example introduced in section 2. The monotonicity violation between t_5 and t_6 is resolved by performing the antitonic regression on $\hat{F}_1(t)$, and on $\hat{F}_2(t)$, $t \in \{t_4, t_5, t_6, t_7\}$. Note that we have the total order $t_4 \preceq t_5 \preceq t_6 \preceq t_7$ on the leaf nodes in this particular case. The solution is to average $\hat{F}_1(t_5)$ with $\hat{F}_1(t_6)$ and $\hat{F}_2(t_5)$ with $\hat{F}_2(t_6)$ with $n(t_5)$ and $n(t_6)$ as weights:

$$F_1^*(t_5) = F_1^*(t_6) = \frac{5 + 10}{25 + 35} = \frac{1}{4} \qquad F_2^*(t_5) = F_2^*(t_6) = \frac{10 + 30}{25 + 35} = \frac{2}{3}$$

Using these revised estimates we assign to the median, meaning we assign to class 2 in both t_5 and t_6. Note that the L_1 error of the original nonmonotone tree is: $25+15+15+10=65$. By relabeling leaf node t_5 to class 2, the error increases to 70. This is the best possible monontone relabeling of the leaf nodes; for example, relabeling t_6 to class 3 would increase the error to 90.

4.3   ICT Pruning

After relabeling the leaf nodes, there may be pairs t_ℓ and t_r with common parent t that have been assigned the same class label. In that case, the tree is pruned in t, and we apply the normal allocation rule (5) to t. This may result in a nonmonotone tree, in which case the leaf nodes are relabeled again. Also note that pruning may create a new pair t'_ℓ, t'_r with the same class label and a common parent t' in which case the algorithm will prune in t'.

4.4   Algorithm Outline

The ICT algorithm is summarized in Algorithm 1. The algorithm takes as input a tree T, and returns a monotone tree T' which is a possibly relabeled and pruned version of T.

If the tree is monotone, it is returned unchanged. However, if the tree is nonmonotone, the antitonic regression computes the new probability estimates in line 4-6. In line 7-10 the leaf nodes are subsequently relabeled. Line 11-14 prune away leaf nodes with the same class label and common parent. The resulting tree may be nonmonotone, hence the recursive call to ICT in line 15.

It should be noted that we combine the ICT algorithm with cost-complexity pruning in the following way. Let $T_1 > T_2 > \ldots > \{t_1\}$ denote the tree sequence produced by standard cost-complexity pruning, where t_1 denotes the root node

Algorithm 1. ICT(T)

1: **if** T is monotone **then**
2:     **return** T
3: **else**
4:     **for** $i \in \{1, \ldots, k-1\}$ **do**
5:         $F_i^* \leftarrow$ AntitonicRegression($\tilde{T}, \preceq, n(t) : t \in \tilde{T}, \hat{F}_i(t) : t \in \tilde{T}$)
6:     **end for**
7:     **for all** $t \in \tilde{T}$ **do**
8:         $F_k^*(t) \leftarrow 1$
9:         $c_T(t) \leftarrow \min_i F_i^*(t) \geq 0.5$
10:     **end for**
11:     **while there are** $t_\ell, t_r \in \tilde{T}$ with common parent t **and** $c_T(t_\ell) = c_T(t_r)$ **do**
12:         $T \leftarrow$ prune T in t
13:         $c_T(t) \leftarrow \min_i \hat{F}_i(t) \geq 0.5$
14:     **end while**
15:     $T' \leftarrow$ ICT(T)
16:     **return** T'
17: **end if**

---

of the tree, and $T_j > T_k$ means T_k is obtained by pruning T_j in one or more nodes. We apply ICT pruning to every tree in this sequence (except the root of course) to obtain a sequence of monotone trees $T_1' > T_2' > \ldots > \{t_1\}$. This sequence may be shorter than the original sequence, since sometimes two trees from the original cost-complexity sequence are pruned back to the same tree by ICT.

4.5 Partial Monotonicity

In many applications there will be attributes for which there is no reason to assume that they have a monotone relation with the class label. Therefore we extended the ICT algorithm to be able to handle such cases.

The ICT algorithm for partial monotonicity is largely the same as it is for complete monotonicity. We just need to change the partial order used in the antitonic regression and the check that determines if two leafs are non-monotone.

First we define a partially monotone classification rule. Let $\mathcal{X}$ be defined as before, and let $\mathcal{Z} = \times \mathcal{Z}_i, i = 1, \ldots, q$. The values $\mathcal{Z}_i$ may be either ordered or unordered. A classification rule $c : \mathcal{X} \times \mathcal{Z} \to \mathcal{Y}$ is monotone in $\mathbf{X}$ iff

$$\forall \mathbf{x}, \mathbf{x}' \in \mathcal{X}, \forall \mathbf{z} \in \mathcal{Z} : \mathbf{x} \preceq \mathbf{x}' \Rightarrow c(\mathbf{x}, \mathbf{z}) \leq c(\mathbf{x}', \mathbf{z})$$

Our orginal ordering on $\tilde{T}$ was defined in such a way that $t \preceq t'$ if node t contained elements that were smaller than some elements of t'. This was the case when $\min(t) \prec \max(t')$. Now we have to add the constraint that t and t' should have overlapping values on $\mathbf{Z}$. Hence, we define a new partial order $(\tilde{T}, \preceq)$ with $t \subset \mathcal{X} \times \mathcal{Z}$ as follows:

$$t \preceq t' \Leftrightarrow \min(t_{\mathbf{X}}) \prec \max(t_{\mathbf{X}}') \wedge t_{\mathbf{Z}} \cap t_{\mathbf{Z}}' \neq \varnothing, \qquad t, t' \in \tilde{T}.$$

Here $t_{\mathbf{X}}$ denotes the projection of t on the monotone attributes $\mathbf{X}$.

5   Experiments

In order to evaluate the proposed algorithm, we performed a number of experiments. This section contains information on the datasets, how we pre-processed the data, the experiments and their results. The programs were implemented in R[1].

5.1   Datasets

We selected a number of datasets where monotonicity constraints are likely to apply. We used the KC4, PC3, PC4 and PC5 datasets from the NASA Metrics Data Program [15], the Acceptance/Rejection, Employee Selection, Lecturers Evaluation and Social Workers Decisions from A. Ben-David [4], the Windsor Housing dataset [1], the Den Bosch Housing dataset [9], as well as several datasets from the UCI Machine Learning Repository [2]. All datasets except Den Bosch Housing are publicly available. Table 1 lists all the datasets used.

5.2   Pre-processing of the Data

ICT makes the harmless assumption that all monotone attributes have an increasing relation with the response. This means that if the actual relation is decreasing, the attribute values have to be inverted. We tested this by looking at the correlation between the attribute and the response. In case of a negative correlation between some attribute x and the response, we transformed the values of x as follows:

$$x_i = x_{max} - x_i + x_{min}, \qquad i = 1, \ldots, n \qquad (8)$$

with $x_{max} = \max(x)$, and $x_{min} = \min(x)$.

For datasets with a numeric response that is not a count (Auto MPG, Boston Housing, CPU Performance, Windsor Housing and Den Bosch Housing) we discretized the response values into four separate intervals, each interval containing roughly the same number of observations.

For all datasets from the NASA Metrics Data Program the attribute ERROR_COUNT was used as the response. All attributes that contained missing values were removed. Furthermore, the attribute MODULE was removed because it is a unique identifier of the module and the ERROR_DENSITY was removed because it is a function of the response variable. On the remaining attributes we used the function stepAIC with backward elimination in R to fit a linear model; attributes that did not occur in the final model were removed from the dataset. Since the distribution of ERROR_COUNT was highly skewed (most modules have zero errors) we sampled the modules with zero errors to create a more balanced distribution. High error counts are less frequent than low error counts. In order to increase frequencies, the higher counts were merged into a single class. For example, for KC4, all class labels greater than five were set to five.

---

[1] http://www.r-project.org/

For the CPU Performance dataset the machine cycle time in nanoseconds was converted to clock speed in Khz, in order to make it positively correlated with the class label. From this dataset the attributes Vendor Name, Model Name and ERP were removed.

From the Den Bosch Housing dataset the independent attributes year, x-coordinate and y-coordinate were removed.

5.3   Relabeling toward Monotonicity

Besides enforcing a monotone model, one can also use prior knowledge about monotonicity by relabeling the dataset to make it monotone. As shown in [22,13], models learned on relabeled datasets on average perform better than models learned with the original class labels.

Therefore, we also tested ICT on relabeled versions of the original datasets. We computed y^* as the relabeling of the observations that minimizes

$$\sum_{i=1}^{\text{ntrain}} |y_i - y_i'|$$

within the class of monotone relabelings y'. Here ntrain denotes the number of observations in the training sample. The test data was not relabeled in the experiments.

Table 1 summarizes for all datasets the cardinality, the number of attributes after pre-processing, the number of distinct class labels and the L_1 distance between y and y^*. For example, to make the Australian Credit data monotone we have to relabel for a total absolute error of 14. Since Australian Credit has a binary class label, this means that 14 observations have to be relabeled.

Table 1. Dataset charasterics and relabeling information

Dataset	cardinality	#attributes	#labels	$\sum |y_i - y_i^*|$
Australian Credit	690	14	2	14
Auto MPG	392	7	4	23
Boston Housing	506	13	4	28
Car Evaluation	1728	6	4	21
Den Bosch Housing	119	8	4	10
Empoyee Rej/Acc	1000	4	9	1161
Employee selection	488	4	9	104
Haberman survival	306	3	2	55
Lecturers evaluation	1000	4	5	364
CPU Performance	209	6	4	26
Pima Indians	768	8	2	53
Social Workers Decisions	1000	10	4	375
Windsor Housing	546	11	4	134
KC4	122	4	6	80
PC3	320	15	5	1
PC4	356	16	6	6
PC5	1032	21	6	141

5.4   Experimental Results

Each of the datasets was randomly divided one hundred times into a training set consisting of four fifth of the data and a test set consisting of the remaining one fifth of the data. On every training set a tree was grown, after which cost complexity pruning was applied to obtain a sequence of trees. The ICT algorithm was applied to each tree in this sequence to obtain a sequence of monotone trees. Subsequently, the test set was used to select the best tree from the original sequence and to select the best tree from the monotone sequence. The test errors of the best standard trees and the best monotone trees were averaged over the one hundred repititions of the experiment.

Table 2 shows the results of the experiments on all datasets. The errors are indicated as the mean absolute error on the test sample. For each column the mean error and the standard deviation of this mean error are indicated, separated by a ± sign. The lowest error and the lowest number of leafs for each dataset are printed in boldface.

First we consider the results with the original class labels. In that case ICT almost always has a slightly lower error than the standard tree. The two exceptions

Table 2. Results of monotone trees (ICT) and standard trees

Dataset	Label	Error ICT	Error Standard	#Leafs ICT	#Leafs Standard
Australian	y	0.1426±0.0070	0.1431±0.0068	3.4300±1.9553	**3.3500**±2.2490
Credit	y^*	**0.1418**±0.0078	0.1426±0.0071	3.3600±1.9515	3.5700±2.9241
Auto MPG	y	**0.2982**±0.0292	0.3045±0.0282	**9.2800**±2.7746	10.6300±5.3497
	y^*	0.2985±0.0295	**0.2982**±0.0293	10.1100±2.7484	12.3800±4.5964
Boston	y	0.3966±0.0370	0.4050±0.0376	8.3100±3.1065	**7.7100**±4.9222
Housing	y^*	**0.3861**±0.0334	0.3935±0.0339	8.3200±2.7957	8.5700±5.2073
Car	y	0.0871±0.0181	**0.0836**±0.0164	**27.6200**±5.4417	32.4400±8.4271
Evaluation	y^*	0.0897±0.0190	0.0849±0.0184	28.2300±5.3802	32.3100±7.8787
Den Bosch	y	0.4922±0.0832	0.5165±0.0852	5.5200±1.5274	5.6600±1.9396
Housing	y^*	**0.4755**±0.0829	0.4856±0.0841	**5.5100**±1.4106	5.5500±1.6840
Employee Rej/Acc	y	1.2764±0.0407	1.2926±0.0415	10.3400±3.9778	**8.2200**±3.4629
	y^*	1.1773±0.0242	**1.1627**±0.0208	17.0600±2.2103	19.3900±1.1538
Employee	y	0.4348±0.0369	0.4590±0.0395	**23.6900**±4.3128	26.9500±8.4452
Selection	y^*	0.3829±0.0265	**0.3822**±0.0304	23.9100±2.9063	27.6300±3.9457
Haberman	y	0.2585±0.0146	0.2605±0.0139	2.6000±2.1742	**1.9900**±1.6112
Survival	y^*	**0.2482**±0.0164	0.2484±0.0164	3.8300±2.0003	4.3200±2.4980
Lecturers	y	0.4764±0.0267	0.4903±0.0267	**19.9000**±5.1981	19.9200±8.9563
Evaluation	y^*	0.4151±0.0233	**0.3832**±0.0157	21.5300±3.5773	28.7900±3.1311
CPU Performance	y	0.4556±0.0540	0.4773±0.0554	**8.3300**±2.3401	9.2900±4.4818
	y^*	**0.4323**±0.0486	0.4324±0.0453	8.5900±1.9700	9.8800±2.7016
Pima	y	0.2586±0.0145	0.2619±0.0144	5.6300±3.1866	4.6700±3.1464
Indians	y^*	0.2580±0.0130	**0.2576**±0.0117	**4.3900**±2.7299	4.5200±3.9936
Social Workers	y	0.4707±0.0209	0.4772±0.0181	9.3300±4.3579	**7.6100**±4.5436
Decisions	y^*	0.4309±0.0163	**0.4045**±0.0137	12.6700±3.6350	27.4500±4.7298
Windsor	y	0.6244±0.0328	0.6619±0.0364	17.0100±5.1198	**15.5400**±12.8860
Housing	y^*	**0.5992**±0.0333	0.6103±0.0377	18.7900±4.2433	24.9500±12.0566
KC4	y	1.1871±0.1349	1.2358±0.1474	**4.4300**±2.2031	4.6300±3.2649
	y^*	**1.0153**±0.1298	1.0174±0.1349	4.8400±1.5681	5.0900±1.7529
PC3	y	0.5357±0.0440	0.5363±0.0394	2.6000±1.3780	**2.4900**±1.5986
	y^*	**0.5351**±0.0443	0.5359±0.0399	2.6100±1.4695	**2.4900**±1.5986
PC4	y	**0.5735**±0.0492	0.5835±0.0543	4.9500±2.9418	**4.5100**±3.0600
	y^*	0.5886±0.0617	0.5928±0.0632	5.4200±3.2167	5.1900±4.0394
PC5	y	0.4960±0.0188	0.4948±0.0242	6.1600±4.1310	**5.8400**±4.5543
	y^*	0.4939±0.0189	**0.4904**±0.0229	6.1111±4.4006	6.8800±5.1410

are the Car evaluation data and dataset PC5. There is no clear winner on the tree size criterion.

On the relabeled data the conclusions are quite different. Now there is no clear winner in terms of the error but ICT clearly has the smaller trees.

Comparing the error on the relabeled and original data, we can conclude that it is beneficial to relabel the training data, since it tends to reduce the error.

It should be noted that the differences found were small, and nowhere significant. Nevertheless it is safe to conclude that in the datasets studied, enforcing a monotone model does not lead to a degradation of the predicitive performance. Hence, when a monotone model is required, or just preferred, such a model can be obtained without loss of predicitive accuracy.

Finally, we discuss our experiments with partially monotone trees. One of the important problems is how to determine which attributes are to be constrained, and which are not. In practice such information may be obtained from domain experts. Here we used a data-based test proposed by [21]. This sometimes resulted in the removal of the constraint for a particular attribute, but the results did not improve compared to complete monotonicity, and are therefore not reported.

6 Conclusions

We have presented a new algorithm, called ICT, for learning monotone classification trees from data. ICT differs from existing monotone tree algorithms in that it relabels the leaf nodes of the tree in case of monotonicity violations: ICT produces the monotone relabeling that minimizes absolute error on the training sample. Furthermore, in contrast to existing monotone tree algorithms, ICT can also be applied to partially monotone problems.

Our experiments have shown that ICT usually performed slightly better than standard trees on the original data. After relabeling, the performance of ICT and the standard tree algorithm was virtually identical. It should be noted however that a standard tree algorithm applied to monotone data does not necessarily produce a monotone tree. Therefore, if a monotone model is required, application of a standard algorithm to relabeled data may not be sufficient. Furthermore, on the relabeled data ICT on average produced smaller trees than the standard algorithm. This warrants the conclusion that ICT trees are easier to understand than their somewhat larger and possibly non-monotone counterparts.

References

1. Anglin, P.M., Gençay, R.: Semiparametric estimation of a hedonic price function. Journal of Applied Econometrics 11(6), 633–648 (1996)
2. Asuncion, A., Newman, D.J.: UCI machine learning repository (2007)
3. Barile, N., Feelders, A.: Nonparametric monotone classification with MOCA. In: Giannotti, F. (ed.) Proceedings of the Eighth IEEE International Conference on Data Mining (ICDM 2008), pp. 731–736. IEEE Computer Society, Los Alamitos (2008)

4. Ben-David, A., Sterling, L., Pao, Y.: Learning and classification of monotonic ordinal concepts. Computational Intelligence 5, 45–49 (1989)
5. Ben-David, A.: Monotonicity maintenance in information-theoretic machine learning algorithms. Machine Learning 19, 29–43 (1995)
6. Bloch, D.A., Silverman, B.W.: Monotone discriminant functions and their applications in rheumatology. Journal of the American Statistical Association 92(437), 144–153 (1997)
7. Breiman, L., Friedman, J.H., Olshen, R.A., Stone, C.J.: Classification And Regression Trees. Chapman and Hall, Boca Raton (1984)
8. Brunk, H.D.: Conditional expectation given a σ-lattice and applications. Annals of Mathematical Statistics 36, 1339–1350 (1965)
9. Daniels, H.A.M., Kamp, B.: Application of MLP networks to bond rating and house pricing. Neural Computing & Applications 8(3), 226–234 (1999)
10. Dykstra, R., Hewett, J., Robertson, T.: Nonparametric, isotonic discriminant procedures. Biometrika 86(2), 429–438 (1999)
11. Feelders, A., Pardoel, M.: Pruning for monotone classification trees. In: Berthold, M.R., Lenz, H.-J., Bradley, E., Kruse, R., Borgelt, C. (eds.) IDA 2003. LNCS, vol. 2810, pp. 1–12. Springer, Heidelberg (2003)
12. Karpf, J.: Inductive modelling in law: example based expert systems in administrative law. In: Proceedings of the third international conference on artificial intelligence in law, pp. 297–306. ACM Press, New York (1991)
13. Kotlowski, W., Slowinski, R.: Statistical approach to ordinal classification with monotonicity constraints. In: ECML PKDD 2008 Workshop on Preference Learning (2008)
14. Lievens, S., De Baets, B., Cao-Van, K.: A probabilistic framework for the design of instance-based supervised ranking algorithms in an ordinal setting. Annals of Operations Research 163, 115–142 (2008)
15. Long, J.: NASA metrics data program (2008),
    http://mdp.ivv.nasa.gov/repository.html
16. Maxwell, W.L., Muckstadt, J.A.: Establishing consistent and realistic reorder intervals in production-distribution systems. Operations Research 33(6), 1316–1341 (1985)
17. Pazzani, M.J., Mani, S., Shankle, W.R.: Acceptance of rules generated by machine learning among medical experts. Methods of Information in Medicine 40, 380–385 (2001)
18. Potharst, R., Bioch, J.C.: Decision trees for ordinal classification. Intelligent Data Analysis 4(2), 97–112 (2000)
19. Potharst, R.: Classification using Decision Trees and Neural Nets. PhD thesis, Erasmus University Rotterdam (1999)
20. Royston, P.: A useful monotonic non-linear model with applications in medicine and epidemiology. Statistics in Medicine 19(15), 2053–2066 (2000)
21. Velikova, M.: Monotone Models for Prediction in Data Mining. PhD thesis, Tilburg University (2006)
22. Velikova, M., Daniels, H.: Decision trees for monotone price models. Computational Management Science 1(3-4), 231–244 (2004)

Author Index

Adams, Niall   1
Aussem, Alex   35, 202

Balakrishnan, Rajesh   143
Barile, Nicola   405
Berlingerio, Michele   237
Bifet, Albert   249
Böhm, Klemens   309
Bonnevay, Stéphane   202
Borisov, Alexander   225
Bradley, Elizabeth   321

Cain, James   261
Carroll, J. Douglas   47
Clyne, John   321
Cohen, Paul   1
Coscia, Michele   237
Counsell, Steve   261
Crémilleux, Bruno   155

Dasu, Tamraparni   21
De Las Rivas, Javier   107
Di-Jorio, Lisa   297
Diallo, Alpha   273
Diamantini, Claudia   285
Douzal-Chouakria, Ahlame   273

Eichinger, Frank   309

Feelders, Ad   405
France, Stephen L.   47

Gavaldà, Ricard   249
Giannotti, Fosca   237
Giroud, Francoise   273
Godin, Robert   393
Gruchalla, Kenny   321

Habich, Dirk   59
Hahmann, Martin   59
Hollmén, Jaakko   213
Höppner, Frank   71

Ienco, Dino   83

Jensen, Pablo   10
Jiang, Eric P.   95

Kamp, Rémon van de   405
Kaski, Samuel   178, 381
Klawonn, Frank   71
Kopanakis, Ioannis   131
Krajca, Petr   333
Kramer, Stefan   119
Krishnan, Shankar   21
Krishnan, Sriram   119
Kumar, Abhishek   143

Laurent, Anne   297
Lehner, Wolfgang   59
Li, Jian   345
Lin, Dongyu   21
Liu, Xiaohui   345
Lurponglukana-Strand, Nuttha   225

Martín-Merino, Manuel   107
Martínez-Álvarez, Francisco   357
Meo, Rosa   83
Mininni, Pablo   321
Mueller, Marianne   119

Napoli, Amedeo   393

Panagiotakis, Costas   131
Parvathy, Anju G.   143
Pelekis, Nikos   131
Pensa, Ruggero G.   83
Plantevit, Marc   155
Potena, Domenico   285
Puolamäki, Kai   178, 381

Rao, Bharat   119
Rast, Mark   321
Riquelme, José C.   357
Rodrigues de Morais, Sérgio   35
Rosales, Rómer   119
Rosenthal, Frank   59
Runger, George   225

Sapozhnikova, Elena P.   167
Savia, Eerika   178
Steck, Harald   119

Steele, Emma   190
Storti, Emanuele   285
Swift, Stephen   261, 345
Szathmary, Laszlo   393

Teisseire, Maguelonne   297
Thibault, Grégory   202
Toivola, Janne   213
Troncoso, Alicia   357
Tucker, Allan   190, 261
Tuv, Eugene   225

Valtchev, Petko   393
Vasudevan, Bintu G.   143
Venkatasubramanian, Suresh   21
Volk, Peter B.   59
Vychodil, Vilem   333

Yi, Kevin   21

Zagoruiko, Nikolay   369